Handbook of Nanophase and Nanostructured Materials

Handbook of Nanophase and Nanostructured Materials
Volume II: Characterization

Edited by

Zhong Lin Wang
Center for Nanoscience and Nanotechnology
School of Materials Science and Engineering
Georgia Institute of Technology
Atlanta, Georgia

Yi Liu
Department of Mechanical Engineering
University of Michigan
Ann Arbor, Michigan

and

Ze Zhang
Beijing Laboratory of Electron Microscopy
Institute of Physics
Chinese Academy of Sciences
Beijing, China

Kluwer Academic / Plenum Publishers
New York, Boston, Dordrecht, London, Moscow

Tsinghua University
Press

Library of Congress Cataloging-in-Publication Data

Handbook of nanophase and nanostructured materials/edited by Zhong Lin Wang, Yi Liu, Ze Zhang.
   p.   cm.
   Includes bibliographical references and index.
   Contents: v. 1. Synthesis—v. 2. Characterization—v. 3. Materials systems and applications I—v. 4. Materials systems and applications II.
   ISBN 0-306-46737-2 (vol. 1)—ISBN 0-306-46738-0 (vol. 2)—ISBN 0-306-46739-9 (vol. 3)—ISBN 0-306-46740-2 (vol. 4)
   1. Nanostructure materials—Handbooks, manuals, etc. I. Wang, Zhong Lin. II. Liu, Yi, 1967– III. Zhang, Ze, 1953–

TA418.9.N35 H358 2002
620.1′1299—dc21

2002075492

Cover image: STM image of self-assembled CuPcOC8 on HOPG surface [Courtesy Dr. Chunli Bai and Dr. Chen Wang (Institute of Chemistry, CAS, Beijing)]

ISBN 7-302-05442-8 (in People's Republic of China)
ISBN 0-306-46738-0 (Rest of World)
ISBN 0-306-47249-X (Four-Volume Set)

©2003 Kluwer Academic/Plenum Publishers, New York
233 Spring Street, New York, N.Y. 10013

http://www.wkap.nl/

10   9   8   7   6   5   4   3   2   1

A C.I.P. record for this book is available from the Library of Congress

All rights reserved

No part of this book may be reproduced, stored in a retrieval system, or transmitted in any form or by any means, electronic, mechanical, photocopying, microfilming, recording, or otherwise, without written permission from the Publisher, with the exception of any material supplied specifically for the purpose of being entered and executed on a computer system, for exclusive use by the purchaser of the work

Printed in the United States of America

List of Contributors

Xunli Wang
Oak Ridge National Laboratory, Oak Ridge, TN 37831.
WangXl@ornl.gov

J. L. Robertson
Oak Ridge National Laboratory, Oak Ridge, TN 37831.

Zhonglin Wang
School of Materials Science and Engineering, Georgia Institute of Technology, Atlanta GA 30332-0245 USA.
Zhong.Wang@mse.gatech.edu

Ze Zhang
Beijing Laboratory of Electron Microscopy, Institute of Physics, Chinese Academia of Sciences, P.O. Box. 2724, Beijing 100080 China.
Zhang@image.blem.ac.cn

Yi Liu
Department of Mechanical Engineering and Center for Materials Research and Analysis, University of Nebraska-Lincoln, NE 68588-0656 USA.
Yliu@unlserve.unl.edu

Alexander J. Shapiro
Metallurgy Division, National Institute of Standards and Technology, Gaithersburg, MD USA.
alexander.shapiro@nist.gov

Chunli Bai
Institute of Chemistry, Chinese Academy of Sciences, Beijing 100080 China.
Clbai@infoc3.icas.ac.cn

Chen Wang
Institute of Chemistry, Chinese Academy of Sciences, Beijing 100080 China.

Tiejin Li
Department of Chemistry, Jilin University, Changchun 130023, China.
tjli@mail.jlu.edu.cn

Jin Z. Zhang
Department of Chemistry, University of California, Santa Cruz CA 95064 USA.
Zhang@Chemistry.ucsc.edu

Adam J. Rondinone
School of Chemistry and Biochemistry, Georgia Institute of Technology, Atlanta GA 30332-0400 USA.

Z. John Zhang
School of Chemistry and Biochemistry, Georgia Institute of Technology, Atlanta GA 30332-0400 USA.
john.Zhang@Chemistry.gatech.edu

Zhong Shi
School of Materials Science and Engineering, Georgia Institute of Technology, Atlanta GA 30332-0245 USA.

Meilin Liu
School of Materials Science and Engineering, Georgia Institute of Technology, Atlanta GA 30332-0245 USA.
meilin.liu@mse.gatech.edu

H. Mizubayashi
Institute of Materials Science, University of Tsukuba, Japan.
mizuh@ims.tsukuba.ac.jp

H. Tanimoto
Institute of Materials Science, University of Tsukuba, Japan.

M. Suganuma
Industrial Research Institute of Aichi Prefectural Government, Japan.

A. Shimatani
Department of Quantum Engineering, Nagoya University, Nagoya, Japan.

H. Saka
Department of Quantum Engineering, Nagoya University, Furo-cho, Nagoya, 464-8603, Japan.
Saka@numse.nagoya-u.ac.jp

Qing Jiang
Department of Materials Science and Engineering, Jilin University of Technology, Changchun 130025, China.

Ke Lu
State Key Laboratory for RSA, Institute of Metal Research, Chinese Academy of Sciences, Shenyang 110015, China.
kelu@imr.ac.cn

Foreword I

Materials are the base and forerunner of modern civilization. The progress of society usually takes the widely used material as a milestone in history, for examples the Stone Age, the Pottery Age, the Bronze Age, etc. modern civilization is symbolized by the application of steel and different metals in the 19th century. The present information age is based on the discovery and wide application of semiconducting materials, and it is sometimes called the Silicon Age.

Materials used by mankind were obtained from nature in ancient times, and then followed by artificial synthesis, and most important materials used at present are prepared artificially. The process of preparation may be macroscopic, microscopic or even atomic. Material made with atomic accuracy is called nanomaterial or nanostructured material. Nanomaterials usually show unique properties through nanoscale size confinement, predominance of interfacial phenomena and the quantum effect. Therefore, by reducing the dimension of a structure to nanosize, many inconceivable properties will appear and may lead to different applications.

In February 2000, a proposal, the "National Nanotechnology Initiative Leading to the Next Industrial Revolution," was published. This was produced by agencies of the U. S. government under the auspices of the National Science and Technology Council. They considered that nanotechnology could have a profound impact on the economy and society in the early 21st century, which might be comparable to that of information or biological technology. Hence, they suggested that materials science and technology should be given top priority.

Based on the government's concern, it can be predicted that an upsurge of R & D on nanotechnology will soon come, and *Nanophase and Nanostructured Materials*, edited by Z. L. Wang, Y. Liu and Z. Zhang, is just in time. This book is a complete and up-to-date treatise dealing with nanostructured materials. It starts from materials synthesis and preparation with different methods, followed by characterization of a nanostructure and properties with different types of sophisticated instruments and novel experimental methods. About half of the book concentrate on materials systems and applications, which is considered a major part of the exploration of nanoscience and technology. Several chapters deal with information technology of the next generation, such as materials related to computer, storage and display. Several chapters are deal with superhard and superhigh strength materials, which are vital for future space and transportation technology. Biomaterials and biotechnology are hot topics for the next century, and several chapters focus on this aspect too.

The second feature of this book is that most authors are young scientists from different countries: United States, China, Japan and Singapore. They are active on the frontier of the field of nanoscience and technology and their works are stimulating and suggestive.

In all, nanoscience and technology is a rapidly growing field. I believe that the book will give a general survey for newcomers and a better understanding of the whole field to those who are working in specific areas.

83 Shuangqing Road	Chang Xu Shi
Haidian District	August, 2000
Beijing 100085, China	

Foreword II

Nanoscale science and technology are experiencing a rapid development, and they are likely to have profound impact on every field of research in the first decade of the 21st century. The technologies for real-space atomic-scale imaging, atomic and molecular manipulation and nano-fabrication have been developed since the discovery of the Scanning Tunneling Microscope by G. Binning and H. Rohrer, who were awarded the Nobel Prize in physics in 1986. This invention has played a key role in promoting the development of nanoscale science and technology. The governments and enterprises of developed countries, such as the United States, Japan, Germany, United Kingdom, France, etc., are investing a lot of funds for nanotechnology research.

Nanoscale science and technology link many fields. The former chief editor of Nature, John Maddox, said: "The idea that physics has almost been worked out as a field for interesting investigations is widely current but mistaken. There is as much left to do as has been done." There are large quantities of unknown phenomena in the fields of mesoscopic physics, nanochemistry and nanobiology. Nanochemistry and nanoelectronics should be related to mesoscopic physics. Nanochemistry, for example, will face to the systems with limited atoms, molecules and ultrafine particles(i.e., clusters), and is related to chemical synthesis, self-assembled systems, self-organization growth, and so on.

Small microbes with life phenomenon belong to nanoscale. The diameter of a DNA molecule is less than 3 nm, and the diameter of a protein molecule is about a few nanometers. Therefore, nanobiology and nanomedicine have many virgin fields that remain to be explored. A scientist said: "When some matter can be designed and manufactured in biomolecular scales, the prospects will be wonderful. This is the excitement of nanoscience and nanotechnology."

Due to the development of integrated circuits, microelectronics should advance into nanoelectronics and further, to molecular electronics: But the new materials, device design and manipulation, and manufacturing are the key challenges that have to be conquered step-by-step. Nanophase and nanostructured materials are fundamental to nanoscale science and technology.

The four volumes edited by Zhong Lin Wang, Yi Liu and Ze Zhang are very timely. There are 40 chapters in these volumes. Each chapter was written by specialists. These books will serve as excellent textbooks and references for researchers and graduate students in the field. This is the first set of the books published domestically in the field, and they will be very useful in promoting the development of nanoscience and nanotechnology in China as well as in the world.

Dr. Zhong Lin Wang, the main editor of these volumes, is a young professor at Georgia Institute of Technology, who has published over 200 scientific papers. He is famous for his work on "nanobalance". Dr. Wang is also a visiting professor at the Center for Nanoscale Science and Technology, Peking University. It is a privilege for me to write a foreword for these volumes. I am confident that their publication will be a great success.

  Center for Nanoscale Science and Technology   Quan De Wu
  Peking University   August, 2000
  Member of Chinese Academy of Sciences

Preface

Advances in science and technology in the 21st century are likely to focus on four directions: information science, life science, environmental science and nanotechnology. Wireless telecommunication, super-fast computers, and advanced computing technology have impacted everyone's daily life. The information highway has revolutionized international communication and business operation. Life science is expected to have many breakthroughs in fields such as genetic engineering, biomedical sciences, disease prevention, control and curing. Preservation of natural resources and the environment is a challenge to human civilization and social progress because we only have one earth. Nanotechnology is based on the smallest unit of matter to engineering new materials and devices atombyatom, aiming at achieving superior properties and performance through atomic-scale architecture.

Advanced materials and advanced manufacturing will be the basis of the technological revolution. Take microelectronics as an example. Silicon technology is nearlly reaching its limit and the device size is in the sub-micro range; the era of nanoelectronics is arriving. As a result of reduced device sizes, many interesting quantum mechanical phenomena become dominant. A totally new and different approach is needed in device fabrication and system assembly. Therefore, nanophase and nanostructured materials, as a field of advanced materials, are the basis of nanoscience and nanotechnology. Research in nanoscience and nanotechnology faces four main challenges:
- Synthesis of structurally controlled nanomaterials, with well defined atomic-scale structure, high purity and large yield;
- Characterization of the structures and properties, especially the properties of individual nanostructures;
- Device fabrication, nanomanipulation and inter-device interconnection;
- System integration and large-scale manufacturing.

Research in nanomaterials is a multidisciplinary effort that involves physicists, chemists, materials scientists, electrical engineers, biological and, possibly, medical scientists. A rapid development in the field requires a book that covers the forefront research in a wide range, including materials synthesis, structure and property characterizations, theoretical modeling and applications. After editing these four volumes, we are convinced that nanomaterials will be the basis of nanotechnology. These books are about the synthesis, characterization and applications of nanophase and nanostructured materials, which are referred to as nanomaterials. By nanophase materials we mean dispersive nanoparticles. Nanostructured materials are solid materials made of nanocrystallites. Nanomaterials are attractive because of their unique and superior electrical, mechanical and chemical properties, where the size of the grains approaches a few to a few hundreds of nanometers. Many of the outstanding properties are strongly enhanced when the size of the object is smaller than the electron mean-free-path length. There is no clear cut size smaller than that for which the materials are called nanomaterials, but a common understanding is that nanomaterials must exhibit some unique size-dependent properties that are minimal or vanish for large bulk crystals.

The contents of the volumes are classified into three parts. The first part emphasizes the synthesis of nanocrystal materials, aiming at describing the principles and approaches of the synthesis techniques, processing controls and the outcoming quality of the nanomaterials using chemical and physical techniques. The second part emphasizes the techniques used for characterizing nanomaterials, aiming at describing the physical mechanism, data interpretation and detailed applications for characterizing nanophase materials to understand the morphology, surface and the atomic level microstructures of nanophase materials and their associated properties. The final part focuses on the systems of different nanomaterials. The objective is to show their characteristics, unique properties and applications.

These volumes are intended as textbooks which not only reflect the state-of-the-art and give a sound review of the literature, but delineate the underlying concepts and bearing of this interdisciplinary field. The book is aimed at being a handbook which is the standard reference in the field for years to come. Our goal is to provide a comprehensive and complete introduction about nanomaterials to graduate students and researchers, whose background can be in chemistry, physics, materials science, chemical engineering, electrical engineering or even biomedical science.

We express our gratitude and appreciation to all of the authors for their hard work and contributions.

Zhong Lin Wang (Editor in Chief)
School of Materials Science and Engineering
Georgia Institute of Technology,
Atlanta GA 30332-0245 USA
zhong.wang@mse.gatech.edu

Yi Liu (Editor)
Department of Mechanical Engineering
and Center for Materials Research and Analysis
University of Nebraska-Lincoln, Lincoln, NE 68588-0656 USA

yliu@unlserve.unl.edu

Ze Zhang (Editor)
Beijing Laboratory of Electron Microscopy
Institute of Physics
Chinese Academy of Sciences, P.O. Box. 2724, Beijing 100080, China
zhang@image.blem.ac.cn

Introduction

The era of nanophase and nanostructured materials has dawned. An exciting future of developments in fundamental science and fascinating discoveries in many fields lies before us. A broad range of applications in applied science and in industry is already emerging and new applications will appear at an accelerating rate.

The publication of these four volumes provides a valuable and timely contribution to the development of the science of materials on a nanometer scale. Already the scientific and popular literature abounds with thousands of articles on the many aspects of the subject and new ideas and results are reported at an increasing number of conferences. It is very difficult to keep track of the progress: what has been established and what things are mere projections. The Editors have provided an important service to the scientific community in bringing together in these four volumes a collection of authoritative reviews of the status of the many lines of investigation in a well-organized and coherent presentation.

The explosive growth of the science of nano-systems in recent years has followed the many years of development of the means for characterizing such materials on the scale of nanometers or smaller. The accrued store of knowledge derived by high-resolution electron microscopy, electron nanodiffraction, scanning electron microscopy and the various forms of scanning probe microscopy provided the background on the detailed structures and morphologies of a large number of solid state systems. The more traditional methods for study of materials in bulk, such as the various optical spectroscopies and X-ray diffraction, have made important contributions, giving the averaged structures which are appropriate for the understanding of many physical and chemical properties. But the possibility of seeing directly the forms and perturbations of individual nanometer-size particles and structural features was critical in providing the inspiration for much of the new approach to the solid state. The high-resolution techniques give us the means for rapid and efficient means for testing new methods for preparing nanostructures, for evaluating modifications of the structures and for correlating the structures with the physical properties. Carbon nanotubes, for instance, were discovered by high-resolution electron microscopy and, as in the case of other nano-structured materials, the high resolution techniques are essential for their manufacture, modification, manipulation and measurement.

The Editors of these volumes are scientists who have had intense personal experience in research on nanostructured materials and are well known for their contributions to the literature on relevant topics. Readers will benefit from their choices of topics and contributors, based on their intimate knowledge of the fields covered. The books should serve not only as a complete collection of scientific reference material but also as text books for graduate students and advanced undergraduates.

John M. Cowley, FRS
Regents' Professor Emeritus
Arizona State University

Contents

1 **X-ray and Neutron Scattering** 1
  1.1 Introduction 1
  1.2 X-ray and Neutron Diffraction 4
  1.3 Inelastic Neutron Scattering 14
  1.4 Small Angle Scattering 21
  1.5 Concluding Remarks 24
  References 25

2 **Transmission Electron Microscopy and Spectroscopy** 29
  2.1 Major Components of a Transmission Electron Microscope 29
  2.2 Atomic Resolution Lattice Imaging of Crystalline Specimens 31
    2.2.1 Phase Contrast 31
    2.2.2 Abbe's Imaging Theory 32
    2.2.3 Image Interpretation of Very Thin Samples 34
    2.2.4 Image Simulation 34
  2.3 Faceted Shapes of Nanocrystals 37
    2.3.1 Polyhedral Shapes of Nanoparticles 37
    2.3.2 Twinning Structure and Stacking Faults 41
    2.3.3 Decahedral and Icosahedral Particles 42
    2.3.4 Nucleation and Growth of Nanoparticles 43
  2.4 Electron Holography 46
  2.5 Lorentz Microscopy 48
    2.5.1 Principle of Lorentz Microscopy 48
    2.5.2 Elimination/Reduction of Magnetic Field from Objective Lens 49
    2.5.3 Fresnel Lorentz Microscopy 49
    2.5.4 Foucault Lorentz Microscopy 50
    2.5.5 Differential Phase Contrast Mode of Lorentz Microscopy in STEM 51
  2.6 Nanodiffraction 52
    2.6.1 Optics for Nanodiffraction 53
    2.6.2 Experimental Procedures to Obtain a Nanodiffraction Pattern 53
    2.6.3 Some Applications 54

2.7	In situ TEM and Nanomeasurements		61
	2.7.1	Thermodynamic Properties of Nanocrystals	62
	2.7.2	Nanomeasurement of Electrical Transport in Quantum Wires	68
	2.7.3	Nanomeasurement of Mechanical Properties of Fiber-Like Structures	70
	2.7.4	Femtogram Nanobalance of a Single Fine Particle	71
	2.7.5	Electron Field Emission from a Single Carbon Nanotube	72
2.8	Electron Energy Loss Spectroscopy of Nanoparticles		75
	2.8.1	Valence Excitation Spectroscopy	75
	2.8.2	Quantitative Nanoanalysis	77
	2.8.3	Near Edge Fine Structure and Bonding in Transition Metal Oxides	79
	2.8.4	Doping of Light Elements in Nanostructures	81
2.9	Energy-Filtered Electron Imaging		85
	2.9.1	Chemical Imaging of Giant Magnetoresistive Multilayers	85
	2.9.2	Imaging of Spin Valves	89
	2.9.3	Mapping Valence States of Transition Metals	91
2.10	Energy Dispersive X-ray Microanalysis (EDS)		93
2.11	Summary		94
References			95

3 Scanning Electron Microscopy — 99

3.1	Introduction		99
3.2	Basic Principals of Scanning Electron Microscopy		100
	3.2.1	Main Parameters of Electron Optics	101
	3.2.2	The Minimum Attainable Beam Diameter	102
3.3	Contrast Formation and Interpretation		103
3.4	Secondary Electron Detectors		111
	3.4.1	Everhart-Thornley Detector	111
	3.4.2	In-iens Secondary Electron Detector	112
3.5	Dedicated Detectors		114
	3.5.1	Solid-State Diode Detector	114
	3.5.2	Scintillator Backscattered Electron Detector	115
	3.5.3	BSE-to-SE Conversion Detectors	115
	3.5.4	Multi-detector System	115
	3.5.5	Electron Backscattered Diffraction (EBSD)	116
	3.5.6	Magnetic Contrast	117
	3.5.7	X-ray Spectrometers	118
3.6	Conclusions		120
References			121

4 Scanning Probe Microscopy ... 124
4.1 Overview ... 124
4.2 Scanning Tunneling Microscopy ... 125
4.2.1 Introduction ... 125
4.2.2 STM Studies on Metals ... 127
4.2.3 STM Studies on Semiconducting Surfaces ... 130
4.2.4 Organic Molecules Studied by STM ... 135
4.3 Atomic Force Microscopy ... 138
4.3.1 Introduction ... 138
4.3.2 The Force Sensor ... 139
4.3.3 Illustration of AFM Applications ... 141
4.3.4 Force Spectrum Analysis ... 144
4.3.5 Lateral Force Microscopy ... 146
4.3.6 Force Microscope Operating in Non-contact Mode ... 147
4.3.7 Force Microscope Operating in Tapping Mode ... 148
4.3.8 Magnetic Force Microscopy ... 150
4.4 Ballistic-Electron-Emission Microscopy ... 152
4.4.1 The Principle of BEEM ... 152
4.4.2 BEEM Experiments ... 154
4.4.3 Ballistic-Hole Spectroscopy of Interfaces ... 156
4.5 Applications of STM and BEEM in Surface and Interface Modifications ... 159
4.5.1 Surface Nanofabrication with STM ... 160
4.5.2 Single Atom Manipulation ... 164
4.5.3 Interfacial Modification with BEEM ... 166
4.6 Concluding Remarks ... 167
References ... 168

5 Optical Spectroscopy ... 172
5.1 Introduction ... 172
5.2 Nanoclusters and Nanocrystals ... 173
5.2.1 Absorption and Photoluminescence Spectroscopic Evidence for Quantum Confinement ... 174
5.2.2 Raman and FTIR Studies on the QDs and Its Supramolecular Assemblies ... 181
5.2.3 High Resolution Spectroscopy of Individual Quantum Dots ... 184
5.2.4 Ultrafast Spectroscopy in Quantum Confined Structures ... 192
5.3 The Control of Nanostructures by Spectroscopic Diagnosis ... 197
5.3.1 Processing on the Nanostructures ... 197
5.3.2 Spectroscopic Diagnosis ... 201

	5.3.3	Photovoltage Spectroscopy of Surface and Interface	212
	References		215

6 Dynamic Properties of Nanoparticles — 219
- 6.1 Introduction — 219
- 6.2 Experimental Techniques — 220
  - 6.2.1 Synthesis of Semiconductor Nanoparticles — 220
  - 6.2.2 Synthesis of Metal Nanoparticles — 222
  - 6.2.3 Characterization of Nanoparticles — 222
  - 6.2.4 Dynamics Measurements with Time-Resolved Techniques — 223
- 6.3 Dynamic Properties of Semiconductor Nanoparticles — 225
  - 6.3.1 Theoretical Considerations — 225
  - 6.3.2 CdS, CdSe and Related Systems — 228
  - 6.3.3 Metal Oxide Nanoparticles: TiO_2, Fe_2O_3, ZnO, SnO_2 — 231
  - 6.3.4 Other Semiconductor Nanoparticle Systems: Si, AgI, Ag_2S, PbS — 234
  - 6.3.5 Nanoparticles of Layered Semiconductors: MoS_2, PbI_2 — 235
  - 6.3.6 Effects of Particle Surface, Size and Shape — 237
- 6.4 Dynamic Properties of Metal Nanoparticles — 238
  - 6.4.1 Background and Theoretical Considerations — 238
  - 6.4.2 Gold (Au) Nanoparticles — 240
  - 6.4.3 Other Metal Nanoparticles: Ag, Cu, Sn, Ga and Pt — 242
  - 6.4.4 Effects of Surface, Size and Shape — 242
- 6.5 Summary and Prospects — 243
- References — 244

7 Magnetic Characterization — 252
- 7.1 Introduction — 252
- 7.2 SQUID Magnetometry — 255
- 7.3 Mössbauer Spectroscopy — 262
- 7.4 Neutron Powder Diffraction — 273
- 7.5 Lorentz Microscopy — 277
- 7.6 Summary — 281
- References — 281

8 Electrochemical Characterization — 283
- 8.1 Introduction — 283
- 8.2 Preparation of Nanostructured Electrode — 285
  - 8.2.1 Electrodeposition and Electrophoretic Deposition — 285

		8.2.2 Formation of Nanoparticles in Polymers	287
8.2.3 Electrochemical Self-Assembly	288
8.2.4 Mesoporous Electrodes	289
8.2.5 Composite Electrodes Consisting of Nanoparticles	291
8.2.6 Powder Microelectrode	291

8.3 Principles of Electrochemical Techniques 293
    8.3.1 Impedance Spectroscopy 293
    8.3.2 Potential Sweep Method 300
    8.3.3 Potential Step Method 304
    8.3.4 Controlled-Current Techniques 307
    8.3.5 Electrochemical Quartz Crystal Microbalance 312
8.4 Application to Nanostructured Electrodes 316
    8.4.1 Characterizing the Reversibility of Battery Electrode Materials 316
    8.4.2 Characterizing the Transport Properties 319
8.5 Summary 320
References 321

9 Mechanical Property Characterization 326
9.1 Elasticity Study of Metal Nanometer Films 326
    9.1.1 Vibrating Reed Method 326
    9.1.2 Elasticity Measurements on Ag and Al Films 328
    9.1.3 Supermodulus Effect in Ag/Pd Multilayers 332
9.2 Mechanical Behavior of High-Density Nanocrystalline Gold 336
9.3 FIB/TEM Observation of Defect Structure Underneath an Indentation 348
    9.3.1 Introduction 348
    9.3.2 FIB Milling 349
    9.3.3 Experimental Procedures 349
    9.3.4 Load-Displacement Curve 349
    9.3.5 TEM Observation 352
    9.3.6 Conclusion 355
References 355

10 Thermal Analysis 358
10.1 Introduction 358
10.2 Fundamental Techniques 359
10.3 Experimental Approach 364
    10.3.1 Melting of Nanophases and Nanostructured Materials 365
    10.3.2 Kinetics of Glass-Nanocrystal Transition and Grain Growth of Nanostructured Materials 366
    10.3.3 Heat Capacity of Nanostructured Materials 369
    10.3.4 Interface Enthalpy of Nanostructured Materials 370

10.4	Data Interpretation		372
	10.4.1	Size-Dependent Melting Thermodynamics of Nanophases	372
	10.4.2	Glass-nanocrystal Transition Thermodynamics	374
10.5	Examples of Applications		374
10.6	Limitations		382
10.7	Prospects		383
References			384

Index ········ 386
Appendix ········ 389

Handbook of Nanophase and Nanostructured Materials

1 X-ray and Neutron Scattering

Xun li Wang and J. L. Robertson

1.1 Introduction

X-ray and neutron scattering are an important class of experimental tools for materials characterization. The scattering of X-ray and neutron beams by crystalline materials can be divided into two categories: elastic and inelastic. Elastic scattering, or diffraction, is referred to as a scattering process in which the incident X-ray or neutrons change direction without losing energy when they interact with the material. During inelastic scattering, the incident X-ray or neutrons change both direction and energy. The loss (or gain) of energy by the X-ray or neutron beams excites (or annihilates) fundamental excitations that describe the collective motion of atoms within the material. As will be seen below, diffraction by X-ray or neutrons allows us to determine the structure (i.e., how atoms are arranged), whereas inelastic scattering allows us to investigate the dynamic behavior (i.e., how atoms move) of materials.

It is perhaps an understatement that much of our understanding and use of advanced materials were made possible by advances in X-ray and neutron scattering techniques. For example, an important aspect of modern materials design is based on the establishment of a relationship between structure and physical properties. The structure referred to here includes the crystal, magnetic, and microstructure of the material. X-ray and neutron diffraction are ideal, in some cases the only, tools for the determination of these structures. In addition, study of lattice dynamics with inelastic neutron scattering elucidates in detail how a structural transition takes place (Stassis, 1986). In recognition of these and other important contributions by X-ray and neutron scattering, four Nobel prizes in physics were awarded to the inventors of X-ray and neutron scattering techniques:

- W. C. Röntgen (1891) for the extraordinary services he has rendered by the discovery of the remarkable rays subsequently named after him.
- M. von Laue (1914) for his discovery of the diffraction of X-rays by crystals.
- W. H. Bragg and W. L. Bragg (1915) for their services in the analysis of crystal structure by means of X-rays.

- B. N. Brockhouse and C. G. Shull (1994) for pioneering contributions to the development of neutron scattering techniques for studies of condensed matter.

Moreover, the 1962 Nobel Prize in medicine went to F. H. C. Crick, J. D. Watson, and M. H. F. Wilkins for their discoveries concerning the structure of DNA molecules using X-ray diffraction and its significance for information transfer in living material. Based on their discovery, an entire new branch of science, i. e., genetic engineering, has emerged, which will forever change our lives.

X-ray and neutron scattering are complementary techniques. X-rays interact with matter via electromagnetic interaction, which scales with the electron density. While this strong interaction between X-rays and samples means a strong scattering intensity, it also dictates that X-ray scattering is essentially a surface technique. A principal advantage of neutrons versus X-rays is that neutrons are highly penetrating. This is because neutrons have no charge and interact with matter via the relatively week neutron-nucleon interaction. This highly penetrating capability ensures that the results obtained from neutron scattering measurements are representative of the bulk, rather than surface layers. In addition, with neutron diffraction in situ studies can be readily realized as neutrons easily penetrate through furnaces, pressure cells, and other special environments. As an example, more than 95% of a neutron beam is transmitted through a 5 mm thick aluminum plate, compared with less than 0.1% of an X-ray beam. This example also illustrates the basis of an important practical application of neutron diffraction-mapping residual stress within engineering components (Krawitz and Holden, 1990; Withers and Holden, 1999).

In addition to its highly penetrating capability, there are several other factors that make neutron scattering a powerful experimental probe. First, the neutron scattering cross section for a given element is not simply related to the atomic number. This means that neutron scattering is ideally suited for study of materials containing light elements, such as hydrogen and oxygen. Second, although having no charge, neutrons possess a magnetic moment. As a result, neutron scattering continues to dominate the study of magnetic materials. Third, inelastic neutron scattering remains a superior tool for the study of dynamic behavior, as the energies of thermal neutrons are of comparable magnitude with the fundamental excitations in crystalline materials. There are a number of neutron facilities around the world where neutron scattering experiments can be carried out (see Appendix). Access is gained by submitting an experiment proposal to one of these facilities.

X-ray and neutron scattering techniques have been used to characterize nanomaterials for a long time. A search in the Science Citation Index indicates that there has been a steady increase in the number of publications on nanomaterials that involve either X-ray or neutron scattering. Figure 1.1 shows a tally of these publications for each year between 1995 and 1999. As can be seen, the total number of publications that involved either X-ray or neutron scattering has increased by a factor of 3 over a span of five years, showing the increased level of interest in nanomaterials. The number of publications that used X-rays is approximately 10 times those that used neutrons, reflecting the fact that X-ray instruments are more readily available than neutron instruments.

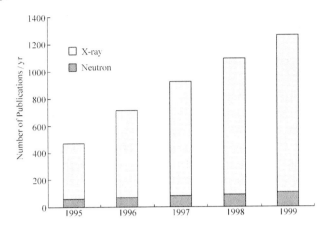

Figure 1.1 Number of publications on nanomaterials that involved the use of either X-ray or neutron scattering techniques from 1994 to 1999

The primary use of X-ray or neutron diffraction in the characterization of nanomaterials has been the determination of grain size, as many of the physical properties of nanomaterials are grain size dependent. For example, when the grain size of a metal is reduced to the nanometer range, its yield strength decreases with decreasing grain size, a phenomenon that has been referred to as the inverse Hall-Petch relationship (Schiotz, 1998). The determination of grain size is usually achieved by analyzing the peak shape profile of one or more diffraction peaks. Materials chemistry is also of interest, especially if the nanomaterial is a mixture of several phases (e.g., a composite). In such a case, it is important to know the crystal structure and fraction of each phase. This information can be obtained though Rietveld refinement of the collected diffraction patterns. In some cases, both X-ray and neutron diffraction patterns are needed in order to resolve the positions of certain atoms. Details of these analysis methods will be discussed in Section 1.2.

Another area that has received considerable interest is the vibrational dynamics of nanocrystalline materials, because they can help us understand how the physical properties of these materials differ from those of bulk specimens. For nanocrystalline materials with grain sizes less than 10 nm, a large fraction of the atoms lie within or adjacent to grain boundaries, and it is not surprising that these grain boundary atoms have a substantial effect on the vibrational spectra. Indeed, measurements of the Debye-Waller factors [see eq. (1.2)] have shown that the atoms in nanocrystals have larger thermal mean-square displacements than those found in large-grained materials (Von Eynatten, et al., 1986; Hayashi, et al., 1990; Eastman and Fitzsimmons, 1995; Long, et al., 1995). While specific aspects of the vibrational spectra can be studied using inelastic X-ray scattering (Fultz, et al., 1997) or Raman spectroscopy (Mlayah, et al., 1997; Siu, et al., 1999), the dynamical behavior as a whole is often best characterized by measuring the vibrational density of states using inelastic neutron scattering.

The vibrational modes of atoms on a crystal lattice are referred to as *phonons* and their energies are quantized. Consequently, when a neutron passing through the specimen is inelastically scattered, it exchanges quanta of energy with the crystal lattice by creating or annihilating phonons. Because the kinetic energy of a thermal neutron (10—100 meV) is of comparable magnitude to that of a phonon, neutrons are well suited for studying the vibrational spectra of materials. Thus, there is a significant change in the neutron's energy when a phonon is either created or destroyed and this change is relatively simple to measure. The vibrational density of states is simply the distribution of phonon energies given by $g(E)$, where $g(E)dE$ is the number of phonons with energies between E and $E+dE$. In addition to the vibrational density of states, there is also the phonon dispersion that describes the relationship between the phonon energy and its wave vector [see Eq. (1.17)] for any direction in the crystal. Fundamental aspects of phonon dispersion curves and how they are determined are discussed in great detail in several textbooks (Marshall and Lovesey, 1971; Bacon, 1975; Kittel, 1986; Stassis, 1986)

1.2 X-ray and Neutron Diffraction

When an X-ray or neutron beam is incident on a crystalline material, reflections occur due to the interaction between the incident beam and the periodically aligned atoms. This scattering process is schematically illustrated in Fig. 1.2(a). Experimentally, a reflection or diffraction peak is characterized by three sets of parameters: peak position, peak intensity, and peak shape profile.

The condition for a reflection to occur is dictated by the Bragg's law,

$$n\lambda = 2d\sin\theta \qquad (1.1)$$

where λ is the wavelength of the X-ray or neutron beam, d is the lattice spacing of the reflecting plane, and θ is the diffraction angle, measured with respect to the incident beam. The integer n is known as the order of the

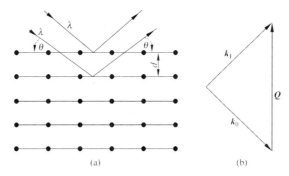

Figure 1.2 Illustration of a reflection from crystalline materials. (a) Bragg reflection from a particular family of planes, separated by a distance d. (b) Laue representation of a Bragg reflection

corresponding reflection. In a diffraction experiment, (1.1) determines the position of the peak. The intensity of the peak, on the other hand, is determined by the coherent interference of the individual atoms that decorate the crystal lattice, and is usually discussed in terms of the unit cell structure factor

$$F(Q) = \sum_m b_j e^{iQ \cdot r_j} e^{-w_j} \qquad (1.2)$$

where r_j is the position of atom j with respect to the origin of the unit cell and w_j is the Debye-Waller factor, which is a measure of the thermal vibration of atom j. Q is the scattering vector defined as

$$Q = k_f - k_i \qquad (1.3)$$

where k_f and k_i are the wave vectors of the incident and scattered beams, respectively. It can be shown that the Bragg's law as described in (1.1) requires that

$$Q = k_f - k_i = \tau \qquad (1.4)$$

where τ is the reciprocal lattice vector. Equation (1.4) is usually referred to as Laue's representation of Bragg's law, which is illustrated in Fig. 1.2(b). The quantity b_j specifies the scattering amplitude of atom j. In the field of neutron scattering, b_j is known as the neutron scattering length, whereas in the field of X-ray scattering, b_j is known as the atomic scattering factor and is more commonly denoted as f_j. Note that b_j and f_j have rather different characteristics due to the differences in the scattering mechanism. The neutron scattering length, b_j, has no Q dependence and appears to vary randomly with the atomic number, Z. The X-ray atomic scattering factor, f_j, on the other hand, is proportional to Z and decreases rapidly with increasing Q.

The measured diffraction peak always exhibits a finite peak profile, even for perfectly ordered samples. This peak profile depends on the type of radiation beams used and the exact experimental setup, and is commonly referred to as the instrument resolution function. A variety of factors contribute to the instrument resolution function, including the inherent divergence of the incident beam, sample size, and detector width.

If the sample is not perfectly ordered, additional peak broadening occurs. The resulting peak shape profile is given by a convolution of the instrument resolution function and a sample broadening function (Klug and Alexander, 1974a),

$$h(x) = \int g(x - x')f(x')dx' \qquad (1.5)$$

Here $h(x)$ is the experimentally observed peak shape profile at point x, $g(x)$ is the instrument resolution function, and $f(x)$ is known as the "pure" peak shape profile. To see the physical meaning of $f(x)$, consider an ideal instrument with a negligible peak width, i.e.,

$$g(x) = \begin{cases} 1 & \text{if } x = 0 \\ 0 & \text{otherwise} \end{cases} \qquad (1.6)$$

Substituting (1.6) into (1.5), one obtains $h(x) = f(x)$. Thus $f(x)$ is simply the peak shape profile measured with an ideal instrument.

Two types of peak broadening most commonly observed are due to finite crystallite size and lattice microstrains. Here a crystallite is referred to as a coherently diffracting domain or grain. It is now generally accepted that crystallite-size broadening can often be approximated by a Lorentzian function (Halder and Wagner, 1966; Nandi and Sen Gupta, 1978; de Keiser, et al., 1982),

$$f(x) = \frac{1}{1 + \left(\frac{2x}{\beta_L}\right)^2} \qquad (1.7)$$

where β_L is the full-width-at-half-maximum (FWHM) of $f(x)$. The parameter β_L is inversely proportional to the average crystallite or grain size, D. However, the proportionality depends on the experimental method and the type of instrument used. Sherrer showed that for constant-wavelength diffractometers ($x = 2\theta$), for example, β_L is related to D by

$$D = \frac{K\lambda}{\beta_L \cos\theta} \qquad (1.8)$$

where the coefficient K has been found to be in the neighborhood of 0.9 by most theoretical calculations (Klug and Alexander, 1974b). Microstrain broadening, on the other hand, is better described with a Gaussian function (Nandi and Sen Gupta, 1978; de Keiser, et al., 1982; Nandi, et al., 1984).

$$f(x) = \exp\left[-4 \ln 2 \cdot \left(\frac{x}{\beta_G}\right)^2\right] \qquad (1.9)$$

where β_G is the FWHM of $f(x)$. β_G is proportional to the root-mean-square strain of the microstrain field, but here again, the proportionality depends on the experimental methods. In the case of constant-wavelength diffractometers, it can be shown that

$$\beta_G = 4\sqrt{2 \ln 2} \cdot \mathrm{tg}\theta \cdot \bar{e} \qquad (1.10)$$

where $\bar{e}$ is the root mean square (rms) strain. Thus, by analyzing the profile shape of $f(x)$, it is possible to determine the nature of peak broadening. A more decisive method to further distinguish these two types of broadening is by examining the order (n)-dependence of $f(x)$. In the case of constant-wavelength diffractometers, for example, β_G increases sharply as a function of $\mathrm{tg}\theta$ for 2θ below $100°$, whereas β_G increases at the much slower rate of $1/\cos\theta$.

It is clear from the above discussion that diffraction data provide very detailed information about crystalline samples. Measurements of the peak positions and intensity yield the crystal structure or chemistry of the material. Microstructural information such as the grain size and lattice microstrains can be determined by analyzing the broadening of the measured peak with respect to the instrument resolution.

Since by definition nanocrystalline materials consist of grains of nanometer sizes, the relevant instruments for characterizing nanocrystalline materials are powder diffractometers. In terms of experimental methods, X-ray and neutron powder diffractometers can be classified into two categories: constant-wavelength and energy-dispersive types. In the constant-wavelength setup, schematically shown in Fig. 1.3, an incident beam with a well-defined wavelength is selected from the X-ray or neutron source, typically with the use of a monochromating crystal. A diffraction pattern is obtained by recording the intensity of the diffracted beam as a function of the scattering angle, using either a wire or a position-sensitive detector. The use of position-sensitive

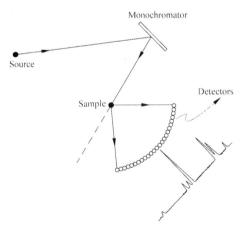

Figure 1.3 Schematic diagram of the experimental setup for a constant-wavelength powder diffractometer

detectors significantly increases the overall data rate, since one or more diffraction peaks are now recorded simultaneously. The recently commissioned D20 neutron powder diffractometer at the Institute Laue-Langevin, for instance, features a large banana-shaped position-sensitive detector covering an angular range of 160°, enabling high-quality diffraction patterns to be collected in a matter of seconds (Convert, et al., 1997).

In the energy-dispersive setup, the entire wavelength spectrum from the X-ray or neutron source is used. The detector is set at a fixed scattering angle and the intensity of the diffracted beam is recorded and analyzed as a function of energy or wavelength. The broad use of energy-dispersive X-ray diffractometry (see, for example, Otto, 1997) is, unfortunately, limited by the energy resolution of available X-ray detectors. Most applications of the energy-dispersive method are with neutron powder diffractometers at a pulsed neutron source, where a diffraction pattern is recorded as a function of the time of flight (from the source to the detector), which is inversely proportional to the neutron wavelength. An advantage of time of flight neutron diffractometers is that the resolution function is nearly independent of d, so peaks at small d values can still be well resolved. This is in contrast to constant-wavelength diffractometers whose resolution deteriorates sharply at sufficiently small d. For more information on neutron time-of-flight diffractometry, see Jorgensen et al. (1989).

A variety of ancillary equipment has been developed for in situ studies. Furnace and cryostat attachments are now routinely used for temperature-dependent studies. Load frames mounted on the sample table have allowed in situ investigations of deformation in crystalline materials (Pang, et al., 1998; Clausen, et al., 1998; Bourke, et al., 1999). Electric and magnetic fields have also been applied to study materials' responses to applied fields (Misture, et al., 1997). Because neutrons are highly penetrating, complex special environments, such as pressure cells, can also be realized with neutron diffractometers. In fact, much of the progress in high-pressure physics was associated with development of high-pressure cells for neutron diffractometers (see, for example, Hull, et al., 1998).

Precise knowledge of the peak shape profile function is essential to all analyses of X-ray and neutron diffraction data. As stated earlier, the observed peak shape profile is a convolution of the instrument resolution function, $g(x)$, and the "pure" peak shape profile, $f(x)$. The instrument resolution function is instrument specific. For example, the resolution function of a constant-wavelength neutron powder diffractometer is well characterized with a Gaussian distribution function, i.e.,

$$g(x) = \exp\left[-4 \ln 2 \cdot \left(\frac{x - x_0}{\beta}\right)^2\right] \quad (1.11)$$

where $x = 2\theta$ is the point of measurement, x_0 is the center of the peak, and β is the FWHM. The resolution function for a constant-wavelength X-ray powder diffractometer is not as simple, but in general can be described as a Voigt function, which is a convolution of a Gaussian and a Lorentzian function. The "pure" peak shape profile due to the presence of both small crystallites and lattice microstrain is also a Voigt function, as can be seen from (1.7) and (1.9). Thus, for simple diffractometers such as those discussed above, the expected peak shape profile is given by the convolution of two Voigt functions. In practice, various analytical expressions approximating the convolution integral are used to facilitate data analysis. For example, the pseudo Voigt function (Wertheim, et al., 1974) which is a linear combination of a Gaussian and a Lorentzian has been introduced to describe the resolution function of X-ray diffractometers.

The first step of analyzing a diffraction data set is to identify all observed diffraction peaks with known phases. Next, a few well-defined peaks should be identified and fitted to the peak shape profile functions. The fitted peak positions are used to calculate the lattice parameters for the identified phases. The fitted peak shape profile parameters can be used to determine the average grain size and lattice microstrain, after correcting for the instrument resolution effects.

The determination of crystallite size and lattice microstrain from diffraction peak shape has been and continues to be a subject of research. Many of the practical methods that are used today are well documented in a special chapter in the classical book by Klug and Alexander (1974c), while more recent advances can be found in review articles by Lanford (1992) and Le Bail (1992). A simplified analysis procedure exists if $h(x)$, $g(x)$ and $f(x)$ are each characterized by a Voigt function, a valid assumption for constant-wavelength diffractometers. In this case, the following relationship holds (Lanford, 1992):

$$\beta_L^f = \beta_L^h - \beta_L^g$$
$$(\beta_G^f)^2 = (\beta_G^h)^2 - (\beta_G^g)^2 \quad (1.12)$$

From β_L^f and β_G^f, the average grain size (D) and rms strain ($\bar{e}$) are obtained through the use of (1.7) and (1.9) for constant-wavelength diffractometers or similar equations for other types of diffractometers. This procedure is applicable for materials containing uniform sized grains. More sophisticated methods are available to deal with complex material systems, where for example the peak broadening is anisotropic along different (hkl) or the peak shape profile cannot be described by a simple Voigt function. For further reading on this subject, the readers are referred to Lanford (1992) and Le Bail (1992) and the references therein.

Analysis of a few well-defined peaks provides a quick estimate of the chemistry and microstructure of the material under study. However, the results thus obtained can be strongly influenced by the quality of the sample. For example, if the sample exhibits preferred orientation for certain crystallographic planes, the lattice parameters determined by fitting to the reflections off these planes can be in substantial error. More precise and complete information about the sample can be obtained, however, if the diffraction data are analyzed together using the so-called Rietveld pattern refinement technique (Rietveld, 1969). In this approach, the entire observed diffraction pattern is fitted with a pattern calculated from a structural model. Since a large number of peaks are fitted simultaneously, the statistical errors associated with fitting of individual peaks are reduced. Moreover, by fitting to the whole pattern, any effects of preferred orientation, extinction, and other systematic aberrations, if present, are also minimized. The main use of Rietveld analysis is for determining the crystal structure or chemistry of the material. Parameters in the structural model are determined through iterative refinement of the fit. Another useful application of Rietveld analysis is with multiphase materials such as ceramic composites, where the volume or weight fraction of each phase can be readily determined using the scale factors obtained from the refinement (Hill and Howard, 1987). This feature is quite useful for studying phase transformation, where the evolution of different phases can be monitored continuously throughout the transformation (see, for example, Alexander, et al., 1995). Full-pattern Rietveld analysis also leads to a more precise determination of the microstructural parameters (Le Bail, 1992), but to carry out the refinement a microstructure model (in addition to the crystal structure model) is required. The study of La_2NiO_4 gave an excellent example in this aspect, where the particular ($h\ k\ l$) dependence on the peak broadening is successfully accounted for by using a one-parameter model assuming that strain broadening occurs only in the a-b planes.

As stated in the introduction, thousands of publications in the literature that have used either X-ray or neutron scattering techniques to characterize nanomaterials and the number of publications is increasing with time. It is thus impossible to do a complete survey of published work. In light of this situation, we have decided to concentrate on illustrating the use of these techniques to characterize some of the most important properties of nanomaterials.

The first and foremost issue in the development of nanomaterials is processing, as many of the properties vary with grain size. Control of grain size is commonly achieved with heat treatment or annealing. Loffler et al. (1998) reported a study on the effect of heat-treatment temperature for nanostructured Fe and Ni. In this study, each sample was heat-treated for 24 h at various temperatures. The grain size and lattice microstrain were derived from X-ray data using the simplified procedure outlined above. The experimental results, shown in Fig. 1.4, indicate that at low temperatures (up to 200°C), the grain size increases only slightly but the lattice microstrain is significantly relieved. Rapid grain growth was seen only at higher heat-treatment temperatures. Thus, there exists a heat-treatment window in which the quality of the sample can be improved without significantly changing the grain size. The results of grain size measurements are often used to understand the influence of grain size on the properties of nanomaterials. For example, Loffler et al. (1998) have found that for nanostructured Fe the coercive field shows a pronounced variation with grain size, with the maximum around 30 nm and a steep decrease at smaller grain sizes. Also, a study by Siu et al. (1999) on ZrO_2 reveals a critical grain size of about 15 nm, above which the Raman (phonon) spectra are essentially similar to those of the bulk. Below the critical grain size, the Raman spectra changed significantly, as evidenced by line broadening, reduced intensity, and shift in line positions. The authors found that the changes in Raman spectra are associated with increasing defects in the nanograins, which result in mode softening and the generation of surface modes. These observations are consistent with inelastic neutron measurement results (see next section).

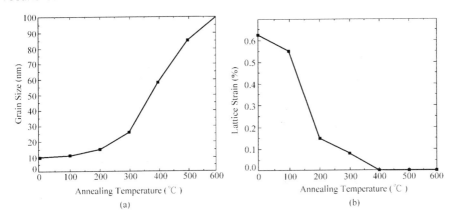

Figure 1.4 Change of average grain size and lattice microstrain as a function of heat-treatment temperature for nanostructured Fe. The treatment time is 24 h in each case

In general nanomaterials retain the crystal structure of the bulk, but certain processing conditions can lead to a change of materials chemistry. For example, it has been shown that high-energy ball milling of Ti-Ru-Fe, an electrocatalytic material, incorporates oxygen into the compound. The added oxygen significantly improves the mechanical stability of the material without affecting the electrocatalytic activity. To understand how oxygen could improve the structural integrity, it is important to first determine, how the crystal structure, phase composition, and nanocrystalline material are affected by the presence of oxygen. A combined Rietveld analysis of X-ray and neutron diffraction data reveals that during high-energy ball milling the oxygen atoms readily oxidized Ti to form various types of titanium oxides depending on the oxygen content. This depletes Ti on the (1a) site of the simple cubic structure of Ti_2RuFe, which is preferentially filled with Fe. At high oxygen concentrations, the alloy is actually a multiphase material containing

$Ti_{2-x}Ru_{1+y}Fe_{1+z}$, Ti oxides, Ru and Fe. Figure 1.5 shows the X-ray and neutron diffraction data obtained for a compound of Ti : Ru : Fe : O (2 : 1 : 1 : 0.5) milled for 40 h. Multiphase Rietveld analysis reveals that the final product contains 85 wt% of Ti_2RuFe, 6 wt% of TiO, 3 wt% of Ru, and 6 wt% of Fe. The average grain sizes are approximately 8 nm for all phases. Having knowledge of the chemistry, the authors argue that the formation of the TiO phase increased the resistance to stress-corrosion cracking, which leads to improved mechanical stability.

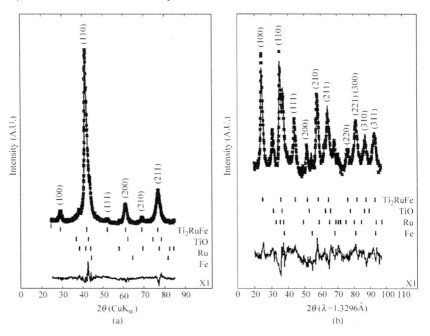

Figure 1.5 X-ray (a) and neutron (b) diffraction patterns of a powder mixture of Ti : Ru : Fe : O (2 : 1 : 1 : 0.5) milled for 40 h. The peaks are indexed for the Ti_2RuFe (cP2-CsCl) phase. The experimental data are indicated by the symbols. The solid lines are fits from Rietveld analysis. The bottom plot is the difference between the experimental data and the fit. The tickmarks are positions of diffraction peaks from different phases

In this study, X-ray and neutron diffraction were used to selectively enhance the chemical contrast (difference in scattering amplitudes) between various elements. For scattering by X-ray, the chemical contrast between Ru and Fe, Ti is quite good due to the large differences in atomic number, while that between Fe and Ti is poor. The use of neutron diffraction solves this problem, as the neutron scattering length of Fe is significantly different from that of Ti. In addition, neutron scattering is more sensitive to the presence of oxygen atoms than X-ray scattering. The success of this study demonstrates the power of the combined use of X-ray and neutron diffraction techniques in solving the chemistry of complex multiphase materials.

For some materials, the crystal structure undergoes a phase transition as the grain size varies. Oh et al. (1999) reported one such observation for nanometer-sized W clusters, where the X-ray pattern revealed a structural change from amorphous to face-centered cubic (FCC) to body-centered cubic (BCC) as the cluster size increased. Their results indicate that W clusters do not simply approach the bulk BCC structure when the cluster becomes sufficiently large. Instead, at an intermediate size, the cluster is in a metastable FCC phase.

The presence of residual stress has a pronounced impact on the structural integrity and mechanical properties of the material, since tensile residual stresses tend to assist both the initiation and propagation of cracks. Quantitative knowledge of these residual stresses is, therefore, essential in the design of materials for structural applications. Neutron diffraction has proven to be an effective technique for accurate determination of residual stresses in composite materials (Krawitz and Holden, 1990; Kuperman, et al., 1991; Wang, et al., 1994; Sergo, et al., 1995) in that the measurement results are representative of the entire irradiated volume. The principle for residual stress measurements with neutron diffraction has been well documented. Basically, one measures the shift in the peak position of one or more reflections, from which the lattice strain is calculated using

$$\varepsilon = \frac{d - d_0}{d_0} \quad (1.13)$$

where d_0 is the stress-free lattice parameter. From the lattice strains, residual stresses are calculated using Hooke's law. Full-pattern Rietveld analysis further improves the measurement precision (Wang, et al. 1994; Daymond, et al., 1997). Todd, et al. (1997) recently reported a neutron diffraction study of residual stresses in Al_2O_3/SiC nanocomposites. The grain size of the SiC reinforcement particles is 200 nm. The experimentally determined residual stresses for each phase are shown in Fig. 1.6. As a result of the thermal expansion mismatch between the two phases, the SiC nanoparticles are in

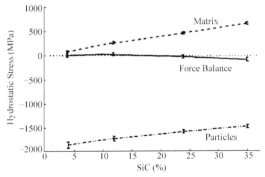

Figure 1.6 Residual stresses in Al_2O_3/SiC nanocomposites, where Al_2O_3 is the matrix and SiC is reinforcing particles. The grain size of SiC particles is 200 nm

compression while the Al_2O_3 matrix is in tension. The condition of force equilibrium requires that the volume-averaged stresses vanish, i.e.,

$$\sigma_{Al_2O_3} f_{Al_2O_3} + \sigma_{SiC} f_{SiC} = 0 \qquad (1.14)$$

where σ is the residual stress and f is the volume fraction of each phase. The calculated values with (1.14) are also plotted in Fig. 1.6 and, as can be seen, they are indeed very close to zero for all samples, demonstrating the validity of the experimental results. These residual stress data were later used to explain the observed fracture mode transition from intergranular in Al_2O_3 ceramics to transgranular in Al_2O_3/SiC nanocomposites.

1.3 Inelastic Neutron Scattering

The inelastic scattering process involves an energy transfer between the neutrons and the sample by either creating or annihilating the fundamental excitations (e.g., phonons) in the crystalline solid. The fundamental equations governing inelastic neutron scattering are conservation of energy and momentum. The energy and momentum transfer during inelastic scattering is given by

$$\begin{aligned} E &= E_f - E_i \\ Q &= k_f - k_i \end{aligned} \qquad (1.15)$$

where E_i and E_f are the energies of incident and scattered neutrons. As discussed in 1.2 Section: elastic scattering ($E=0$) occurs only if $Q=\tau$. For one-phonon scattering, the following conditions must be met,

$$E = E_f - E_i = \hbar\omega(q) \qquad (1.16)$$
$$Q = k_f - k_i = \tau + q \qquad (1.17)$$

Here ω and q are the frequency and wave vector, respectively, of the phonon involved in the inelastic scattering process and $\hbar$ is Plank's constant. Equations (1.16) and (1.17) are the basis for experimental determination of phonon dispersion curves, which typically require large single crystals. When large single crystals are unavailable, as is the case for nanomaterials, one could still learn a great deal about the lattice dynamics by determining the vibrational density of states.

One method for measuring the vibrational density of states using inelastic neutron scattering is with a triple-axis spectrometer, which is schematically shown in Fig. 1.7 In this case, an incident neutron energy, E_i, is selected from a "white" neutron beam by using a monochromating crystal to Bragg diffract neutrons with the desired energy in a new direction. The neutron energy is related to its wavelength by $\lambda^2 \approx 81.8/E_{fi}$, where λ is in Å (1 Å = 10^{-10} m) and E is in meV. These neutrons interact with the sample and the energy transfer is determined by using a second (analyzer) crystal to select a final energy, E_f, from the scattered neutron beam. The scattering vector Q is determined from the values of E_i, E_f, and the scattering angle. Thus, by scanning either E_i or E_f the energy transfer as a function of the scattering vector Q can be mapped out.

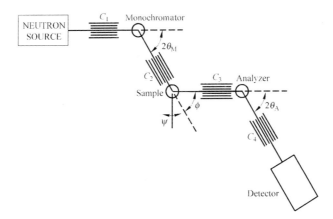

Figure 1.7 A schematic diagram of a triple-axis spectrometer for measurements of inelastic neutron scattering spectra. C_1 to C_4 are Soller collimators

In general, there are both a coherent, σ_c, and an incoherent, σ_i, scattering cross section for each isotope of a chemical element. The total scattering cross section, σ_t, is then given by

$$\sigma_t = \sigma_c + \sigma_i \tag{1.18}$$

The coherent scattering is the scattering process that is coherent amongst nuclei and is most useful for diffraction (see above section) experiments and measurement of the phonon dispersion from large single crystals. The complicated Q dependence of the coherent inelastic scattering from polycrystalline materials makes it difficult to determine the vibrational spectrum from measurements at a small number of Q values. Thus we must rely on incoherent scattering to measure the vibrational density of states. Incoherent scattering arises from the lack of either isotope or spin coherence or both and is isotropic, i.e., Q independent. For a given element, each isotope has its own characteristic scattering length, b, or more generally for isotopes that possess a nuclear spin, b_+ and b_- for the spin-up and spin-down nuclear states, respectively. Since all the isotopes of a given element are chemically equivalent, the isotopes will be randomly distributed among the atomic sites for that element. The relative fractions of each isotope are, in the absence of isotopic enrichment, given by the isotopic abundance ratios for each element and by the weighting factors for each spin state, w_+ and w_-. Thus, the incoherent scattering cross section is given by

$$\sigma_i = 4\pi w_+ w_- (b_+ - b_-)^2 \tag{1.19}$$

Because of the random nature of the isotopic distribution, the observed vibration spectrum is, aside from the intensity factor, independent of Q. Therefore, for material systems where the scattering is dominated by incoherent scattering, the vibrational density of states can be measured at a fixed value of Q by simply varying the energy transfer. For material systems where there is a small but significant coherent component to the inelastic scattering, measurements at several values of Q, away from the Bragg peaks, can often be averaged together to obtain satisfactory results.

As an example, consider the study of the phonons in nanocrystalline Ni_3Fe by Frase et al. (1998). Here, inelastic neutron scattering spectra were measured in order to obtain the vibrational density of states of nanocrystalline FCC Ni_3Fe produced by mechanical alloying. The nanocrystalline material was subjected to a variety of heat treatments to alter its crystallite sizes and internal strains. Inelastic neutron scattering measurements were made on the HB3 triple-axis spectrometer at the high flux isotope reactor at the Oak Ridge National Laboratory. Typical energy loss spectra (raw data) for both

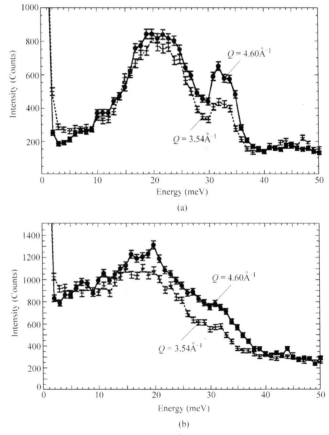

Figure 1.8 Neutron energy loss spectra obtained with $Q = 3.54$ and 4.60 Å^{-1}. Upper panel: powder annealed at 600°C (average crystallite size is 50 nm). Lower panel: as milled nanocrystalline material (average crystallite size is 6 nm)

coarse-grained (50 nm) and nanocrystalline (6 nm) Ni_3Fe are shown in Fig. 1.8. Note that inelastic neutron scattering data were collected at two Q values, 3.54 and 4.60 $Å^{-1}$, because there is a coherent contribution to the inelastic spectra from Ni_3Fe. These Q values were chosen so that when the spectra from the coarse-grained material are combined, the vibrational density of states of bulk Ni_3Fe (Hallman and Brockhouse, 1969) is reproduced. To obtain the vibrational density of states from the observed inelastic spectra, the measurements at the two values of Q were summed and a constant background subtracted for each sample. The conventional multi-phonon expansion (Marshall and Lovesey, 1971; Sears, 1973) was used to calculate and subtract away the multi-phonon contribution at room temperature. Assuming that the remaining inelastic scattering, $I(E)$, is all due to one-phonon scattering, $g(E)$ can be obtained with the relation

$$g(E) = E \cdot [1 - \exp(-E/k_B T)] \cdot I(E) \qquad (1.20)$$

where k_B is the Boltzmann constant and T is the temperature in kelvins. The experimentally determined vibrational density of states for the various crystallite sizes is shown in Fig. 1.9.

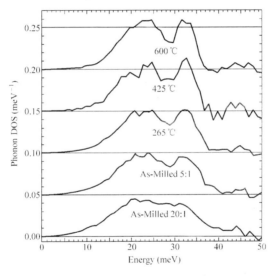

Figure 1.9 Phonon vibrational density of states extracted from the inelastic neutron scattering data for Ni_3Fe powders with crystallite sizes of (top to bottom) 50, 15, 12, 10 and 6 nm

In comparison to coarse-grained Ni_3Fe, nanocrystalline material shows two distinct features in its vibrational density of states. First, the nanocrystalline vibrational density of states is more than twice as large at low energies (below ~15 meV) and the increase is approximately proportional to the fraction of atoms in the grain boundary material. Second, the features of the nanocrystalline vibrational density of states are broadened substantially.

This broadening does not depend in a simple way on the crystallite size, suggesting that it has a different physical origin than the enhancement of the vibrational density of states at low energies.

A second example to illustrate this technique is the study of the vibrational excitations of hydrogen in nanocrystalline palladium by Stuhr, et al. (1995). From previous studies of the solubility and chemical diffusion of hydrogen in single-crystal, coarse-grained, and nanocrystalline palladium, it is known that for low hydrogen concentrations, the hydrogen solubility in nanocrystalline palladium is greater than that found in coarse-grained or single-crystal palladium (Mütschele, 1987a; 1987b; Eastman, et al., 1993). It was unclear, though, where these additional hydrogen atoms are located—whether they are within the grain, at the grain boundaries, or at the surface of the grains.

To answer this question, nanocrystalline palladium was produced by the inert gas condensation method and investigated with inelastic neutron scattering (Stuhr, et al., 1995). The average grain size was determined using X-ray diffraction to be less than 23 nm. The nanocrystalline palladium was then charged with hydrogen as needed. The neutron scattering spectra were collected before and after doping with H on the BT4 triple-axis spectrometer at the National Institute of Standards and Technology. In this case a cold (~80 K) Be filter rather than a second crystal was used to analyze the energy of the scattered neutrons, E_f (a Be filter cooled to 80 K is transparent only to neutrons whose energy is below 5.257 meV). Two samples with 2.4 and 4.8 at% H were studied in detail. The spectra of the H modes obtained at 10 and 300 K are shown in Fig. 1.10. The most striking feature is that at higher doping levels (4.8 at%), the spectra exhibit an intense peak in

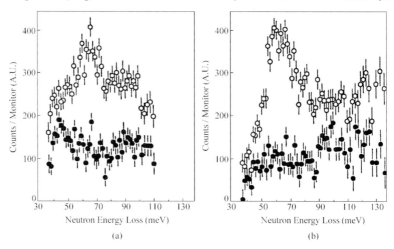

Figure 1.10 Spectra of the H modes in Pd-H nanocrystalline materials at 10 and 300 K. The spectra of the undoped samples have been subtracted from that of the H-doped samples. The filled circles represent the spectra of the sample with low H concentration (2.4%) and the open circles the spectra of the higher H concentration (4.8%)

the energy range of 50—80 meV. A comparison with inelastic neutron scattering data obtained for coarse-grained samples as well as with the optical modes of H in Pd indicates that these modes are attributable to the H in the crystalline regions of the grains. The spectra for the 2.4 at% sample, on the other hand, show hardly any structure between 50 and 80 meV. Beyond 80 meV, the spectra for both samples (especially the 10 K data set) exhibit features that are consistent with the surface modes found in H absorbed on the surface of Pd. Based on these observations, Stuhr, et al. conclude that, up to 2—3 at% H concentration, the H atoms occupy sites in the grain boundaries and/or at the inner surfaces of the sample. If the H concentration is increased to beyond 3 at%, the additional H atoms occupy sites in the crystalline region of the grains.

While the triple-axis spectrometer provides a direct measurement of the vibrational density of states for samples from which the scattering is dominated by incoherent scattering, the most general inelastic neutron scattering technique for determining the vibrational density of states is time-of-flight spectrometry. This method takes advantage of the pulses of neutrons, either from a spallation neutron source or a reactor source where the neutron beam is mechanically switched on and off. Typical pulse rates employed are 20—60 Hz and a large array of neutron detectors is used to cover as much reciprocal space as possible. The energy transfer due to the interaction of the neutron pulse with the sample is determined from the time it takes the neutrons to travel from the sample to the detector. Neutrons that gain energy by annihilating vibrational modes in the sample have a larger velocity and will in turn arrive at the detector array before those that lose energy by creating vibrational waves.

A good example of this technique can be seen in the continuing work by Stuhr, et al. (1998). They studied the vibrational modes of the H-doped nanocrystalline Pd using the small H content as a probe for a separate determination of the vibration of the Pd atoms located at surfaces and/or grain boundaries. The vibrations of a given atom are described by the spectral density $<u_2(E,T)>$, of its temperature-dependent vibrational atomic displacements or by its local vibrational density of states, $g(E)$. The relation between $<u_2(\omega,T)>$ and $g(E)$ is given by (Lovesey, 1984)

$$<u_2(E,T)> = g(E) \cdot \left[\frac{3\hbar^2 \coth(E/k_B T)}{2mE}\right] \qquad (1.21)$$

where m is the mass of the atom. The total $g(E)$ of the samples as a whole is given by the average of the local $g(E)$ of all the individual atoms. The light H atoms follow the vibrational motion of the surrounding much more massive Pd atoms, so that their low-energy spectral density $<u_2(E,T)>$ is almost identical to that of their Pd neighbors. It has already been established in the previous example that at low H concentration the H atoms are located predominately at the surface and grain boundaries. Thus, the H atoms can be used as a probe of the local $g(E)$ of the surface/grain boundary Pd atoms.

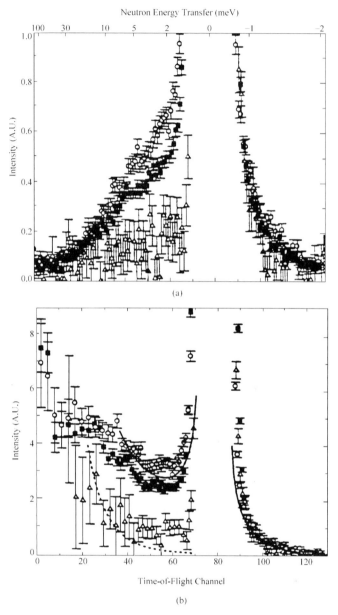

Figure 1.11 Spin-incoherent time-of-flight spectra of the inert-gas-condensed Pd (open circles, 180 K), Pd-black (solid squares, 260 K), and the coarse-grained Pd (open triangles, 270 K) samples. The solid and dashed lines show for a local vibration density of states proportional to E (vibrational frequency) and E^2, respectively. Note that the energy scale is not linear

To separate the spectra of H from that of the Pd, Stuhr et al. took advantage of the fact that Pd has no spin-incoherent cross section [see Eq. (1.19)], whereas scattering by H is dominated by incoherent scattering. Therefore, the incoherent spectra obtained for H doped Pd represent the band modes of H, despite its low concentration. Combining with the arguments presented earlier, it can be seen that for this nanocrystalline system, incoherent scattering will map out the local $g(E)$ of the Pd atoms located at surfaces and/or in grain boundaries.

Experimentally, the spin-incoherent part of the spectra can be separated from the coherent part by using neutron spin-polarization analysis (Squires, 1978). Figure 1.11 shows the spin-incoherent time-of-flight spectra for coarse-grained and nanocrystalline H-doped Pd as a function of temperature with and without the instrumental corrections. The spectra were collected on the D7 time-of-flight spectrometer at the Institute Laue-Langevin in Grenoble, France. From these results, it is clear that there is an enhancement of the low-energy modes in the nanocrystalline material that is much the same as that observed by Frase et al. (1998), and that the local vibrational density of states of the surface/gain boundary Pd is essentially linear in vibrational frequency. By comparing the incoherent and coherent scattering spectra, Stuhr et al. (1998) also note that the Pd atoms from the interior of the grains do not noticeably contribute to the additional low-energy modes.

1.4 Small Angle Scattering

When a highly collimated beam of either X-rays or neutrons is passed through a material, the width of the beam is increased by diffraction and refraction of the beam by the particles that make up the material. These extra intensities are commonly referred as small angle scattering. Theoretically, the intensity of small angle scattering is related to the Fourier transform of the distribution of scattering length density in the material. This feature has particular application to nanocrystalline materials because it is sensitive to the difference in atomic density between the atoms in the (ordered) interior of the grains and those located in the interfacial region. Thus, both the variation in atomic density within the particles and the particle morphology can be studied using this technique.

A good example can be seen in a paper by Sanders et al. (1996). They used small angle neutron scattering (SANS) to investigate the structure of nanocrystalline palladium and copper produced by inert gas condensation. These compacted powders still exhibit traditional powder processing problems such as pores between the grains and impurities on the particle surfaces even though the grains are extremely small. The goal of this particular study was to quantify those aspects of the morphology that are associated with the processing of the material. The small angle neutron scattering data were collected on the 30 m NSF SANS instrument at the National Institute of Standards and Technology. The differential scattering cross sections for the palladium and copper samples are shown in Fig. 1. 12 along with the results from a maximum entropy analysis (for MaxSAS software package, see URL http://www.mrl.uiuc.edu/%7Ejemian/sas.html). In general, small angle scattering profiles can be analyzed using the Porod relation (Porod, 1982) between the characteristics of the particles and the Q dependence of the small angle scattering intensity:

$$\frac{d\Sigma}{d\Omega}(Q) = \frac{2\pi(\Delta\rho)^2}{Q^4} \cdot \frac{S}{V} + \frac{n\sigma_{inc}}{4\pi} \quad (1.22)$$

where $\Delta\rho$ is the difference in the scattering length density between the matrix and the scattering entity, S/V is the total surface area of the scattering particles per unit volume, n is the number of atoms per unit volume, and σ_{inc} is the incoherent scattering cross section. Analysis of the data using maximum entropy methods indicated a high concentration of hydrogen impurities. The pore sizes and number densities scaled with the grain size for the room temperature compacted nanocrystalline Pd and Cu samples. Annealing was found to decrease the hydrogen level and number of pores, but at the same time increased the grain and pore sizes. Warm compaction of the Pd material also dramatically decreased the hydrogen concentration and the number of pores while only slightly increasing the grain size.

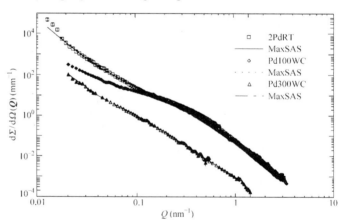

Figure 1. 12 Logarithmic plot of the differential scattering cross section $d\Sigma/d\Omega$ vs. the scattering vector Q for room temperature and warm compacted nanocrystalline Pd. The solid lines are fits from the MaxSAS routine

Another small angle neutron study of interest is the work by Wagner, et al. (1991) on the magnetic microstructure of nanocrystalline Fe in relation to the compositional microstructure of grains and grain boundaries. Here, the fact that neutrons have a magnetic moment is used to study the magnetic aspects of the microstructure. Neutrons combine interactions with compositional as well as magnetic structures of the material and therefore represent a unique probe to simultaneously monitor compositional and magnetic correlation.

Nanometer-sized crystalline α-Fe samples were prepared by compacting 8 nm Fe crystals produced by inert gas condensation into a solid. In this case the small angle measurements were carried out using the D11 instrument at Institute Laue-Langevin in Grenoble, France (Ibel, 1973). Since the neutrons probe both the nuclear and magnetic scattering length density, there are two contributions to the differential scattering cross section:

$$\left(\frac{d\sigma}{d\Omega}\right)_{nucl}(Q) = (1-Q)^2 (\Delta b_{nucl})^2 \cdot i_{nucl}(Q,R) \qquad (1.23)$$

and

$$\left(\frac{d\sigma}{d\Omega}\right)_{magn}(Q) = (\Delta b_{magn})^2 \cdot i_{magn}(Q,R) \qquad (1.24)$$

where b_{nucl} and b_{magn} are the nuclear and magnetic neutron scattering lengths, respectively, and $i_{nucl}(Q,R)$ and $i_{magn}(Q,R)$ represent the square of the Fourier transform of the nuclear and magnetic correlation functions. The measured SANS intensity is a linear superposition of the two components. Figure 1.13 shows the two-dimensional scattering patterns for nanocrystalline Fe in different external magnetic fields. Without an applied field, the scattering would be isotropic because on average all orientations of the grains and their magnetization vectors relative to the scattering vector are equally probable. Since the magnetic cross section depends on the angle between Q and the magnetization vector, M, the magnetic component of the scattering will become anisotropic when an external field is applied and M is no longer random. If α is the angle between the two vectors, projected onto the plane normal to the incident neutron beam, (1.24) can be transformed into the so-called radial cross section:

$$\left(\frac{d\sigma}{d\Omega}\right)_{magn}(Q,\alpha) = (\Delta b_{magn})^2 \cdot i_{magn}(Q,R) \cdot \sin^2\alpha \qquad (1.25)$$

For applied magnetic fields of 5 kOe and greater, the intensity distribution, shown in Fig. 1.13, become elongated in the vertical direction, as expected from the increasing alignment of the sample magnetization, M, toward the direction of the applied field, H. From an analysis of the Q-dependence of the intensity data, Wagner et al. (1991) found that the magnetic fluctuations in the nanocrystalline Fe were not confined to single grains, but extended across the interfaces over several hundreds of grains. These fluctuations are suppressed by external magnetic magnetization, but recover reversibly when the external field is removed.

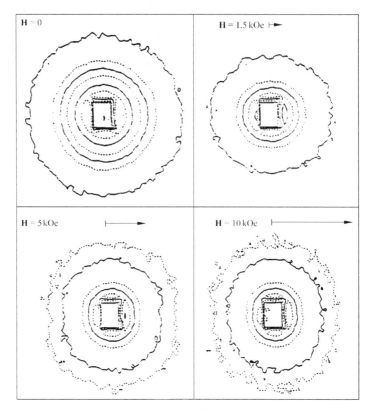

Figure 1.13 Small angle neutron scattering intensity from nanocrystalline Fe for different external magnetic fields applied in the horizontal direction to the sample

1.5 Concluding Remarks

This chapter provides a general background of using X-ray and neutron scattering for characterization of nanocrystalline materials. Selected examples were discussed to illustrate how these techniques were used to understand the chemistry, microstructure, lattice dynamics, and magnetism of nanocrystalline materials. From these examples, it is clear that X-ray and neutron scattering have played and will continue to play an important role in the development of nanoscience and technology. In many instances, the scattering techniques provide essential information that simply could not be obtained by other means. Readers are reminded, however, while the overall strategy of applying these techniques has been presented, an in-depth understanding requires a much more detailed study of the references indicated.

Looking forward, processing remains a key issue in nanomaterials research. In the future, with improved instrumentation, it should be possible to perform real-time in situ measurements that will inevitably lead to a better understanding of the influence of various processing conditions. Also, the application of X-ray and neutron scattering for study of magnetic behaviors is still largely unexplored. There have not been adequate instruments, for instance, for the study of the super-paramagnetic fluctuations in magnetic nano-clusters (Casalta, et al., 1998), a key issue in the development of high-performance data storage media. There is no doubt that challenging materials issues like these will continue to drive advances in X-ray and neutron scattering instrumentation.

References

Alexander, K. B., P. F. Becher, X.-L. Wang, and C. H. Hsueh. J. Amer. Ceram. Soc., 78, 291—296 (1995)

Bacon, G. E., Neutron Diffraction. Oxford University Press, London, United Kingdom (1975)

Bourke, M. A. M., D. W. Brown, B. Clausen, H. Choo, P. Rangaswamy, and R. Vaidyanathan. Neutron News. 10, 24—30 (1999)

Casalta, H., P. Schleger, C. Bellouard, M. Hennion, I. Mirebeau and B. Fargo. Physica B 241—243, 576—578 (1998)

Clausen, B., T. Lorentzen, and T. Loffers. Acta Mater., 46, 3087—3098 (1998)

Convert, P., T. Hansen, A. Oed, and J. Torregrossa. Physica. B 241—243, 195—197 (1997)

Daymond, M. R., M. A. M. Bourke, R. von Dreele, B. Clausen, and T. Lorentzen. J. Appl. Phys., 82, 1554—1562 (1997)

de Keiser, Th. H., J. I. Lanford, E. J. Mittemeijer, and A. B. P. Vogels. J. Appl. Cryst., 15, 308—314 (1982)

Eastman, J. A., and M. R. Fitzsimmons. J. Appl. Phys., 77, 522 (1995)

Eastman, J. A., L. J. Thompson, and B. J. Kestel. Phys. Rev. B. 48, 84 (1993)

Frase, H., B. Fultz, and J. L. Robertson. Phys. Rev. B, 57, 898 (1998)

Futlz, B., C. C. Ahn, E. E. Alp, W. Sturhahn, and T. S. Toellner. Phys.

Acknowledgment This work was supported by Office of Science, U. S. Department of Energy. Oak Ridge National Laboratory is managed by Lockheed Martin Energy Research Corporation for the U. S. Department of Energy under contract number DE-AC05-96OR22464.

Rev. Lett.. **79**, 937—940 (1997)
Halder, N. C., and C. N. J. Wagner. Adv. X-ray Anal.. **9**, 91—102 (1966)
Hallman, E. D., and B. Brockhouse. Canadian J. Phys.. **47**, 1117 (1969)
Hayashi, M., E. Gerkema, A. M. van der Kraan, and I. Tamura. Phys. Rev.. B. **42**, 98771 (1990)
Hill, R. J. and C. J. Howard. J. Appl. Cryst.. **20**, 467—474 (1987)
Hull, S., D. A. Keen, W. Hayes, and N. Gardner. J. of Physics-Condesed Matt.. **10**, 10941—10954 (1998)
Ibel K.. J. Appl. Cryst.. **9**, 296 (1973)
Jorgensen, J. D., J. Faber, J. M. Carpenter, R. K. Crawford, J. Hanman, R. L. Hitterman, R. Kleb, F. J. Rottella, and T. G. Worlton. J. Appl. Cryst.. **22**, 321 (1989)
Kittel, C. Introduction to Solid State Physics. 6th Edn. John Wiley and Sons, New York. (1986)
Klug, H. P. and L. E. Alexander. in X-ray diffraction procedures for polycrystalline and amorphous materials. 2nd Edn.. John Wiley & Sons Inc., New York p. 291 (1974a)
Klug, H. P. and L. E. Alexander. in X-ray diffraction procedures for polycrystalline and amorphous materials. 2nd Edn.. John Wiley & Sons Inc., New York 687—689 (1974b)
Klug, H. P. and L. E. Alexander. in X-ray diffraction procedures for polycrystalline and amorphous materials. 2nd Edn.. John Wiley & Sons Inc., New York 618—708 (1974c)
Krawitz, A. D. and T. M. Holden. MRS Bull.. November, 57—64 (1990)
Kuperman, D. S., S. Majumdar, and J. P. Singh. Neutron News. 2, 15—18 (1991)
Lanford, J. L.. in Accuracy in Powder Diffraction II, National Institute of Science and Technology Special Edition 846 National Institute of Science and Technology, Gaithersburg. Maryland. 110—126 (1992)
Le Bail, A.. in Accuracy in Powder Diffraction II, National Institute of Science and Technology Special Edition 846 National Institute of Science and Technology, Gaithersburg, Maryland. 142—153(1992)
Loffler, J. F., J. P. Meier, B. Doudin, J.-P. Ansermet, and W. Wagner. Phys. Rev.. B **57**, 2915—2924 (1998)
Long, B. L., C. C. Ahn, and B. Fultz. J. Mater. Res.. 10, 2408 (1995)
Lovesey, S. Theory of Neutron Scattering from Condensed Matter. Vol. 1. Clarendon Press, Oxford (1984)
Marshall, W. and S. W. Lovesey. Theory of Thermal NeutronScattering. Oxford University Press, London, United Kingdom (1971)
Misture, S. T., S. M. Pilgrim, J. C. Hicks, C. T. Blue, E. A. Payzant, and C. R. Hubbard. Appl. Phys. Lett.. 72, 1042—1044 (1997)
Mlayah, A., A. M. Brugman, R. Carles, J. B. Renucci, M. Ya. Valakh,

and A. V. Pogorelov. Solid State Comm.. 199, 50 (1997)
Mütschele, T. and R. Kirchheim. Scripta Metall.. 21, 1101 (1987a)
Mütschele, T. and R. Kirchheim. Scripta Metall.. 21, 135 (1987b)
Nandi, R. K., and S. P. Sen Gupta. J. Appl. Cryst.. 11, 6—9 (1978)
Nandi, R. K., H. K. Kuo, W. Scholsberg, G. Wissler, J. B. Cohen, and B. Christ, Jr.. J. Appl. Cryst.. 17, 22—26 (1984)
Oh, S. J., S. H. Huh, H. K. Kim, J. W. Park, and G. H. Lee. J. Chem. Phys.. 111, 7402—7404 (1999)
Otto, J. W.. J. Appl. Cryst.. 30, 1008—1015 (1997)
Pang, J. W. L., T. M. Holden, and T. E. Mason. Acta Mater.. 46, 1503—1518 (1998)
Porod G.. in Small Angle X-ray Scattering. Edited by O. Glatterand and O. Kratky Academic Press, London. 17 (1982)
Rietveld, H. M.. J. Appl. Cryst.. 2, 65—71 (1969)
Rodriguez-Carvajal, J., M. T. Fernandez-Diaz, and J. L. Martinez. J. Phys.: Condens. Matter. 3, 3215—3234 (1991)
Sanders, P. G., J. A. Eastman, and J. R. Weertman. Acta Mat.. 46, 4195—4202 (1998)
Sanders, P. G., J. R. Weertman, and J. G. Barker. J. Mater Res.. 11, 3110 (1996)
Schiotz, J., F. D. Di Tolla, K. W. Jacobsen. Nature. 391, 561—563 (1998)
Sears, V. F.. Phys. Rev. A 7 340 (1973)
Sergo, V., X.-L. Wang, D. R. Clarke, and P. F. Becher. J. Am. Ceram. Soc.. 78, 2213—2214 (1995)
Siu, C. G., M. J. Stokes, and Y. L. Liu. Phys. Rev.. B 59, 3173—3179 (1999)
Squires, G. L.. Introduction to Thermal Neutron Scattering. Cambridge University Press, Cambridge, United Kindom. 171—194 (1978)
Stassis, C.. Methods of Experimental Physics. Vol. 23A, Neutron Scattering. Academic Press, Amsterdam, The Netherlands, 369—439 (1986)
Stuhr, U., H. Wipf, K. H. Anderson, and H. Hahn. Phys. Rev. Lett.. 81, 1449 (1998)
Stuhr, U., H. Wipf, T. J. Udovic, J. Weißmüller, and H. Gleiter. J. Phys.: Condens. Matter. 7, 219 (1995)
Todd, R. I., M. A. M. Bourke, C. E. Borsa, R. J. Brook. Acta Meter.. 45, 1791—1800 (1997)
Wang, X.-L., C. R. Hubbard, K. B. Alexander, P. F. Becher, J. A. Fernandez-Baca, and S. Spooner. J. Am. Ceram. Soc.. 77, 1569—1575 (1994)
Wertheim, G. K., M. A. Butler, K. W. West, and D. N. E. Buchanan. Rev. Sci. Inst.. 45, 1369—1371 (1974)
Withers, P. J. and T. M. Holden. MRS Bull.. December. 17—23 (1999)

Appendix
List of Neutron Scattering Facilities

For contact information of these facilities, please visit

http://www.neutron.anl.gov/Neutronf.htm

Australian Reactors
Chalk River Neutron Scattering (Chalk River, Canada)
IRI Reactor (Delft, Netherlands)
DR3 Reactor (Risoe, Denmark)
High Flux Beam Reactor——HFBR (Brookhaven, New York, USA)
High Flux Isotope Reactor——HFIR (Oak Ridge, Tennessee, USA)
Berlin Neutron Scattering Center——HMI, BENSC Hahn-Meitner-Institut, (Germany)
IBR Fast Pulsed Reactors (Dubna, Russia)
GKSS/FRG-1 Reactor + HASYLAB (Hamburg, Germany)
Institute of Laue-Langevin (Grenoble, France)
NFL, The Studsvik Neutron Research Laboratory (Studsvik, Sweden)
Intense Pulsed Neutron Source——IPNS (Argonne, Illinois, USA)
ISIS at Rutherford-Appleton Laboratory (United Kingdom)
KENS Ibaraki (KEK, Japan)
KURRI (Kyoto, Japan)
KFA-Juelich-1 (Juelich, Germany)
Laboratoire Leon de Brillouin (Saclay, France)
Los Alamos Neutron Scattering Center——LANSCE (Los Alamos, New Mexico, USA)
NIST Research Reactor (Gaithersburg, Maryland, USA)
SINQ Spallation Source (Villigen, Switzerland)
Spallation Neutron Source (Oak Ridge, Tennessee, USA)
University of Illinois Triga Reactor (Urbana-Champaign, Illinois, USA)
University of Missouri Reactor (Columbus, Missouri, USA)

2 Transmission Electron Microscopy and Spectroscopy

Zhong Lin Wang, Ze Zhang and Yi Liu

Nanomaterials are dominated by particles with sizes in nanometer range. Imaging atoms and quantitative analysis of surface chemistry are vitally important for nanomaterials research. Transmission electron microscopy (TEM) is a powerful and unique technique for structure characterization, which not only can provide real space imaging at a resolution of 1—2 Å (1 Å = 1 × 10^{-10} m), but also an analytical tool for quantitative structure and chemical analysis. By forming a nanometer size electron probe, TEM is unique in identifying and quantifying the chemical and electronic structures of individual nanocrystals. In situ TEM is demonstrated for characterizing and measuring the properties of individual nanostructures, such as the bending modulus of a carbon nanotube, from which the structure-property relationship can be registered with the structure of the nanoparticle. This is a unique application of TEM and it is a new direction in nanomeasurements (Wang, 1999). The objective of this chapter is to introduce the fundamentals of TEM and its applications in structural determination of nanomaterials.

2.1 Major Components of a Transmission Electron Microscope

A modern TEM is schematically shown in Fig. 2.1. It is composed of an illumination system, a specimen stage, an objective lens system, the magnification system, the data recording system(s), and the chemical analysis system. The electron gun typically uses an LaB6 thermionic emission source or a field emission source. The LaB6 gun gives a high illumination current, but the current density and the beam coherence are not as high as those of a field emission source. A field emission source is unique for performing high coherence lattice imaging, electron holography and high spatial resolution microanalysis. The illumination system also includes the condenser lenses that are vitally important for forming a fine electron probe. The specimen stage is a key for carrying out structure analysis, because it can be used to perform in situ observations of phenomena induced by annealing, electric field, magnetic field or mechanical stress, giving the possibility of characterizing the physical properties of individual nanostructures. An elegantly designed specimen stage can change a TEM into a nanomeasurement machine (see Section 2.7 for details). The objective lens determines the limit of image resolution. The magnification system consists of intermediate lenses and projection lenses, and it gives a magnification up to 1.5 million. The data recording system tends to be digital with the use of a charge coupled device (CCD), allowing quantitative data processing and quantification. Finally, the chemical analysis system is the energy dispersive X-ray spectroscopy (EDS) and electron energy-loss spectroscopy (EELS). Both can be used complementarity to quantify the chemical composition of the specimen. EELS can also provide information about the electronic structure of the specimen.

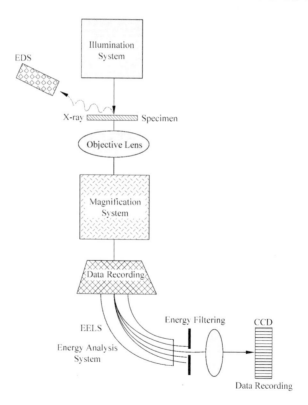

Figure 2.1 Schematic structure of a transmission electron microscope

2.2 Atomic Resolution Lattice Imaging of Crystalline Specimens

2.2.1 Phase Contrast

Images in TEM are usually dominated by three types of contrast. First, diffraction contrast (Hirsch, et al., 1977), which is produced due to a local distortion in the orientation of the crystal (by dislocations, for example), so that the diffracted intensity of the incident electron beam is perturbed, leads to contrast observed in the bright-field image. The nanocrystals oriented with their low-index zone axis parallel or nearly parallel to the incident beam direction usually exhibit dark contrast in the bright-field image that is formed by selecting the central transmitted beam. Since the diffraction intensities of the Bragg reflected beams are strongly related to the crystal orientations, this type of image is ideally suited for imaging defects and dislocations. For nanocrystals, most of the grains are defect-free in volume, while a high density of defects is localized at the surface or grain boundary. Diffraction contrast can be useful for capturing strain distribution in nanocrystals whose sizes are larger than 15 nm. Since the resolution of diffraction contrast for smaller size nanocrystals is about 1—2 nm, its application is limited.

Secondly, phase contrast is produced by the phase modulation of the incident electron wave which transmits through a crystal potential. This type of contrast is sensitive to the atom distribution in the specimen and is the basis of high-resolution TEM. To illustrate the physics of phase contrast, we consider the modulation of a crystal potential to the electron wavelength. From the de Broglie relation, the wavelength λ of an electron is related to its momentum, p, by

$$\lambda = \frac{h}{p} \quad (2.1)$$

When the electron goes through a crystal potential field, its kinetic energy is perturbed by the variation of the potential field, resulting in a phase shift with respect to the electron wave that travels in a space free of potential field. For a specimen of thickness d, the phase shift is

$$\phi = \sigma V_p(b) = \sigma \int_0^d dz V(r) \quad (2.2)$$

where $\sigma = \pi/\lambda U_0$, $b = (x, y)$, U_0 is the acceleration voltage, and $V_p(b)$ is the thickness-projected potential of the crystal. Therefore, from the phase point of view, the electron wave is modulated by a phase factor

$$Q(b) = \exp[i\sigma V_p(b)] \quad (2.3)$$

This is known to be the phase object approximation (POA), in which the crystal acts as a phase grating filter. If the incident beam travels along a low-index zone axis, the variation of $V_p(b)$ across atom rows is a sharp function because an atom can be approximated by a narrow potential well and its width is about 0.2—0.3 Å (1 Å = 1 × 10^{-10} m). This sharp phase variation is the basis of phase contrast, the fundamental of atomic resolution imaging in TEM.

Finally, mass thickness or atomic number produced contrast. Atoms with different atomic numbers exhibit different powers of scattering. If the image is formed by collecting the electrons scattered to high angles, the image contrast would be sensitive to the average atomic number along the beam direction.

2.2.2 Abbe's Imaging Theory

Image formation in TEM is analogous to that in an optical microscope (Cowley, 1995). For easy illustration, a TEM is simplified into a single lens microscope, as given in Fig. 2.2, in which only a single objective lens is considered for imaging and the intermediate lenses and projection lenses are omitted. This is because the resolution of the TEM is mainly determined by the objective lens. The entrance surface of a thin foil specimen is illuminated by a parallel or nearly parallel electron beam. The electron beam is diffracted by the lattices of the crystal, forming the diffracted beams which are propagating along different directions. The electron-specimen interaction results in phase and amplitude changes in the electron wave that are determined by quantum mechanical diffraction theory. For a thin specimen and high-energy electrons, the transmitted wave function $\Psi(x, y)$ at the exit face of the specimen can be assumed to be composed of a forward-scattered wave.

The non-near-axis propagation through the objective lens is the main source of non-linear information transfer in TEM. The diffracted beams will be focused in the back-focal plane, where an objective aperture could be applied. An ideal thin lens brings the parallel transmitted waves to a focus on the axis in the back focal plane. Waves leaving the specimen in the same direction (or angle θ with the optic axis) are brought together at a point on the back focal plane, forming a diffraction pattern. The electrons scattered to angle θ experience a phase shift introduced by the chromatic and spherical aberrations of the lens, and this phase shift is a function of the scattering angle, thus, the diffraction amplitude at the back-focal plane is modified by

$$\psi'(u) = \psi(u)\exp[i\chi(u)] \quad (2.4)$$

where $\psi(u)$ is the Fourier transform of the wave $\Psi(r)$ at the exit face of the specimen, u is the reciprocal space vector that is related to the scattering angle by $u = 2\sin\theta/\lambda$, and $\chi(u)$ is determined by the spherical aberration

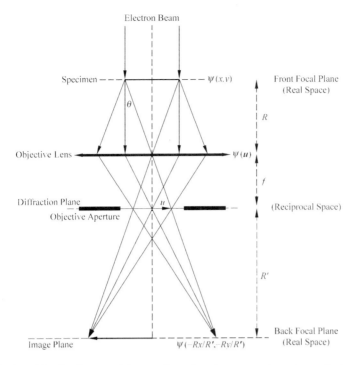

Figure 2.2 Abbe's theory of image formation in an one-lens transmission electron microscope. This theory is for a general optical system in TEM

coefficient C_s of the objective lens and the lens defocus Δf (Baï bich, et al., 1988)

$$\chi(u) = \frac{\pi}{2} C_s \lambda^3 u^4 - \pi \Delta f \lambda u^2 \qquad (2.5)$$

where λ is the electron wavelength. The chromatic aberration arises from the change in focal length as a result of energy fluctuation of the incident electron beam. The spherical aberration is due to dependence of the focal length on the electron scattering angle θ. From simple optics, a change in focal length changes the phase of the electron beam. Thus, the aberration and defocus of the lens are to modulate the phases of the Bragg beams scattered at different angles.

The electron image is the interference result of the beams scattered at different angles, and this interference pattern is affected by the phase modulation introduced by the aberration of the objective lens. The image is calculated according to

$$I(x,y) = |\Psi(r) \otimes t_{obj}(x,y)|^2 \qquad (2.6)$$

where $\otimes$ indicates a convolution calculation of (x, y), and $t_{obj}(x,y)$ is the inverse Fourier transform of the phase function $\exp[i\chi(u)]$. The convolution of the lens transfer function introduces the non-linear information transfer characteristics of the objective lens, leading to complexity in image interpretation.

2.2.3 Image Interpretation of Very Thin Samples

In high-resolution TEM (HRTEM) images, one usually wonders if the atoms are dark or bright. To answer this question one must examine the imaging conditions. For clarity in following discussion, the weak scattering object approximation (WPOA) is made. If the specimen is so thin that the projected potential satisfies $|\sigma V_p(b)| \ll 1$, the phase grating function is approximated by

$$\Psi(b) \approx 1 + i\sigma V_p(b) \qquad (2.7)$$

From (2.3) and ignoring the σ^2 term, the image intensity is calculated by

$$I(x,y) \approx 1 - 2\sigma V_p(b) \otimes t_s(b) \qquad (2.8)$$

where $t_s(b) = \text{Im}[t_{obj}(b)]$. The second term in (2.8) is the interference result of the central transmitted beam with the Bragg reflected beams. Any phase modulation introduced by the lens would result in contrast variation in the observed image. Under the Scherzer defocus, $t_s(b)$ is approximated to be a negative Gaussian-like function with a small oscillating tail, thus, the image contrast, under the WPOA, is directly related to the two-dimensional thickness-projected potential of the crystal, and the image reflects the projected structure of the crystal. This is the basis of structure analysis using HRTEM. On the other hand, the contrast of the atom rows is determined by the sign and real space distribution of $t_s(b)$. Therefore, the image contrast in HRTEM is critically affected by the defocus value. A proper change in defocus could lead to contrast reversal, which makes it difficult to match the observed image directly with the projection of atom rows in the crystal. This is one of the reasons that image simulation is a key step in quantitative analysis of HRTEM images.

The WPOA illustrated above is valid only for a weak scattering object, and, unfortunately fails in most practical cases. Quantitative analysis of high-resolution TEM image requires numerical calculations using full dynamic theory for electron scattering. After many years of research, software for this type of simulation has been well developed and is fairly user friendly.

2.2.4 Image Simulation

Image simulation needs to include two important processes, the dynamic multiple scattering of the electron in the crystal and the information transfer of the objective lens system. The dynamic diffraction process is to solve the Schrodinger equation under given boundary conditions. There are several approaches for performing dynamic calculations (Wang, 1995). For a finite size crystal containing defects and surfaces, the multislice theory is most adequate for numerical calculations. The multislice many-beam dynamic diffraction theory was first developed based on the physical optics approach (Cowley, et al., 1957). The crystal is cut into many slices of equal thickness Δz in the direction of the incident beam (Fig. 2.3). When the slice thickness tends to be very small the scattering of each slice can be approximated as a phase object. The transmission of the electron wave through each slice can be considered separately if the backscattering effect is negligible, which means that the calculation can be made slice by slice. The defect and 3-D crystal shape can be easily accounted for in this approach.

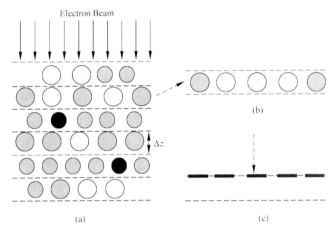

Figure 2.3 (a) A schematic diagram showing the physical approach of the multislice theory for image and diffraction pattern calculations in TEM. (b) Transmission of electron wave through a thin crystal slice. (c) An approximate treatment of the wave transmission through a thin slice

The transmission of the electron wave through a slice can be considered as a two step process—the phase modulation of the wave by the projected atomic potential within the slice and the propagation of the modulated wave in "vacuum" for a distance Δz along the beam direction before striking the next crystal slice. The wave function before and after transmitting a crystal slice is correlated by

$$\Psi(b, z + \Delta z) = [\Psi(b,z) Q(b, z + \Delta z)] \otimes P(b, \Delta z) \quad (2.9)$$

where the phase grating function of the slice is

$$Q(b, z + \Delta z) = \exp\left[i\sigma \int_{z}^{z+\Delta z} dz V(b, z)\right] \quad (2.10a)$$

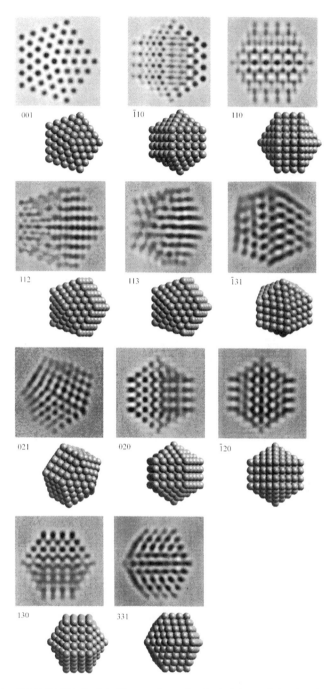

Figure 2.4 Theoretically simulated images for a decahedral Au particle at various orientations and at focuses of (A) $\Delta f = 42$ nm and (B) $\Delta f = 70$ nm. The Fourier transform of the image is also displayed (Courtesy of Drs. Ascencio and M. José-Yacaman)

and the propagation function is

$$P(b, \Delta z) = \frac{\exp(\pi i K |b|^2 / \Delta z)}{i\lambda \Delta z} \quad (2.10b)$$

The most important characteristic of this equation is that no assumption was made regarding the arrangement of atoms in the slices, so that the theory can be applied to calculation of the electron scattering in crystals containing defects and dislocations. This is the most powerful approach for nanocrystals.

The next step is the information transfer through the objective lens system. By taking a Fourier transform of the exit wave function, the diffraction amplitude is multiplied by the lens transfer function $\exp(i\chi)E(u)$, where the defocus is a variable. The defocus values of the objective lens and the specimen thickness are two important parameters which can be adjusted to match the calculated images with the observed ones. Figure 2.4 shows systematic simulations of a decahedral Au particle in different orientations. The particle shape can only be easily identified if the image is recorded along the five-fold axis (Ascencio, et al., 1998). The group A and group B images were calculated for two different defocuses, exhibiting contrast reversal from dark atoms to bright atoms. In practice, with consideration of the effects from the carbon substrate, it is difficult to identify the particle shape if the particle orientation is off the five-fold axis.

2.3 Faceted Shapes of Nanocrystals

2.3.1 Polyhedral Shapes of Nanoparticles

Nanocrystals exhibiting distinctly different properties from the bulk are mainly due to their large portions of surface atoms and the size, and possibly shape, effect as well. A particle constituting a finite number of atoms can have a specific geometrical shape. Surface energies associated with different crystallographic planes are usually different, and a general sequence may hold for face-centered cubic crystals, $\gamma\{111\} < \gamma\{100\} < \gamma\{110\}$. For a spherical single-crystalline particle, its surface must contain high index crystallography planes, which possibly result in a lower surface energy. Facets tend to form on the particle surface to increase the portion of the low index planes. Therefore, for particles smaller than 10—20 nm, the surface is a polyhedron. Figure 2.5(a) shows a group of cubo-octahedral shapes as a function of the ratio, R, of the growth rate in $<100>$ to that of the $<111>$. The fastest growth direction in a cube is $<111>$, the longest direction in the octahedron is the $<100>$ diagonal, and the longest direction in the cubo-octahedron ($R = 0.87$) is the $<110>$ direction. The particles with

$0.87 < R < 1.73$ have the {100} and {111} facets, which are named the truncated octahedral (TO). The other group of particles has a fixed (111) base with exposed {111} and {100} facets (Fig. 2.5(b)). An increase in the area ratio of {111} to {100} results in the evolution of particle shapes from a triangle-based pyramid to a tetrahedron.

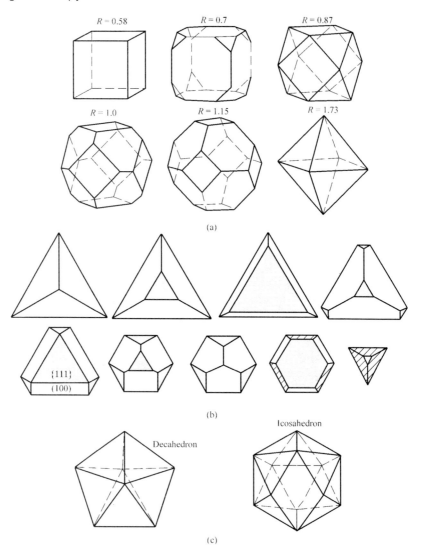

Figure 2.5 (a) Geometrical shapes of cubo-octahedral nanocrystals as a function of the ratio, R, of the growth rate along <100> to that of <111>. (b) Evolution in shapes of a series of (111) based nanoparticles as the ratio of {111} to {100} increases. The beginning particle is bounded by three {100} facets and a (111) base, while the final one is a {111} bounded tetrahedron. (c) Geometrical shapes of multiply twinned decahedral and icosahedral particles

If the particle is oriented along a low index zone axis, the distribution of atoms on the surface can be imaged in profile, and the surface structure is directly seen with the full resolution power of a TEM (Smith, et al., 1981; Yao, et al., 1994). This is a powerful technique for direct imaging the projected shapes of nanoparticles particularly when the particle size is small. With consideration of the symmetry in particle shapes, HRTEM can be used to determine the 3-D shape of small particles, although the image is a 2-D projection of a 3-D object.

Figure 2.6(a) gives a profile HRTEM image of a cubic Pt nanocrystal oriented along [001] (Wang, et al., 1996). The particle is bounded by {100} facets and there is no defect in the bulk of the particle. The distance between the adjacent lattice fringes is the interplanar distance of Pt {200}, which is 0.196 nm, and the bulk structure is face-centered cubic. The surface of the particle may have some steps and ledges particularly at the regions near the corners of the cube. To precisely image the defects and facets on the cubic particles, a particle oriented along [110] is given as shown in Fig. 2.6(b). This is the optimum orientation for imaging cubic structured materials. The {110} facets are rather rough, and the {111} facets are present. These higher energy structural features are present because the particles were prepared at room temperature.

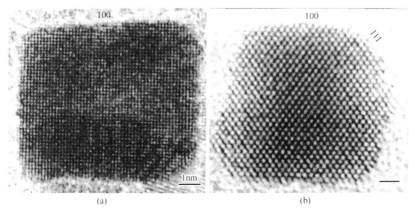

Figure 2.6 HRTEM images of cubic Pt nanocrystals oriented along (a) [001] and (b) [110], showing surface steps/ledges and the thermodynamically nonequilibrium shapes

An octahedron has eight {111} facets, and four {111} facets are edge-on if viewed along [110]. If the particle is a truncated octahedron, six {100} facets are created by cutting the corners of the octahedron, two of which are edge-on when viewed along [110]. Figures 2.7(a) and (b) show the HRTEM images of [110] oriented truncated octahedron and octahedral Pt particles, respectively. A variation in the area ratio of {100} to {111} results in a slight difference in particle shapes.

A tetrahedral particle is defined by four {111} faces and it usually gives a triangular shape in HRTEM. Figure 2.8 gives two HRTEM images of truncated

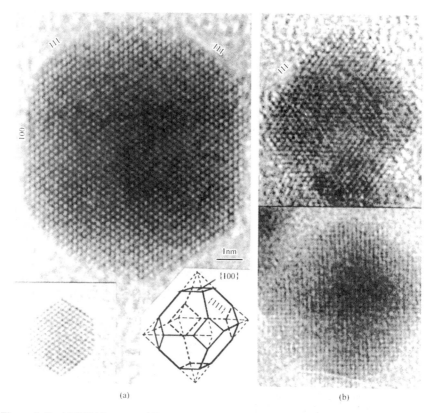

(a) (b)

Figure 2.7 HRTEM images of Pt nanocrystals (a) with a truncated octahedral shape and oriented along [110], and (b) with a octahedral shape and oriented along [110] and [001]. The inset in (a) is a model of the particle shape

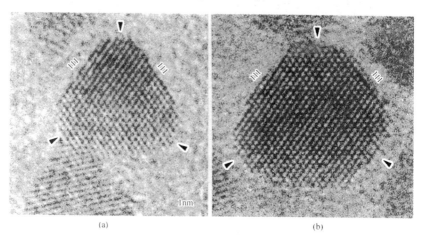

(a) (b)

Figure 2.8 HRTEM images of truncated tetrahedral Pt nanocrystals oriented along [110]. The surface steps and ledges at the truncated corners are clearly resolved

tetrahedral particles oriented along [110]. Two {111} facets and one {001} facet (at the top of the image as a result of truncation) are imaged edge-on. There are some atom-high surface steps on the {111} surfaces and the corners. These atomic-scale structures are likely important for enhancing the catalysis activities of the nanocrystals.

2.3.2 Twinning Structure and Stacking Faults

Twinning is one of the most popular planar defects in nanocrystals, and it is frequently observed for fcc structured metallic nanocrystals. Twinning is the result of two subgrains sharing a common crystallographic plane, thus, the structure of one subgrain is the mirror reflection of the other by the twin plane. Figure 2.9(a) gives a HRTEM image of a Pt particle oriented along [110], which is composed of two grains connected at a twin relation. The twin plan is indicated by a dotted line. The fcc structured metallic nanocrystals usually have {111} twins, which is the fundamental defect mechanism for the growth of spherical-like particles.

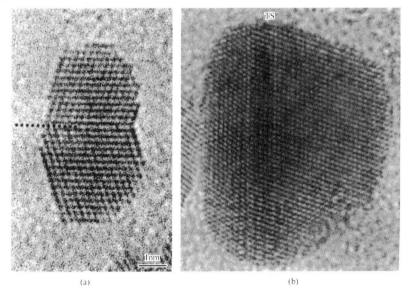

(a) (b)

Figure 2.9 HRTEM images of (a) Pt and (b) Au nanocrystals having twin (T) and stacking fault (S) structures, respectively. The nanocrystals are oriented along [110] and the twin plane and stacking fault are parallel to the direction of the electron beam

Stacking faults are typical planar defects. Stacking faults are produced by a distortion on the stacking sequence of atom planes. The (111) plane stacking sequence of a FCC structure follows A-B-C-A-B-C-A-B-C-. If the stacking sequence is changed to A-B-C-A-B-A-B-C-, a stacking fault is created. Figure 2.9(b) shows a Au particle that contains a twin and a stacking fault. It is known that nanocrystals usually contain no dislocation but strain. Twins and stacking faults are probably created from the high strain energy in the volume.

2.3.3 Decahedral and Icosahedral Particles

The two most typical examples of multiple twinned particles (MTP) are the decahedron and icosahedron (Ino, 1966; Allpress, et al., 1967). Starting from a FCC structured tetrahedron, a decahedron is assembled from five tetrahedrals sharing an edge (Fig. 2.10(a)). If the observation direction is along the five-fold axis and in an ideal situation, each tetrahedron shares an angle of 70.5°, five of them can only occupy a total of 352.6°, leaving a 7.4° gap. Therefore, strain must be introduced in the particle to fill the gap. An icosahedron is assembled using 20 tetrahedra via sharing an apex (Fig. 2.10 (b) and (c)). The icosahedral and decahedral particles are the most extensively studied twinned nanocrystals (Marks, 1994; Yang, 1979a, 1979b, Buffat, et al., 1991; Ino, 1969).

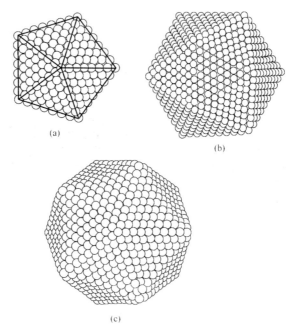

Figure 2.10 Atomic models of (a) decahedral, and (b, c) icosahedral nanocrystals

The simplest orientation for identifying the MTPs is along the five-fold symmetry axis. Figure 2.11(a) shows a large decahedral Au particle, and the five-fold twin structure is clearly seen. The strain introduced is also visible. High-resolution TEM image for a smaller one clearly shows the crystal structure of the particle (Fig. 2.11(b)). Identification of the shape of small MTPs in TEM is not trivial because the image depends sensitively on the orientation of the particle. With the presence of a thin amorphous carbon substrate, it is even more difficult to identify the particle shape directly.

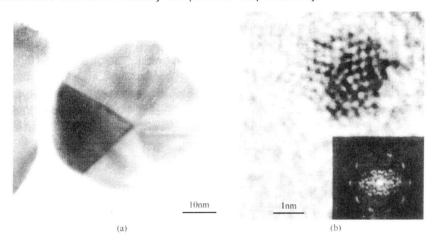

(a) (b)

Figure 2.11 TEM images of decahedral Au nanocrystals when the incident electron beam is parallel or nearly parallel to the five-fold symmetry axis. The strain field introduced by the five-fold twins is apparent in (a). A Fourier transform of the image clearly reveals the five-fold symmetry

2.3.4 Nucleation and Growth of Nanoparticles

Nucleation and growth are usually important for understanding physical and chemical characteristics of nanoparticles. Carbon, probably the most versatile element in nature, is a typical example. Carbon can form a variety of structures distinctly different in either crystallography or surface morphology. In addition to the commonly known amorphous carbon, partially disordered carbon blacks, graphite and diamond structures, carbon fullerene (Kroto, et al., 1985) and tubes (Iijima, 1991) have attracted a great deal of research interest in recent years. The versatility of carbon is due to its unique hybridization, sp^1, sp^2 and sp^3 bonding, allowing the formation of pentagonal, hexagonal and heptagonal carbon rings with positive, zero and negative surface curvatures, respectively. These geometrical modules serve as the fundamental units for constructing a variety of structures that have potential applications in scientific research and industrial technology (Dresselhaus, et al., 1996; Ebbesen, 1996).

TEM has played an irreplaceable role in the study of carbon structures. The carbon nanotube, for example, was first discovered using TEM. Carbon spheres, comprised of concentric graphitic shells, have been synthesized by a mixed-valent oxide-catalytic carbonization (MVOCC) process (Kang, et al., 1996; Wang, et al., 1996). TEM has been applied to image the very center of the carbon sphere to explore its growth mechanism (Fig. 2.12). It is apparent that the image contrast decreases dramatically as the center is approachal, but the graphitic structure is still visible. The weak contrast at the center is accounted for by two reasons: the strong background effect from the outer layers, and fewer and smaller size graphitic layers being parallel to the incident beam. The image contrast suggests that the graphitic layers form a polyhedral shell structure, where the corners of the polyhedron are indicated by arrowheads. The HRTEM image recorded from another sphere (Fig. 2.12(b))

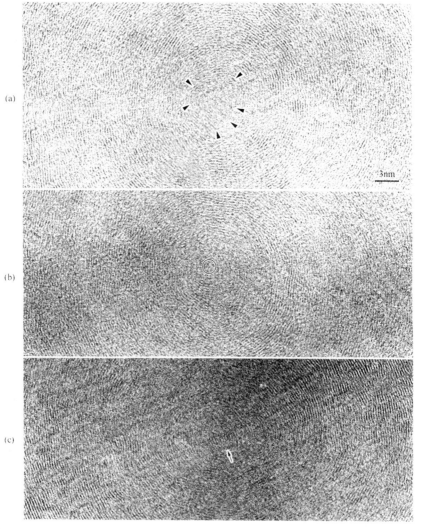

Figure 2.12 HRTEM images recorded from the centers of carbon spheres, showing the spiral growth near the centers

suggests that the distribution of the graphitic layers near the center is not spherical symmetric and the graphitic layers may exhibit a clockwise spiral structure. An HRTEM image recorded from a third carbon sphere (Fig. 2.12 (c)) shows that there are a few parallel aligned graphitic flakes being enclosed by an elliptical graphitic shell at the center. This is possible only with the presence of pentagonal carbon rings, as described below.

Carbon atoms can form three different types of graphitic carbon rings (Fig. 2.13(a)). The flat graphite layer is composed of hexagonal carbon rings. The pentagonal carbon ring causes the hexagonal network to curve inward (with $+60°$ inclinations), forming a surface with positive curvature. The heptagonal carbon ring forces the hexagonal lattice to be curved outward (with $-60°$ inclinations), forming a negative curvature. A curling graphitic carbon particle is believed to be nucleated from pentagonal carbon rings, and its growth forms a quasi-polyhedral spiral shell particle (Fig. 2.13(b)) (Kroto, et al., 1988). A pentagonal carbon ring produces an inward surface with positive curvature, resulting in spiral growth. At the beginning, the particle appears in a quasi-polyhedral shape, and the spiral shell forms the core of the sphere. When the particle size is sufficiently large, it is approximately a sphere. With the increase of the particle surface area, newly created pentagons and heptagons in the reaction chamber fall continuously on the surface of the sphere. These pentagons are the new nucleation sites for growing curved graphitic flakes with orientations that may not have any relationship with the initial spiral layer. In this case, the growth rates of these graphitic layers might be much faster than that of the in-plane spiral growth. With the arrival of many pentagonal and heptagonal carbon rings, the particle surface is covered by a stack of randomly twisted graphitic flakes nucleated at

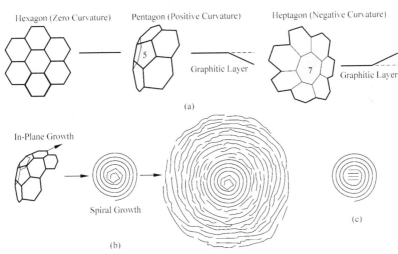

Figure 2.13 (a) The hexagon, pentagon and heptagon carbon rings. (b) The spiral nucleation model of a fine graphitic particle. (c) The growth model of a graphitic carbon sphere

different depths.

An alternative nucleation mechanism follows the same spiral growth, except the initial spiraling starts with a stack of flat graphitic flakes, as shown in Fig. 2.13(c), and the particle is solid. The beginning of the spiral growth is due to the creation of pentagonal carbon rings at the end of a hexagonal layer. The circulation of the layer around the stacked flakes produces an elliptical shape, as observed experimentally (Fig. 2.12(c)).

2.4 Electron Holography

In conventional HRTEM, the phase and amplitude images of the exit electron wave are mixed non-linearly and convoluted with the phase contrast transfer function. It is usually a rather sophisticated and complex process to retrieve the phase and amplitude images. Electron holography can recover the true object image. There are numerous ways of performing electron holography in TEM and STEM (Cowley, 1992). The most popular technique is off-axis holography. Holography is based on the interference and diffraction properties of waves, thereby producing a true image of an object (including amplitude and phase) without any distortion of the lenses (Fig. 2.14). The development of high-brightness high-coherence electron sources has made it possible to obtain holograms using electron waves in TEM (Gabor, 1949; Tonomura, 1993; Lichte, 1991; Tonomura, et al., 1995).

The phase image can provide the distribution of an electrostatic field around a charged particle as well as the thickness projected potential image of a nanocrystal. The former can be used to extract the charge distribution in the particle (Frost, et al., 1995), and the latter is useful to determine if the particle contains a hard core or an empty cavity (Datye, et al., 1995). This measurement is based on the electron phase shift after transmission through the electrostatic potential of the specimen. The perturbation of the field to the electron wave results in a change in the electron wavelength. The relative phase shift of the electron wave with respect to the reference wave is given by Eq. (2.5). For a homogeneous specimen with constant thickness, in addition to the average potential introduced by the atoms in the crystal, a charge barrier can be created at interfaces and defects. If the former contributes only a background, the latter can be retrieved experimentally.

Figure 2.15 shows a phase image of the MgO cubic crystal oriented along [210]. The MgO cube is bounded by four {100} faces. A thickness projection of the cube along [210] results in rapid variation in the projected mass thickness towards the edge of the cube, while there is no change in the projected thickness at the center (see the inset). This result is clearly revealed by the phase image reconstructed from the hologram.

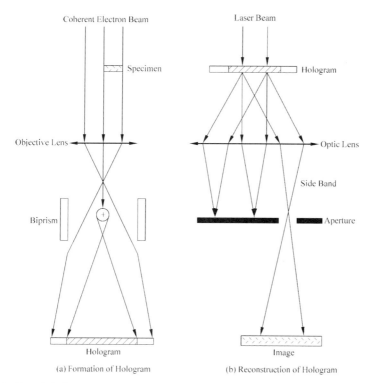

Figure 2.14 (a) Formation of off-axis electron hologram in TEM using an electrostatic biprism, and (b) Reconstruction of electron hologram using laser diffraction

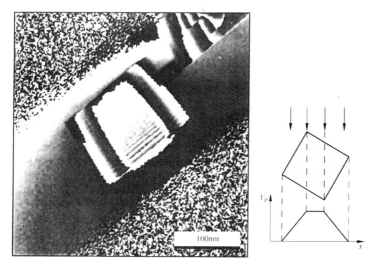

Figure 2.15 A phase image of a MgO cube oriented along [210], showing the phase variation resulting from the change in the projected potential. A schematic model of the projection geometry is given, where the profile of the projected potential across the particle is illustrated (Courtesy of Dr. B. Frost)

Imaging of magnetic domains and interdomain interaction is vitally important for understanding the behavior and improving the properties of magnetic materials. Imaging domains in nanoparticles is really a challenge. The key is to create a magnetic-field-free zone in which the specimen is placed. This can be achieved either by switching off the objective lens or by using a Lorentz objective lens. In the former, experiments can be performed using a conventional TEM equipped with an FEG at the expense of image magnification, which is usually reduced by 50—80 times. The image resolution is also greatly reduced because of the aberration effects from the intermediate lenses. Using a Lorentz objective lens is the optimum choice. Electron holography can reveal the phase created by the presence of the magnetic domain, a quantitative analysis of the phase image gives the domain structure and the strength of the local magnetism. This has been demonstrated for magnetic particles with sizes of 100 nm (Hirayama, et al., 1995; Mankos, et al., 1995).

2.5 Lorentz Microscopy

Magnetic nanocrystals and nanostructured magnetic materials occupy very important positions in cutting edge technologies. The applications of magnetic materials include permanent magnets, magnetic recording, magnetic memory, magnetic sensors, etc. Lorentz microscopy in TEM is a convenient way to image the magnetic structure with the highest achievable resolution.

2.5.1 Principle of Lorentz Microscopy

Electrons passing though a magnetic domain experience Lorentz deflection given by

$$\beta = eB\lambda d/h$$

where e is the electron charge, B is the magnetic induction, λ is the electron wavelength, d is the thickness of the domain and h is the Plank's constant. This deflection angle for 100 kV electrons passing a 200 nm Fe specimen is in the order of 0.4 milliradians. This deflection can be used to image the magnetic domain in either conventional TEM or in a scanning TEM (STEM). The deflection of the electron beam results in diffraction contrast as observed by bright-field TEM imaging using a very small objective aperture, in which the contrast is introduced due to the decrease in number of electrons propagating along the incident beam direction. This allows a direct imaging of local domain structure.

2.5.2 Elimination/Reduction of Magnetic Field from Objective Lens

TEM is based on magnetic lenses. To image magnetic specimens, it is necessary to avoid any change of the magnetic structure caused by the magnetic field of the objective lens. A simple solution is to switch off the objective lens, and a lens for magnetic imaging can be mounted at a lower position. In the JEOL 2010 the objective mini-lens can be used as the magnetic lens. Because the objective lens is now far away from the specimen, magnification loss needs to be compensated by intermediate lenses and projector lenses. Chapman et al. (1994) has incorporated a large gap objective lens with two additional Lorentz lenses above and below it. The Lorentz lenses can take over the functions of the objective lens to create a magnetic-field-free zone, while the objective lens can be used to generate magnetic field for magnetic reversal experiment.

2.5.3 Fresnel Lorentz Microscopy

This mode is easily realized in a conventional TEM, as shown in Fig. 2.16 (a).

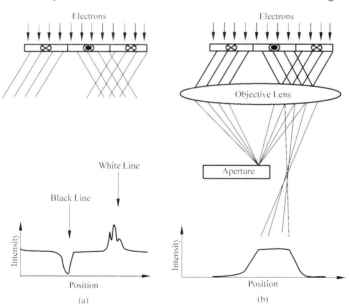

Figure 2.16 Comparison of contrast formation by Lorentz imaging (a) Fresnel mode, (b) Foucault mode

By defocusing, the domain wall can be revealed as dark lines or white lines. The white lines (fringes) are results of interference between the electrons from the two domains. Such fringes can be observed only when the electrons are highly coherent.

2.5.4 Foucault Lorentz Microscopy

This method deliberately makes use of the off-centering of the objective aperture to generate magnetic contrast as illustrated in Fig. 2.16(b). The domains from which the electrons are blocked by the aperture will appear dark while others appear bright. The contrast can be reversed by placing the aperture at the other side. Is advantage over the Fresnel mode is that it is an in-focus image.

Figure 2.17 compares the Fresnel and Foucault images of a Co-Sm magnetic thin film. In the Fresnel image, the black or white lines corresponds to the domain walls, while in the Foucault image the black or white contrast correspond to domains with opposite magnetization. Combining the bright-field TEM image and the Foucault mode image, the mechanism for the magnetic reversal switching volume in Co-Sm has been identified (Liu, et al., 1996).

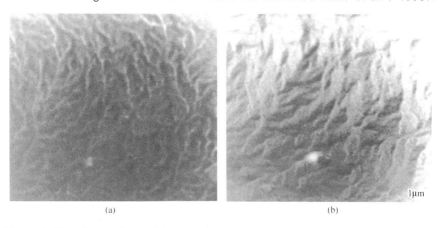

(a) (b)

Figure 2.17 Comparison of Lorentz images recorded in (a) Fresnel mode and (b) Foucault mode

The disadvantage of the Foucault image is that it requires that the deflection angle be large enough so that a split of the central spot is formed. As the deflection angle is proportional to the specimen thickness, the deflection angle becomes small for very thin specimens. In the defocus image, the small deflection angle can be compensated by going to a larger defocus.

By doing so, magnetic contrast characterized by a black-white pair in 25 nm Fe particles has been reported (Tanji, et al., 1999). Figure 2.18 shows the TEM image and simulated images. The Fe particle is supported by MgO. Because of the coherent electron source from the field emission gun, the Fresnel fringes due to the particle dominate the contrast. The magnetic contrast is revealed by the asymmetry of the Fresnel fringes. The Fresnel fringes would be perfect circles if there were no magnetic contrast. The shifting of the fringes is consistent with a single domain particle.

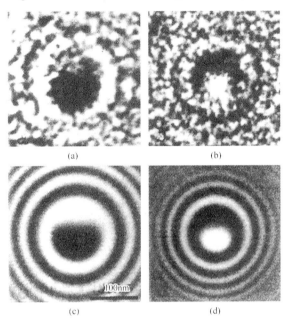

Figure 2.18 Fresnel image of a Fe particle by (a) TEM and (b) simulation (after Tanji, et al., 1999)

2.5.5 Differential Phase Contrast Mode of Lorentz Microscopy in STEM

To overcome the disadvantages in conventional TEM for Lorentz imaging, STEM has been utilized to obtain the optimum magnetic contrast (Chapman, et al., 1990). Figure 2.19 shows the experimental setup in STEM. A quadrant detector is used to collect the signal. The magnetic contrast due to deflection of electrons is retrieved by proper differentiation of signals collected by detectors. For example, an image can be obtained by $(A+B)-(C+D)$ or

(A+D)−(B+C) to characterize magnetic induction in two perpendicular directions. The resolution of the image is determined by the probe size which is at the level of 0.2 nm in the current generation of 200 kV FEG TEMs. This technique is also sensitive for imaging very thin films.

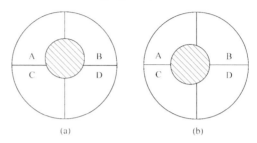

Figure 2.19 Schematic illustration of quadrant detectors in STEM

Lorentz microscopy offers the highest resolution for imaging the magnetic structure. Combined with various TEM techniques, a relationship between magnetic structure and physical structures can be established. The future challenge will be the in situ observation of magnetic reversal in nano-magnetic crystals. This will require high resolution at the 1—5 nm level and a proper magnetic in situ device to generate the magnetic field. These two criteria can be achieved in STEM. However, as the magnetic reversals in nanocrystals are extremely fast and sensitive to the magnetic field, the success of such an endeavor remains unfulfilled.

2.6 Nanodiffraction

Electron diffraction is a powerful tool for materials research ever since the invention of TEM. A popular experimental setup for electron diffraction is selected area diffraction (SAD). Using parallel illumination the area of interest is selected by placing an aperture in the image plane of the objective lens. Because of the physical size of the aperture, the lens aberration and diffraction, the practical selected area by the aperture is limited to micrometer level. For conventional materials with a grain size of 10 to 100 microns, the diffraction pattern is a spot pattern. However, for nanostructured materials with a grain size of, say 20 nm, the diffraction pattern by the SAD method becomes a ring pattern (also called powder pattern) because so many grains are included in the aperture. Such a ring pattern contains rich information about the structure of the specimen and can be used to identify uniquely a known phase (Liu, et al., 1998). However, it is occasionally necessary to obtain a spot pattern from a single grain of nanometer dimension for such applications as determination of crystal orientation, crystal structure of modulated phase, epitaxy, etc. This is the nanodiffraction technique, e. g., electron diffraction from a nano-sized area of a specimen.

2.6.1 Optics for Nanodiffraction

Unlike SAD, nanodiffraction is realized by focusing the electron beam on the specimen. The experimental setup for nanodiffraction is schematically shown in Fig. 2.20 (b), in comparison with SAD in Fig. 2.20 (a). The area contributing to diffraction is controlled by the probe size. Since the probe size can be as small as 0.2 nm, nanodiffraction can therefore be readily achieved from an area smaller than the size of the unit cell. When the probe is focused, minimum probe size can be obtained. The diffraction spots in the diffraction pattern are disks instead of sharp spots. The diameter of the disk is defined by the semi-convergent angle of the electron beam. This could cause overlapping of spots if the unit cell is large. The sharp spots pattern can be obtained by defocusing the probe slightly using the condenser lens as shown in Fig. 2.20(c). The probe size on the specimen in this setup may be larger than the available minimum probe size.

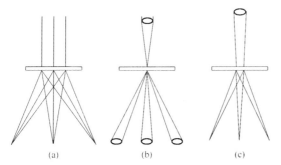

Figure 2.20 Schematic illustration of experimental setup for (a) SAD, (b) conventional nanodiffraction and (c) focused nanodiffraction

2.6.2 Experimental Procedures to Obtain a Nanodiffraction Pattern

The procedures for carrying out nanodiffraction are:
(1) Choose a proper probe size: A probe size smaller than the half grain size under study is preferred; the smaller the probe size, the smaller the

chance of encountering a defect. However, if the probe is too small, the diffraction pattern would be too dark to be seen. Also the probe size must be a few times larger than the unit cell in order to Preserve the symmetry of the crystal in the diffraction pattern.

(2) Align the high voltage center and current center. Both are important for obtaining a good probe.

(3) Correct astigmatism of the condenser lens in the image mode by making the electron beam circular.

(4) Correct astigmatism of the objective lens. This could be done by using a larger spot size. This alignment ensures that a minimum spot size can be obtained.

(5) Focus the image. This is done by setting the objective current at a desired value by adjusting the z control (the vertical translation of the specimen). Fine focus is done by adjusting the objective current.

(6) Find the zone axis. Going through the above procedure will generate a good probe. Now going to the diffraction mode will generate a diffraction pattern. Although tilting can still be used to find the proper zone axis, large angle tilting is limited by the size of the grain and can cause overlapping with adjacent grains. Therefore, finding the zone axis is alternatively done by probing the right grain by scanning the probe over the specimen while watching the diffraction pattern until the one with the desired zone axis is found.

(7) Expose the film. An exposure time of 1—2 seconds is preferred. This can be done by using smaller camera length.

Taking good quality nanodiffraction patterns involves reducing the beam illumination time on the crystal under examination. Nanoprobe placed at crystals can cause very fast contamination. The contaminants are light elements such as C, etc. The clear diffraction pattern can disappear in a few seconds. The contamination can come from poor vacuum or the specimen surface. Using a cold specimen stage is an effective way to reduce specimen contamination. The contaminant atoms can be "frozen" at the specimen surface. Heating the specimen to 100°C is also occasionally helpful to eliminate contaminants from the specimen surface.

2.6.3 Some Applications

2.6.3.1 Crystal Structure Identification of Metastable Phases in Co-Sm Films

Co_5Sm has the highest anisotropy value and is an excellent permanent magnet material. However, when sputtered into film, both amorphous phase and crystalline phase can be formed. X-ray and electron diffraction studies indicate that the diffraction data of the sputtered films produced in our laboratory can not be matched with any equilibrium phases in the Co-Sm system. The nanostructure is made of crystallites of about 5 nm distributed in an amorphous matrix as shown in Fig. 2.21. The crystal structure of the crystallites was identified by nanodiffraction and confirmed by HRTEM (Liu, et al., 1995). It was found that all the nanodiffraction patterns could be indexed by close-packed structure models with different stacking modes. Figure 2.22 compares the experimental and simulated nanodiffraction patterns based on the close-packed structure for different zone axes. The intensity of the diffraction spots in some zone axes such as $[1\ 0\ \bar{1}\ 0]$ are not sensitive to the stacking mode. Quite similar $[1\ 0\ \bar{1}\ 0]$ zone axis patterns can be obtained from all the stacking modes. The $[1\ 1\ \bar{2}\ 0]$ nanodiffraction patterns are characteristics of the stacking mode and lead to a decisive identification of the stacking mode. The stacking modes ABC, ABAC and ABCAB are identified as shown in Fig. 2.22 (g), (i) and (k). In many cases, a pattern with overlapping spots was obtained, suggesting a random stacking. It is realized that different zone axes can be used to retrieve different information. The $[0\ 0\ 0\ 1]$ zone axis pattern is sensitive to ordering and was used to study the Sm solute atom distribution. Figure 2.23 compares the $[0\ 0\ 0\ 1]$ zone axis pattern from different ordered structures. It was found that the Sm atoms were randomly distributed in the Co lattice. The lattice parameters $a = 0.245 - 0.256$ nm and $c_0 = 0.204 - 0.207$ nm (c_0 is the spacing between the close-packed planes, and $c = Nc_0$ for an N layer stacking unit cell) were deduced from the $[1\ 0\ \bar{1}\ 0]$ and $[1\ 1\ \bar{2}\ 0]$ zone axis patterns in Fig. 2.22.

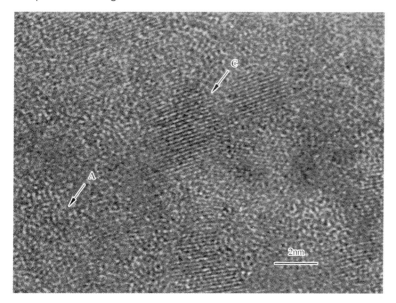

Figure 2.21 HRTEM image of a Co-Sm film

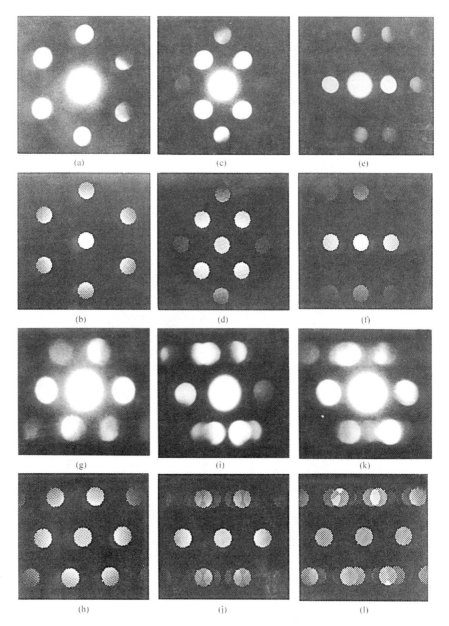

Figure 2.22 Comparison of TEM and simulated nanodiffraction patterns. (a, b): [0 0 0 1] zone axis, three layer stacking ABCABC... (c, d):[2 0 $\bar{2}$ 1] zone axis, ABACABAC... (e, f) :[1 0 $\bar{1}$ 0] zone axis, five layer stacking ABCABABCAB... (g, h):[1 1 $\bar{2}$ 0] zone axis, three layer stacking ABCABC... (i, j) : [1 1 $\bar{2}$ 0] zone axis, four layer stacking ABACABAC... (k, l):[1 1 $\bar{2}$ 0] zone axis, five layer stacking ABCABABCAB...

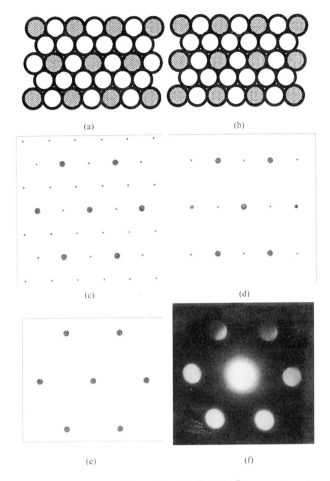

Figure 2.23 Identification of possible ordering by [0 0 0 1] zone axis patterns: (a) Co_3V type ordering in the close packed plane and (b) Co_3Ni type ordering in the close packed plane. (c, d) are corresponding diffraction patterns. (e) Diffraction pattern of a disordered HCP structure. (f) TEM pattern

2.6.3.2 Measurement of Volume Fraction of Crystallite Phase in an Amorphous Phase

Figure 2.24 shows two examples of the nanodiffraction patterns taken from the Co-Sm film. Figure 2.24(a) is a case of crystallites, and (b) is the pattern of the amorphous phase. To measure the volume fraction of the crystallite, each film was probed with a separation of about 2 nm along a straight line, which was scanned in an area for 200 measurements. In most cases, either a spot pattern or an amorphous pattern was observed. For some cases where a mixture of a spot pattern and an amorphous pattern is observed, the pattern is judged as spot or amorphous depending on the intensity distribution. The volume fraction of the crystallites in the film was deduced by the ratio of the number of crystallite patterns against the number of total patterns. The results for different films along the coercivity are listed in Table 2.1. It is seen from Table 2.1 that the volume fraction varies with the processing parameter Ar pressure and the composition. Lower Ar pressure and lower Sm content result in a high volume fraction of the crystallite. Also the Co-Pr as-deposited film has much lower crystallite volume fraction. Annealing of the Co-Pr film resulted in total crystallization of the film. The as-deposited Co-Pt has no amorphous phase.

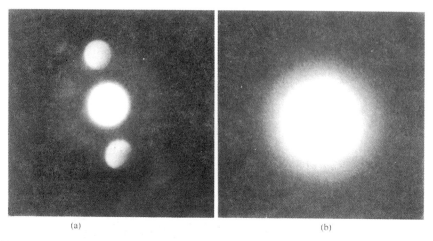

(a) (b)

Figure 2.24 Comparison of nanodiffraction patterns (a) from a crystallite and (b) from the amorphous phase

Table 2.1 Processing, nanostructure and magnetic properties of the Co-X (X= Sm, Pr, Pt) films. V_c is the volume fraction of the crystallite, subscript C indicates crystallite and A indicates amorphous.

Film com. (at%)	Ar pressure (mTorr)	Film thickness (nm)	Coercivity (kOe)	Phases	V_c (%)
Co-19%Sm	5	24	0.61	C+A	91
Co-19%Sm	12	24	2.58	C+A	65
Co-19%Sm	30	24	0.92	C+A	54
Co-22%Sm	5	30	1.2	C+A	81
Co-22%Sm	17	30	4.1	C+A	57
Co-22%Sm	30	30	3.4	C+A	48
Co-22%Pr (as-deposited)	20	35	0.5	C+A	32

Film com. (at%)	Ar pressure (mTorr)	Film thickness (nm)	Coercivity (kOe)	Phases	V_c (%)
Co-22%Pr (400℃ 30 min)	20	35	7.2	C	100
Co-50%Pt (as-deposited)	20	10	0.1	C	100
Co-50%Pt (750℃ 3hrs)	20	10	23	C	100

Note that the grainy contrast without lattice fringe in an HRTEM image does not necessarily indicate an amorphous phase. When a crystallite is aligned along a high-index zone axis, the projection of the structure is beyond the resolution limit and no lattice fringes will be observed. Contamination of the crystallite will also result in an amorphous like contrast. The phase identification of the amorphous phase is most reliably made by nanodiffraction without the limitation of the microscope resolution. No matter what direction is aligned relative to the electron beam, a spots pattern is always generated if a crystal is probed.

2.6.3.3 Identification of the $L1_0$-Structure Ordered Phase in Fe-Pt

Superior high-energy product has been recently found in the Fe-Pt system (Liu, et al., 1999). The magnets were designed based on the theory of exchange coupling between magnetically hard phase and soft phase. Figure 2.25 shows a [110] zone axis HRTEM image of a Fe-Pt film annealed by rapid thermal annealing. Small precipitates of 3—8 nm were revealed. However, the difference in crystal structure between the matrix and the precipitate is not revealed. Nanodiffraction reveals that the matrix is ordered $L1_0$ phase while the precipitate is a disordered FCC phase. Figure 2.26 compares the [001] zone axis nanodiffraction patterns recorded from the matrix and the precipitate. The pattern from the matrix in Fig. 2.26(a) could be matched to the pattern in Fig. 2.26(c), simulated based on the $L1_0$ structure. The pattern from the precipitate in Fig. 2.26(b) does not show the ordering spots, matching well to the pattern in Fig. 2.26(d), simulated using the FCC structure model.

2.6.3.4 Identification of Epitaxy at the Interface

While HRTEM is capable of identifying epitaxy at the interface, it does not work well for thicker specimens. Nanodiffraction, on the other hand, can work comfortably for thicker specimens as long as the image of the interface is visible. Figure 2.27 shows the interface between a nanostructured TiN film and steel and the corresponding nanodiffraction patterns. The epitaxy is found to be steel $(1\bar{1}0)[111]$ // TiN $(\bar{1}1\bar{1})[011]$.

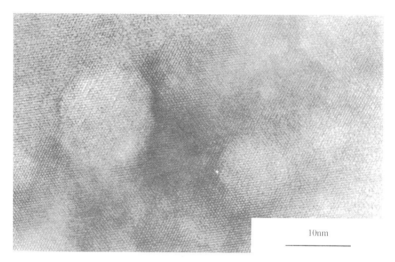

Figure 2.25 HRTEM image of a Fe-Pt film

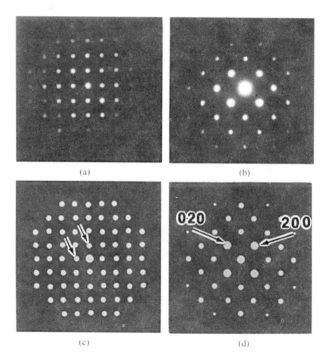

Figure 2.26 Nanodiffraction patterns (a) From $L1_0$ phase. (b) From disordered FCC phase. (c, d) Nanodiffraction patterns

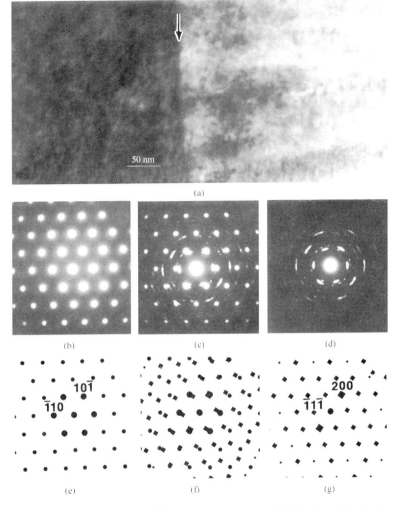

Figure 2.27 Epitaxial growth of TiN on tool steel. (a) TEM bright-field image. The interface is indicated by an arrowhead. (b, c, d) Nanodiffraction patterns recorded from the tool steel, the interface and the TiN, respectively. (e) Calculated [111] nanodiffraction pattern of iron. (g) Calculated [110] pattern of TiN. (f) Superposition of (e) and (g)

2.7 In situ TEM and Nanomeasurements

The presence of a large percentage of surface atoms in nanomaterials results in surface dominated phenomena. The large mobility of the surface atoms and the thermodynamically nonequilibrium faceted shapes of the nanocrystals could have very interesting physical, chemical and electronic structures of fundamental interest and technological importance. TEM is an ideal tool for tracing the in situ bahavior of nanocrystals induced by temperature and/or externally applied fields. The in situ experiments can be carried out on the same nanocrystal at a large range of temperature, giving the possibility of examining the temperature induced shape transformation, structural evolution, melting and surface phenomena. This section focuses on the types of in situ TEM techniques for probing the unique properties of nanocrystals.

2.7.1 Thermodynamic Properties of Nanocrystals

The thermodynamic properties of nanocrystals depend strongly on crystal size and crystal shape (Mohamed, et al., 1999). The thermodynamic properties of Au nanorods are an example. Gold nanorods were prepared by an electrochemical method (Yu, et al., 1997). A series of in situ TEM images recorded for the same specimen region as the temperature was increased continuously from room temperature to 923 K, showing the formation process of Au small nanoparticles on the carbon support in Fig. 2.28(a—h). Nucleation started at 473 K as shown in Fig. 2.28(b) and it was only observed by TEM at 488 K, shown in Fig. 2.28(c). It is clear that as the temperature reached 433 K, at which the structure of the micelles around the rods was found to collapse in solution, no visible change was observed in the morphology of the nanorods. At $T = 488$ K, as shown in Fig. 2.28(c), small nuclei of gold nanoparticles began to form on the surface of the carbon substrate. A small increase in temperature to 493 K (Fig. 2.28(d)) resulted in a rapid growth of these small particles on the carbon substrate. Further increase in temperature to 513 K led to the growth of the particles, and the stable particle size was reached at $T = 600$ K (Fig. 2.31(a)). At $T > 513$ K, the widths of the nanorods increased significantly, while their lengths remained the same (compare the rod marked by an arrow in Fig. 2.28 (a), (e) and (h). This is due to surface melting, which induces shape transformation, so the thickness of the nanorods decreased when placed onto the substrate as a "liquid-like" material. It is noticed that the density of the particles in the regions adjacent to the Au rods is approximately the same as the particles in the regions without rods.

  We propose that the formation of small Au nanoparticles in Fig. 2.28(d)—(h) is the result of surface sublimation from gold nanorods as well as the gold atoms contained in the solution deposited onto the carbon substrate. The sublimation of gold atoms from the surface is supported by the following observation. Shown in Fig. 2.29 is a series of TEM images recorded from the same area, in which the shrinkage of a particle indicated by an arrowhead is

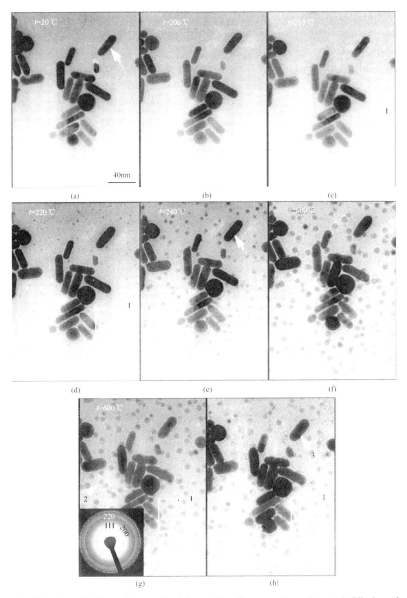

Figure 2.28 (a—f) Show the result of the sublimation, condensation and diffusion of gold atoms from gold nanorods: A series of in situ TEM images recorded for the same specimen region as the temperature was increased continuously from room temperature to 923 K, showing the formation process of Au small nanoparticles on the carbon support. The nucleation started at ∼ 473 K and it was only clearly observed by the TEM at 488 K shown in (b). At $T > 513$ K, the widths of the nanorods increased significantly, while their lengths remained the same (compare the rod marked by an arrow in (a), (e) and (f)). This is due to surface melting, which induces shape transformation, so the thickness of the nanorods decreased when placed onto the substrate as a "liquid-like" material (note an increase in the width does not necessarily mean an increase in the volume)

visible. Such a drastic reduction in size could be the result of surface sublimation, and more interestingly, this phenomenon occurs at a relatively low temperature.

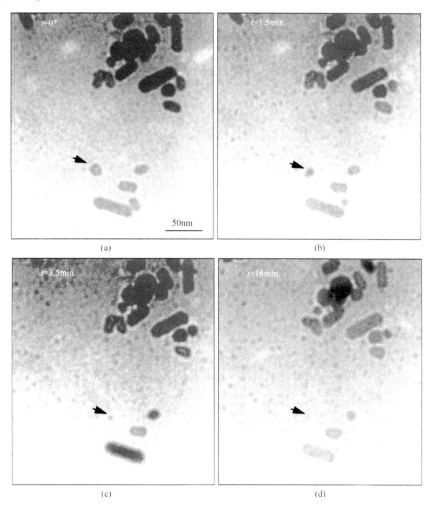

Figure 2. 29 In situ TEM images recorded as a function of time at 225℃, showing the shrinking of an Au particle indicated by an arrowhead at such a low temperature, indicating surface sublimation and diffusion

A model can be proposed to understand the observed phenomena. The initial rapid growth of the Au particles is due to the contribution of the diffusion of atoms on the substrate surface. This process is assumed to follow the hit-and-stick model and is directly related to the intersection length of the particle with the substrate surface. On the other hand, the atoms on the surface of the Au particles can sublime, and their sublimation rate is directly proportional to

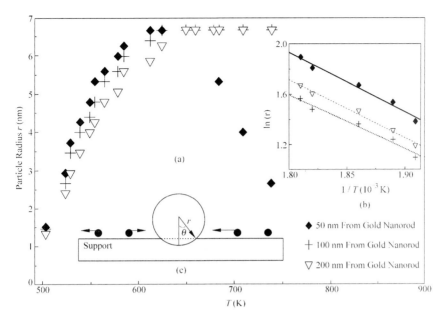

Figure 2.30 (a) The dependence of the steady state particle radii (r_{max}) formed at different specimen temperature for particles located at different distances from the Au rods (diamond shapes are for the closest to the rod and trigonal are for the furthest away). The size of the particles located closest to the rod (~50 nm) drop sharply at 650 K. Considering the specimen drift at higher temperature and the effect of carbon substrate on the image contrast, the newly formed particles were seen in TEM only when their sizes were larger than ~2 nm. ◆, + and ▽ represent the particles at distances of 50, 100 and 200 nm, respectively, from the gold nanorods. (b) Plot of $\ln(r_{max})$ vs. $1/T$ in the low-temperature range, the slopes of which gives the difference between the activation energy of diffusion and the sublimation energy Q of the gold atoms. (c) Sketch showing the growth process of the small gold cluster on the TEM substrate. Growth kinetics of the small particles at a constant temperature of $T = 498$ K

the exposed surface area of the particle. Accordingly, for a spherical particle of radius r, partially embedded in the substrate as shown in Fig. 2.30(c), the net growth of the number of atoms (n_r) in the particle is given by

$$\frac{dn_r}{dt} = (2\pi r \sin\theta) N_0 k_d - 4\pi r^2 \alpha \rho_s k_s \qquad (2.11)$$

where θ is the angle that specifies the degree of contact of the particle with the substrate; k_d is the diffusion rate constant of the atoms adsorbed on the substrate; N_0 is the steady state concentration of the Au atoms on the surface; α is a factor representing the fraction of the exposed surface area, ρ_s is the surface density of the particle, and k_s is the sublimation rate constant. The direct condensation of Au atoms onto the particle surface can be accounted for by a proper choice of the α factor. Initially, the diffusion term is dominant, resulting in a rapid growth of the particle. As the particle grows in size, the sublimation term increases because of the r^2 term. When the two terms are equal, a steady state is reached ($dn_r/dt = 0$), achieving a maximum radius (r_{max}) at this temperature, which is given by

$$r_{max} = N_0 k_d \sin\theta / 2\alpha \rho_s k_s \qquad (2.12)$$

Using the relation $n_r = 4/3\alpha r^3 \rho_0$, where ρ_0 is the volume density of Au atoms, and the boundary condition $r = 0$ at $t = 0$, the solution of Eq. (2.11) relating r to t is given by

$$r_{max} \ln\left(\frac{r_{max}}{r_{max} - r}\right) - r = [\alpha \rho_s k_s / \rho_0] t \qquad (2.13)$$

The kinetics of Au particle growth were determined using in situ TEM at 498 K, as soon as the newly formed particles were observed in TEM. From the kinetic curve given in Fig. 2.31(a), the particles seem to experience a rapid growth in the first 500 s, and then the size reached a constant value. By setting the left-hand side of (2.13) equal to Y, using the experimental data in Fig. 2.31(a), the plot of Y versus time can be used to calculate the activation energy Q for sublimation of gold atoms, which is $Q = (1.39 \pm 0.01)$ eV.

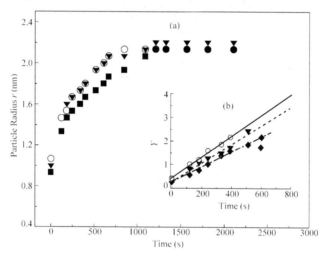

Figure 2.31 (a) Particle radii as a function of time. (b) Comparison of the theoretically calculated Y vs. time curves according to Eq. (2.13) with the experimentally observed data points. From the slopes of the lines, the sublimation energy Q, is calculated

The crystallographic shape of the Au nanorods has been investigated using HRTEM to understand why the gold atoms can sublimate at lower temperature. The nanorods have been found to exhibit the unstable {110} facets (Wang, et al., 1999b) (Fig. 2.32), in addition to the {100} and {111} facets present in the spheres. For the face-centered Au nanocrystals, the surface energy follows the relationship $\gamma\{110\} > \gamma\{100\} > \gamma\{111\}$. The presence of a significant percentage of gold atoms on the {110} surface suggests that the nanorod is relatively thermodynamically unstable. The {111} and {100} faceted Au spheres have lower surface energy. Thus, their atoms may not sublime at relatively low temperatures. Therefore, the sublimation of Au atoms from the nanorods at lower temperatures is due to the shape of the nanorods, which have the unstable {110} surfaces.

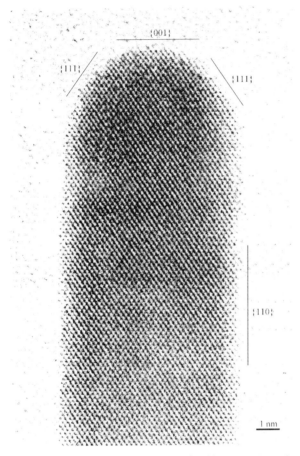

Figure 2.32 HRTEM image showing the {110} and {100} facets of the Au nanorods. The electron beam is parallel to [110]

Studying shape-controlled nanocrystals is of great interest from a catalysis viewpoint. Platinum nanoparticles with a high percentage of cubic-, tetrahedral- and octahedral-like shapes, respectively, have been synthesized by changing the ratio of the concentration of polymer capping material (polyacrylate) to that of Pt^{2+} being reduced by H_2 from K_2PtCl_4 at room temperature (Ahmadi, et al., 1996). The polymer acts not only as the passivative protection layer of the nanocrystals but also as the key factor in controlling their shapes (Whetten, et al., 1996; Wang, 1998). The growth mechanism of the shape controlled Pt nanocrystals is attributed to a kinetic process in which most of the nuclei are tetrahedral, while a shape transformation from tetrahedral to cubic occurs as the particles grow (Petroski, et al., 1998). The in situ behavior of the Pt nanocrystals has also been studied by TEM (Wang, et al., 1998). The surface capping polymer was removed by annealing the specimen to a temperature of 180℃—250℃, while the particle shape showed no change up to ~350℃. In a temperature range of 350℃ to 450℃, a small truncation occurred in the particle shape but no major shape transformation. The particle shape experienced a dramatic transformation into a spherical-like shape when the temperature was higher than ~500℃; macroscopic surface melting occurred at ~600℃, much lower than the melting point of bulk Pt (1773℃).

2.7.2 Nanomeasurement of Electrical Transport in Quantum Wires

It is known that the properties of nanostructures depend strongly on their size and shape. The properties measured from a large quantity of nanomaterials could be an average of the overall properties, so the unique characteristics of individual nanostructure could be embedded. The ballistic quantum conductance of a carbon nanotube (Frank, et al., 1998), for example, was observed only from defect-free carbon nanotubes grown by arc-discharge technique, while such an effect vanishes in the catalytically grown carbon nanotubes because of the high density of defects. Thus, an essential task in nanoscience is the property characterization of an individual nanostructure with well-defined atomic structure. But, this is a challenging task.

The electrical properties of a single multiwalled carbon nanotube have been measured using a setup in an atomic force microscope (AFM). A carbon fiber from the arc-discharge chamber was attached to the tip of the AFM; the carbon tube at the forefront of the fiber was in contact with a liquid mercury bath. The conductance was measured as a function of the depth as the tube was inserted into the mercury. Surprisingly, the conductance shows quantized stairs (Fig. 2.33), and the stair height matches well to the quantum conductance $G_0 = 2e^2/h$. This effect shows up only if the carbon nanotube is defect free, which means the tubes produced by arcdischarge rather than catalytic growth. The experiments were repeated over 100 times in AFM and it was repeated in TEM using an in situ specimen holder (Wang, et al., 2000). The conductance is quantized and is independent of the length of the carbon nanotube. No heat dissipation was observed in the nanotube. This is the result of ballistic conductance, and it is believed to be a result of conductance through a single graphite layer.

Such experiments have also been carried out in TEM using a newly built

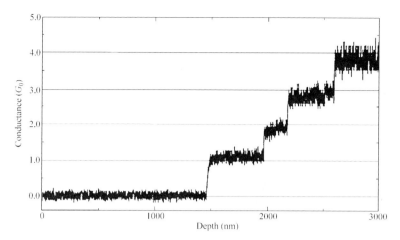

Figure 2.33 Quantized conductance of a multiwalled carbon nanotube as a function of the depth (z) into the liquid mercury in which the nanotfube was inserted in an atomic force microscope. The second stair is introduced as another nanotube touches the mercury (Courtesy of Drs. W. A. de Heer and P. Poncharal)

specimen holder that can apply an electric field across a single carbon nanotube while the tube is in contact with the electrode. Figure 2.34 gives a TEM image of a carbon nanotube which is in electric contact with a liquid mercury droplet. The conductance of the tube was found to be G_0. This type

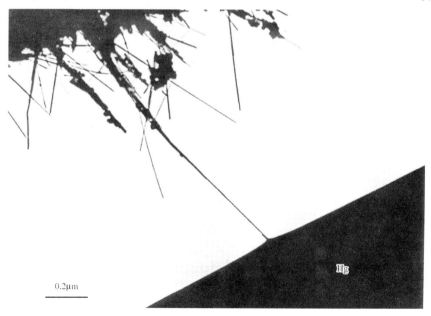

Figure 2.34 Electric contact of a carbon nanotube with a mercury surface imaged by TEM

of measurement ensures the reliability of the experimental result, avoids artifact, and allows a direct view of the nanotube from which the electric transport was measured.

2.7.3 Nanomeasurement of Mechanical Properties of Fiber-Like Structures

One of the significant challenges in nanoscience is the measurement of mechanical properties of individual constituents that comprise the nanosystem. The problem arises due to difficulties in gripping and handling fibers that have nano-size diameters. Mechanical characterization of individual nanofibers was performed using AFM. By deflecting on one end of the nanofiber with an AFM tip and holding the other end, the mechanical strength has been calculated by correlating the lateral displacement of the fiber as a function of the applied force (Salvetat, et al., 1999; Wong, et al., 1997). Another technique that has been previously used involves measurement of the bending modulus of carbon nanotubes by measuring the vibration amplitude resulting from thermal vibrations (Treacy, et al., 1996).

A new approach has recently been demonstrated for measurement of the mechanical strength of single carbon nanotubes using in situ TEM (Poncharal, et al., 1999). The carbon nanotube has a nanofiber-like structure consisting of concentric cylindrical graphitic sheets, with diameters of 5—50 nm and lengths of 1—20 μm. A special TEM specimen holder has been built for applying a voltage across the nanotube and its count electrode. Each nanotube can be clearly observed in the TEM. Thus, the measurements can be done on a specific nanotube whose microstructure is determined by transmission electron imaging and diffraction. If an oscillating voltage is applied on the nanotube with ability to tune the frequency of the applied voltage, resonance can be induced. When the applied voltage frequency equals the natural frequency of the nanotube (Fig. 2.35), resonance is obtained and the frequency can be accurately measured. Resonance is nanotube selective because the natural vibration frequency depends on the tube diameter (D), length (L), density (ρ), and bending modulus (E_b) (Meirovich, 1986):

$$v_i = \frac{\beta_i^2}{8\pi} \frac{D}{L^2} \sqrt{\frac{E_b}{\rho}} \qquad (2.14)$$

where $\beta = 1.875$ and 4.694 for the first and second harmonics. In the above formation, the tube is assured to be a solid and the inner radius of the tube is ignored. The bending modulus of nanotubes has been measured as a function of its diameter (Poncharal, et al., 1999). The bending modulus is as high as 1.2 TPa (as strong as diamond) for nanotubes with diameters smaller than 8 nm, and it drops to as low as 0.2 TPa for those with diameters larger than 30 nm. The technique developed here is being applied to the measurement of the mechanical strength of submicron fibers.

Figure 2.35 Alternating field induced mechanical resonance. A selected carbon nanotube at (a) stationary, (b) the first harmonic resonance ($v_1 = 1.21$ MHz), and (c) the second harmonic resonance ($v_2 = 5.06$ MHz)

2.7.4 Femtogram Nanobalance of a Single Fine Particle

Analogous to a pendulum, the mass of the particle attached at the end of the spring can be determined if the vibration frequency is measured, provided the spring constant is calibrated. This principle can be adopted in the case outlined in Sect 2.6 to determine a very tiny mass attached at the tip of the free end of the nanotube. The resonance frequency drops more than 40% as a result of adding a small mass (Fig. 2.36). The mass of the particle can be thus derived by a simple calculation:

$$v = \left(\frac{D^2}{16\pi L}\right)\left(\frac{3E_b}{Lm_{eff}}\right)^{1/2} \quad (2.15)$$

For the case shown in Fig. 2.36, $D = 42$ nm, $E_b = 90$ GPa, the mass of a particle is 22 ± 6 fg ($1 \text{fg} = 10^{-15}$g). This is the newly discovered "nanobalance" (Poncharal, et al., 1999), the most sensitive and smallest balance in the world. We anticipate that this nanobalance would have application in measuring the mass of large biomolecules and biomedical particles, such as virus, possibly leading to a nano-mass spectrometer.

Figure 2.36 A small particle attached at the end of a carbon nanotube at (a) stationary and (b) the first harmonic resonance ($v = 0.968$ MHz)

2.7.5 Electron Field Emission from a Single Carbon Nanotube

The unique structure of carbon nanotubes clearly indicates they are ideal objects that can be used for producing high field emission current density in a flat panel display (de Heer, et al., 1995). Growth of aligned carbon nanotube films have been reported by several groups (Li, et al., 1996; Pan et al., 1998; Ren, et al., 1998; Fan, et al., 1999). With consideration of the variation in the diameters and lengths of the aligned carbon nanotubes, the measured $I-V$ curve is an averaged contribution from all of the carbon nanotubes. Using the in situ TEM setup we built, the electric field induced field emission characteristics of a single carbon nanotube have been studied. Figure 2.37 shows a TEM image of the carbon nanotubes which are emitting electrons at an applied voltage. The dark contrast near the tips of the nanotube is due to the field induced by the tip charged electrons as well as the emitting electrons. A detailed analysis of the field distribution near the tip of the carbon nanotube by electron holography will provide the threshold field for field emission.

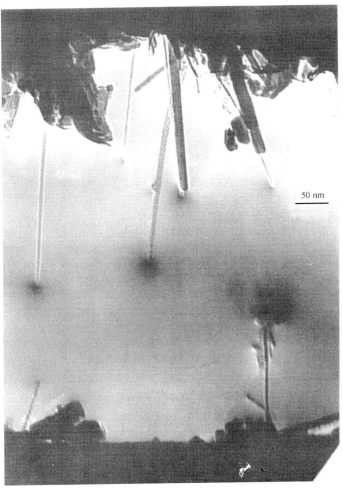

Figure 2.37 in situ TEM observation the field emission from individual carbon nanotubes. The applied voltage across the two electrodes was 80 V

Structure damage can be introduced at the tip of a carbon nanotube if the strength of the externally applied field exceeds some limits. Shown in Fig. 2.38 is a series of TEM images recorded from a carbon nanotube during its field emission. The nanotube is experiencing rapid structural damage starting from the tip. The damage occurs in a form of pilling off, so that part of the segments are cut off abruptly and completely. This type of damage is believed to leave a high density of dangling bonds on the surface of the carbon tube as a result of the broken graphitic sheets. This experimental result disagrees with the model proposed by Rinzler, et al. (1995), who believe that the damage occurs as an unraveling process in which the carbon atoms are removed string by string and layer by layer. More importantly, we found that the carbon nanotube vibrates during field emission, resulting in fluctuation in emission current.

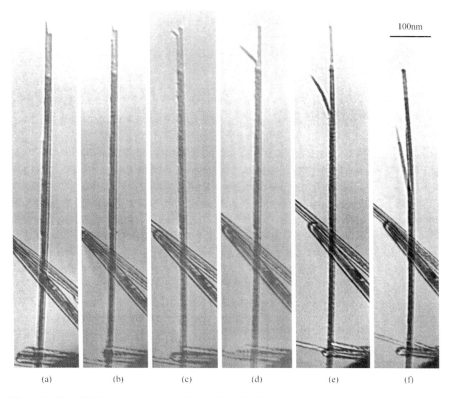

Figure 2.38 TEM images showing the electron field emission induced structure damage at the tip of a carbon nanotube. This damage occurs when the local field exceeds a critical value. The applied voltage is 120 and the distance from the tip to the counter electrode is ~ 3 μm

2.8 Electron Energy Loss Spectroscopy of Nanoparticles

Chemical microanalysis is an important field in TEM. Under the impact of an incident electron beam, the electrons bonded to the atoms in the specimen may be excited either to a free electron state or to an unoccupied energy level with a higher energy. The quantum transitions associated with these excitations will emit photons (or X-rays) and electrons such as secondary electrons, Auger electrons and ionized electrons. These inelastic scattering signals are the fingerprints of the elements that can provide quantitative chemical and electronic structural information.

2.8.1 Valence Excitation Spectroscopy

In studying nanoparticles, it is necessary to probe the electronic structure of a single nanocrystal. This is possible only with the use of a probe that is smaller than the size of the nanocrystal. The valence excitation spectrum of a nanoparticle is most sensitive to its electronic structure and size effects (Wang, 1996, 1997; Schattschneider, et al., 1995). The spectra can be acquired in TEM and STEM using a fine electron probe. The quantification of the spectra relies on theoretical calculations. The valence band excitation of a nanoparticle is most easily and accurately described using dielectric response theory. The impact of an incident electron is equivalent to a time dependent pulse, which causes transitions of valence electrons. In classical dielectric response theory, an incident electron is treated as a particle following a pre-defined trajectory, which is assumed to not be affected by the interaction between the incident electron and the dielectric media, as shown in Fig. 2.39. Electron energy loss is a continuous process in which the electron is decelerated due to the attractive force $F_z = -eE_z$ due to the field of the induced charges, resulting in energy loss. For a general case in which the incident electron is moving along the z-axis and under non-relativistic approximation, if the instantaneous position of the electron is denoted by $r_0 = (x_0, 0, z' = vt)$, where x_0 is called the impact parameter, the energy-loss spectrum of an incidence electron due to surface plasmon excitation of a finite dielectric medium was calculated by (Mohamed, et al., 1999; Naoe, et al., 1994)

$$\frac{dP(\omega)}{d\omega} = \frac{e}{\pi \hbar v^2} \int_{-\infty}^{\infty} dz \int_{-\infty}^{\infty} dz' \omega \text{Im} \left\{ -\exp\left[\frac{i\omega(z'-z)}{v}\right] V_i(r, r_0) \right\} \Bigg|_{r=(x_0, 0, z), r_0=(x_0, 0, z')}$$

(2.16)

where $V_i(r,r_0)$ is the potential due to the induced charge when a "stationary" electron is located at $r_0 = (x_0, 0, z')$, i.e., it is the homogeneous component of V satisfying

$$\nabla^2 V(r,r_0 0) = -\frac{e}{\varepsilon(\omega)\varepsilon_0}\delta(r-r_0) \qquad (2.17)$$

for the dielectric media considered. It is important to note that $V_i(r,r_0)$ is ω-dependent. The potential distribution in space is a quasi-electrostatic potential for each point along the trajectory of the incident electron. The integral over z' sums over the contributions made by all of the points along the trajectory of the incident electron. Therefore, valence-loss spectra are calculated to solve the electrostatic potential for a stationary electron located at r_0 in the dielectric media system.

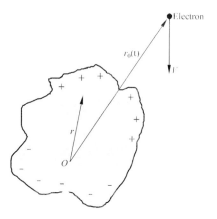

Figure 2.39 Excitation of a dielectric medium by an external incident electron

The electron energy-loss spectrum $dP(\omega)/d\omega$ is experimentally observable using electron energy-loss spectroscopy. In comparison to the theoretically calculated spectra according to (2.16), the dielectric function $\varepsilon(\omega)$ of the nanoparticle can be refined. This is a fine probe analysis on the electronic structure of a single nanoparticle. Figure 2.40(a) shows a comparison of the theoretically calculated inelastic excitation probability as a function of the electron impact parameter and the electron energy-loss from a carbon sphere comprised of concentric graphitic shells (Stockli, et al., 1998). The anisotropic dielectric properties of the graphitic shells were considered. The surface excitation increases significantly as the electron moves towards the surface. The experimental spectra shown in Fig. 2.40(b) agree quantitatively with the simulated results. The calculation has also been extended to carbon nanotubes (Stockli, et al., 1999).

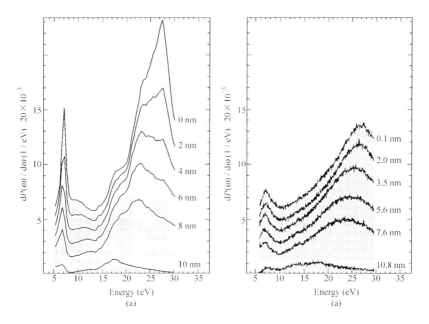

Figure 2.40 (a) Calculated EELS spectra of a carbon sphere (radius = 10 nm) with graphitic onion-like structure as a function of the electron impact parameter x_0. (b) Experimentally observed EELS spectra from a carbon sphere

2.8.2 Quantitative Nanoanalysis

Energy dispersive X-ray spectroscopy (EDS) (Goldstein, et al., 1992) and electron energy-loss spectroscopy (EELS) (Egerton, 1996) in TEM have been demonstrated as powerful techniques for performing microanalysis and studying the electronic structure of materials (Wang and Kang, 1998b). Atomic inner shell excitations are often seen in EELS spectra due to a process in which an atom bounded electron is excited from an inner shell state to a valence state accompanied by incident electron energy loss and momentum transfer. This is a localized inelastic scattering process, which occurs only when the incident electrons are propagating in the crystal. Figure 2.41 shows an EELS spectrum acquired from a superconductor compound $YBa_2Cu_4O_8$, where the ionization edges of O-K, Ba-M and Cu-L are present in the displayed energy-loss range. Since the inner shell energy levels are the unique features of the atom, the intensities of the ionization edges can be used effectively to analyze the chemistry of the specimen.

After subtracting the background, an integration is made to the ionization edge for an energy window of width Δ accounted by the threshold energy. Thus, the intensity of oscillation at the near edge region is flattened, and the

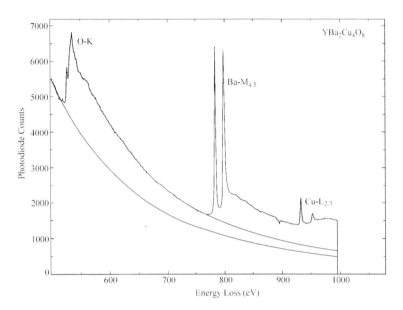

Figure 2.41 An EELS spectrum acquired from $YBa_2Cu_4O_8$ showing the application of EELS for quantitative chemical microanalysis, where the smooth lines are the theoretically simulated background to be subtracted from the ionization regions

integrated intensity is dominated by the properties of single atoms. This type of information is most useful in material analysis and the integrated intensity is given by

$$I_A(\Delta) \approx I_0(\Delta)\sigma_A(\Delta,\beta)n_A d \qquad (2.18)$$

where $I_0(\Delta)$ is the integrated intensity of the low-loss region including the zero-loss peak for an energy window Δ; $\sigma_A(\Delta,\beta)$ is the energy and angular integrated ionization cross section. In imaging mode, β is mainly determined by the size of the objective aperture or the upper cut-off angle, depending on which is smaller. In diffraction mode, the β angle is determined not only by the size of the EELS entrance aperture and the camera length but also by the beam convergence. In general, the width of the energy window is required to be more than 50 eV to ensure the validity of (2.18), and $\Delta = 100$ eV is an optimum choice. If the ionization edges of two elements are observed in the same spectrum, the chemical composition of the specimen is

$$\frac{n_A}{n_B} = \frac{I_A(\Delta)\sigma_B(\Delta,\beta)}{I_B(D)\sigma_A(\Delta,\beta)} \qquad (2.19)$$

This is the most powerful application of EELS because the spatial resolution is almost entirely determined by the size of the electron probe. The key quantity in this analysis is the ionization cross section. For elements with atomic numbers smaller than 14, the K edge ionization cross section can be calculated using the SIGMAK program (Egerton, 1996), in which the atomic wave function is approximated by a single-electron hydrogen-like model. The ionization cross section for elements with $13 < Z < 28$ can be calculated using the SIGMAL program. For a general case, the ionization cross section may need to be measured experimentally using a standard specimen with known chemical composition.

2.8.3 Near Edge Fine Structure and Bonding in Transition Metal Oxides

The energy-loss near edge structure (ELNES) is sensitive to the crystal structure. This is a unique characteristics of EELS and in some cases it can serve as a "fingerprint" to identify a compound. In EELS, the L ionization edges of transition-metal and rare-earth elements usually display sharp peaks at the near edge region (Fig. 2.42), which are known as white lines. For transition metals with unoccupied 3d states, the transition of an electron from 2p state to 3d levels leads to the formation of white lines. The L_3 and L_2 lines are the transitions from $2p^{3/2}$ to $3d^{3/2}d^{5/2}$ and from $2p^{1/2}$ to $3d^{3/2}$, respectively, and their intensities are related to the unoccupied states in the 3d bands. Numerous EELS experiments have shown that a change in valence state of cations introduces a dramatic change in the ratio of the white lines, leading to the possibility of identifying the occupation number of 3d orbital using EELS.

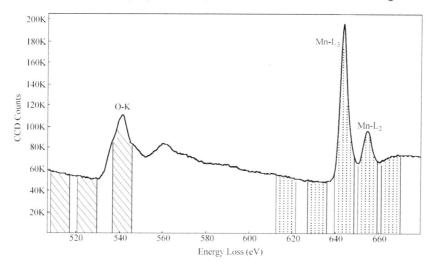

Figure 2.42 EELS spectrum acquired from MnO_2 showing the Mn-L_3 and Mn-L_2 white lines. The five windows pasted in the Mn-L edge are to be used for extracting the image formed by the ratio of white lines

EELS analysis of valence state is carried out in reference to the spectra acquired from standard specimens with known cation valence states. Since the intensity ratio of L_3/L_2 is sensitive to the valence state of the corresponding element, then if a series of EELS spectra are acquired from several standard specimens with known valence states, an empirical plot of these data serves as the reference for determining the valence state of the element present in a new compound (Kurata, et al., 1993; Pearson, et al., 1993; Mansot, et al., 1994; Fortner, et al., 1996; Lloyd, et al., 1995; Yuan, et al., 1994; Wang, et al., 1997a, 1997b). The L_3/L_2 ratios for a few standard Co compounds are plotted in Fig. 2.43(a). EELS spectra of Co-$L_{2,3}$ ionization edges were acquired from $CoSi_2$ (with Co^{4+}), Co_3O_4 (with $Co^{2.67+}$), $CoCO_3$ (with Co^{2+}) and $CoSO_4$ (with Co^{2+}). Figure 2.43(b) shows a plot of the experimentally measured intensity ratios of white lines L_3/L_2 for Mn. The curves clearly show that the ratio L_3/L_2 is very sensitive to the valence states of Co and Mn. This is the basis of our experimental approach for measuring the valence states of Co or Mn in a new material.

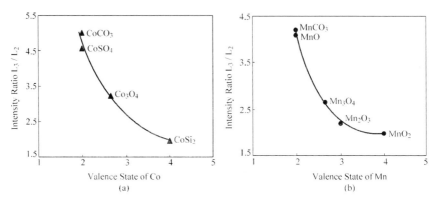

Figure 2.43 Plots of the intensity ratios of L_3/L_2 calculated from the spectra acquired from (a) Co compounds and (b) Mn compounds as a function of the cation valence. A nominal fit of the experimental data is shown by a solid curve

Determining the crystal structure of nanoparticles is a challenge, particularly when the particles are smaller than 5 nm. The intensity maxima observed in the X-ray or electron diffraction patterns of such small particles are broadened due to the crystal shape factor, and greatly reduce the accuracy of structure refinement. The quality of the high-resolution TEM images of the particles is degraded because of the strong effect from the substrate. This difficulty arises in our recent study of CoO nanocrystals whose shape is dominated by tetrahedrals of sizes smaller than 5 nm (Yin, et al., 1997). Electron diffraction indicates the crystal has an fcc-type cubic structure. The synthesized nanocrystals are likely to be CoO. EELS is used to measure the valence state of Co. Figure 2.44 shows a comparison of the spectra acquired from Co_3O_4 and CoO standard specimens and the synthesized nanocrystals. The relative intensity of the Co-L_2 to Co-L_3 for the nanocrystals is almost identical to that for CoO standard, while the Co-L$_2$ line of Co$_3$O$_4$ is significantly higher, indicating that the Co valence in the nanocrystals is 2+, confirmimg the CoO structure of the nanocrystals.

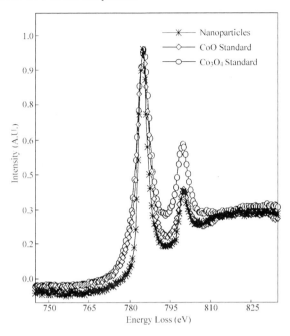

Figure 2.44 A comparison of EELS spectra of Co-L$_{2,3}$ ionization edges acquired from Co$_3$O$_4$ and CoO standard specimens and the synthesized nanocrystals, proving that the valence state of Co is 2+ in the nanocrystals. The full width at half-maximum of the white lines for the Co$_3$O$_4$ and CoO standards is wider than that for the nanocrystals, possibly due to size effect

2.8.4 Doping of Light Elements in Nanostructures

Today's electronics related information industry is mainly based on silicon technology. But owing to its indirect band-gap nature and poor photoluminescence (PL) property, silicon in bulk form has almost no future in opto-electronics devices which are desperately needed for high-speed signal transmission and processing systems. Motivated by great potential applications in the information industry, people have made intensive studies in searching for silicon based opto-electronic materials. Enlightened by quantum size confinement effect from nanomaterials, people have carried out a lot of work on quantum wells, quantum wires, and quantum dots. In these low-dimensional materials, the movements of carriers are restricted in a nano-scale, resulting in a band structure transformation from indirect band gap to a directed version and/or increasing of its band gap, which result in a possible PL from nanosized silicon. Along this line, a promising progress has been achieved recently by porous silicon and nanosilicon crystallites with efficient PL.

Recently, a new form of low-dimensional silicon material-silicon nanowires (Si-NWs) had been successfully fabricated by a high-temperature-laser-ablation method (Yu, et. al., 1998). The uniformity of the as-grown Si-NWs is very high, which provides an opportunity for systematic investigation of their physical properties and microstructures.

As mentioned above, Si-NWs has stimulated intensive interests in their physical properties and potential applications because of their low dimensionality and quantum-confinement effect (Yu, et al., 1999). However, controlled doping of Si-NWs is a prerequisite for the realization and efficiency of many Si-NW based devices. The properties of these nano-structured materials will be modified by doping a large amount of guest ions. Conventional doping techniques have some difficulties in doping Si-NWs, especially for controlled doping. Therefore we use the electrochemical insertion method to dope Si-NWs with Li^+ ions, and the doping level of the Si-NWs can be controlled efficiently by this method.

Figure 2.45 shows a series of selected area electron diffraction (SAED) patterns at different doping levels. The SAED pattern in Fig. 2.45(a) is taken from the undoped Si-NWs, and the sharp diffraction rings reveal the c-silicon

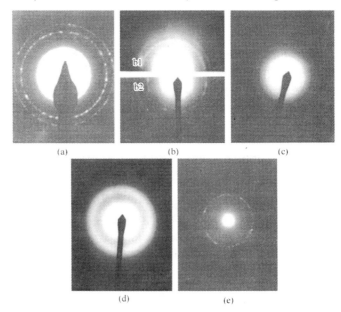

Figure 2.45 SAED patterns of Si-NWs with different doping doses of Li^+ ions: (a) undoped Si-NWs; (b) lightly doped; (c) heavily doped; (d) more heavily doped; and (e) extracted from the state of (d)

structure. Figure 2.45(b) is taken from Si-NWs with a light doping level, which shows two different diffraction patterns. One is with c-Si structure and the other is an unknown structure. With the increasing dose of Li^+ ions, only the weak diffraction pattern of the c-Si structure is found. At the more heavily doped stage, the crystalline diffraction pattern disappears and diffuse diffraction halos are present in Fig. 2.45(d), indicating that the crystalline structure is destroyed completely. However, the crystalline diffraction pattern indexed with the c-Si structure (Fig. 2.45(e)) is visible again when the Li^+ ions are extracted from the Si-NWs.

The HREM images in Fig. 2.46(a)—(d), respectively, correspond to the Li^+ ion doping of the state of the Si-NWs as indicated in Fig. 2.45(b)—(e). A crystalline core and an amorphous outer layer with thickness of 3 nm can be seen in Fig. 2.46(a). The interspacing is 0.31 nm in the core of the nanowire, which corresponds to the {111} plane. However, the spacing of 0.32 nm is also visible at some regions. Furthermore, the amorphous layer at the outer part of the doped Si-NWs is also thicker than that occurring as the surface layer of undoped nanowires, and some completely amorphous regions are also visible as indexed by the black arrows. This nonuniform lattice expansion and destruction of the crystalline structure of the Si-NWs indicate a heterogeneous distribution of Li^+ ions. The HREM image in the inset of Fig. 2.46(a) is taken from the Li^+ ion-doped Si-NWs whose diffraction pattern is shown in Fig. 2.45(b). The interplanar spacing of the image is 0.25 nm. With increasing doping dose, the lattice expansion becomes more obvious and the outer amorphous region develops further towards the core of the Si-NWs (Fig. 2.46(b)). The amorphous layer is about 5 nm, and many regions with a crystalline structure in the core of the Si-NWs are destroyed completely. The spacing of the remaining crystalline regions is about 0.32 nm. When the Si-NWs are doped with a high dose of Li^+ ions, the crystalline structure is destroyed completely and is transformed into amorphous Li-Si nanowires (Fig. 2.46(c)).

The foregoing HREM investigation found that the lattice of the crystalline Si will expand due to the insertion of Li^+ ions. With increasing doping dose, the Si—Si bonds are broken and, at last, the crystalline nanowires transform into the amorphous phase. On the other hand, the kinetic metastable phase will disappear with increasing doping dose. When the Li^+ ions are extracted from the heavily doped Li-Si nanowires, the short range ordered structure with lattice distortion will be formed again in some regions of the nanowires, as shown in Fig. 2.46(d).

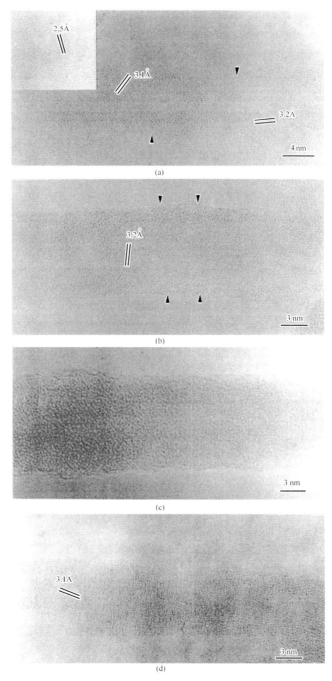

Figure 2.46 HREM images of Si-NWs at different doping states (a—d). Microstructural characteristics of the doping state shown in (b—e) of Fig. 2.45

2.9 Energy-Filtered Electron Imaging

Energy-filtered transmission electron microscopy (EF-TEM) is a rapidly developing field for high spatial-resolution chemical imaging. With an energy filter, images (or diffraction patterns) formed by electrons with specific energy losses can be obtained (Fig. 2.44) (Reimer, 1995). Here we demonstrate a few applications of EF-TEM.

2.9.1 Chemical Imaging of Giant Magnetoresistive Multilayers

Magnetic multilayers (MLs) with giant magnetoresistance (GMR) effect have huge promise for future technology (Baibich, et al., 1988). Among different types of GMR MLs, NiFe/Cu based MLs are regarded as good candidates because their high MR ratios and field sensitivity for magneto-electronic device applications. Since the prominent physical property of GMR can only occur in MLs with sub-layers 1—2 nanometers thick, the microstructure of MLs plays an important role in oscillatory exchange coupling between magnetic layers and in spin-dependent electron scattering. It has been reported that the GMR effect is strongly dependent on the microstructure of the NiFe/Cu films (Miyamoto, et al., 1996; Naoe, et al., 1994; Nakatani, et al., 1993). For example, it is proposed (Pettit, et al., 1997) that pinholes in the Cu spacer layers in NiFe/Cu MLs may be responsible for the ferromagnetic coupling of neighboring magnetic layers and will lead to strong biquadratic coupling. But there is not enough experimental result to support such speculation.

Columnar crystallites (CCs) were found as a prominent structure character in these films (Fig. 2.47). The CC starts from the Fe buffer and penetrates all the sub-layers upward to the surface of the film. Similar CCs were also found in NiFe/Cu based spin valves. If some of the CCs are single crystal phase of NiFe, they would provide local ferromagnetic bridges in the multi-layer film and result in the possible decrease of GMR.

Though the multi-layer structure within the CCs can be seen from diffraction contrast images, but it is relatively hard to recognize from the entire film, due to the complex diffraction contrast resulting from variation of crystallographic orientations, defects in the CCs and Morie fringes from overlapping of two CCs. The difficulty in identifying the NiFe and Cu layers by HREM images arises from the small difference in electron scattering ability between NiFe and Cu and their small lattice mismatch (~2%). A careful examination of the digitized HRTEM images of CCs shown in Fig. 2.48 did not show any change in lattice spacing along growth direction of the film, which

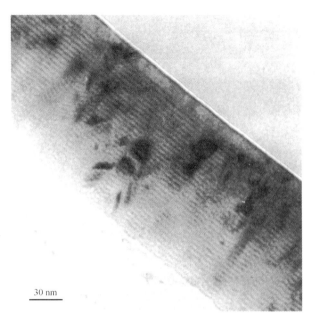

Figure 2.47 Cross-sectional TEM image of $Ni_{80}Fe_{20}/Cu$ multi-layers. Note the pronounced columnar structure. Multi-layer structure inside the columnar crystallites could be identified clearly by electron diffraction contrast images

reveals a two-fold possibility. First, the CC could be a single crystal of NiFe and Cu, which may result in the formation of ferromagnetic "bridge" and non-ferromagnetic pinholes, respectively. Another possibility is to form an alternate coherent growth of NiFe and Cu layers in the CCs. Accordingly, we need a more direct experiment measurement to investigate the structure of the CCs.

EF-TEM can provide chemical information at high spatial resolution of 1 nm in magnetic MLs as well as in spin valves. Figure 2.49 is an EELS spectrum in which electron energy loss peaks corresponding to inner-shell ionization of $Fe-L_{2,3}$, $Ni-L_{2,3}$ and $Cu-L_{2,3}$ corresponding to energy losses of 708 eV, 855 eV and 931 eV are shown, respectively. Figure 2.50 shows a series of elemental mapping of Fe, Ni and Cu by using the three-window method (Gao, et al., 1998; Zhang, et al., 1996) on above EELS spectrum from $Ni_{80}Fe_{20}/Cu$ MLs, which shows, at nanometer level, again the multi-layered structure of the film with CCs structure. In other words, the multi-layer structure is maintained within the CCs, though HREM images showing epitaxy growth and no lattice difference between the two different types of sub-layers.

Comparing with the electron diffraction contrast image shown in Fig. 2.50 (a), the multilayered structure of CCs can be seen more clearly from the elemental mapping by using Ni, Fe and Cu signals as shown in Fig. 2.50(b)—

Figure 2.48 HRTEM image of a columnar crystallite in $Ni_{80}Fe_{20}$/Cu multi-layer showing coherent growth of NiFe and Cu with f.c.c. structure. Twin defects are also visible

(d). This shows the advantage of elemental mapping in chemical analysis on MLs. It should also be noted that Fe concentration in this film is only 20% in each NiFe sub-layer, therefore the signal/noise ratio is not high enough to show a sharp elemental image. Nevertheless, the Fe buffer can be seen more clearly in Fig. 2.50(b). Because the atomic ratio of Ni in NiFe sub-layer is four times higher than Fe, the layered structure of the film appeared more pronounced in Fig. 2.50(c) by Ni mapping. The non-magnetic layer of Cu could also be clearly seen in Fig. 2.50(d).

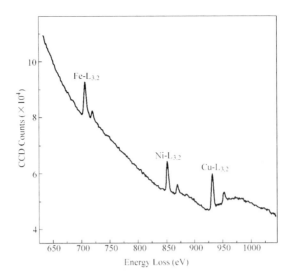

Figure 2.49 Electron energy loss spectrum of $Ni_{80}Fe_{20}/Cu$ multi-layers

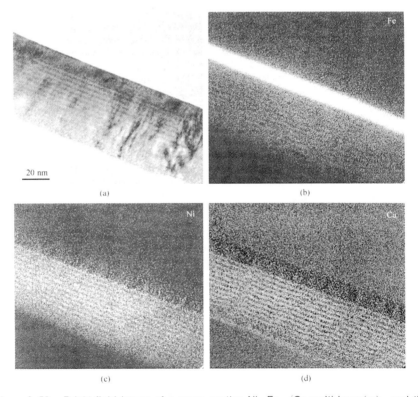

Figure 2.50 Bright-field image of a cross section $Ni_{80}Fe_{20}/Cu$ multi-layer(a), and the corresponding elemental maps of Fe (b), Ni (c) and Cu (d) by using their $L_{2,3}$ edges, respectively. The multilayer structure can be seen clearly from these elemental mappings

2.9.2 Imaging of Spin Valves

Spin valve (SV) is a more promising structure that could be developed for real application as GMR read heads. Transport properties in SV are very closely related to the microstructure of these materials. The density of grain boundaries within each layer as well as the quality of the interfaces between the layers influences markedly the resistivity of the SV film. Figure 2.51(a) is the bright-field TEM image of the SV of Ta/NiFe/Cu/NiFe/FeMn/Ta. The dark area is the Ta layer. Above it, the film exhibits the columnar grains with an average size of about 20 nm. However the sub-layers such as NiFe, Cu and FeMn could not be distinguished clearly. The element mapping of Ni, Fe and Mn was carried out in this area using electron energy loss peak corresponding inner-shell ionization edge of Ni-$L_{2,3}$ (855 eV), Fe-$L_{2,3}$ (708 eV) and Mn-$L_{2,3}$ (640 eV). Figure 2.51(b), (c) and (d) are Ni, Fe and Mn maps, respectively. It shows clearly that the sub-layers of NiFe and FeMn are very smooth and straight. In addition one can see the contrast variation in the maps even if in the same sub-layer. This is owing to the diffraction effects visible in Fig. 2.51(a) and in the energy-filtered images as well (Hofer, et al., 1995). Such effect could be used to identify the columnar grain boundaries in the element map. Figure 2.51(b) shows that Ni in the pinned layer protrudes to the Cu layer at grain boundary 1 (GB1). Figure 2.52 gives the line profiles of Ni, Fe, Mn maps along the direction of film growth. It shows that there is no obvious concentration difference of Ni between the NiFe free layer and pinned layer within the grains. However, at GB1, the concentration of Ni in the pinned layer is markedly lower than that in the free layer, while the concentration of Ni in the Cu layer increased evidently. This indicates that Ni diffuses towards Cu layer along the columnar grain boundary. The profiles of Fe within the grains are almost the same as that at GB1, indicating the diffusion of Fe towards Cu is not significant even if at grain boundaries. The profiles of Mn are almost constant too, indicating the diffusion of Mn in the FeMn layer towards the NiFe pinned layer is not striking both within the grain and at grain boundaries.

These results are not difficult to understand. For Ni and Cu, they can completely dissolve each other and the diffusivity along the grain boundaries is much higher than that in the bulk. So we could observe that Ni diffuses towards the Cu layer along the columnar grain boundaries. However, Fe and Cu are a phase separating system and solubility of Mn in Ni is limited, so the diffusions of Ni towards the Cu layer and Mn towards the NiFe layer are not striking even at columnar grain boundaries in SVs.

Figure 2.51 shows (a) bright-field TEM image of the SV and elemental

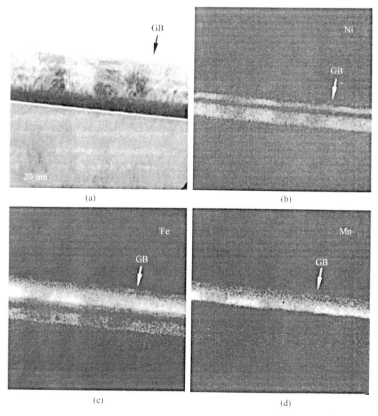

Figure 2.51 (a) Bright-field TEM image of the spin valve, elemental maps of (b) Ni, (c) Fe and (d) Mn, respectively. GB stands for columnar grain boundary

maps of (b) Ni, (c) Fe and (d) Mn, respectively. GB stands for columnar grain boundary.

In Fig. 2.51(b), the diffusion of Ni to the Cu layer along the GB is not obvious. This implies that the structure of grain boundaries may influence the diffusion. At small angle grain boundaries and at large angle grain boundaries with some special orientation angle whose grain boundary energy is lower, the diffusion of Ni to Cu may not be striking compared with that at large angle grain boundaries with high grain boundary energy. So at some columnar grain boundaries the diffusion of Ni to Cu could be observed, however at other grain boundaries the diffusion could not be observed.

The thickness of sub-layers could be calculated from line profiles. We define the width measured at half maximum of the profile as the thickness of one sub-layer. The thicknesses of NiFe free layer, Cu space layer, NiFe pinned layer and FeMn pinning layer are 8.5, 2.8, 3.5 and 9.8 nm, respectively.

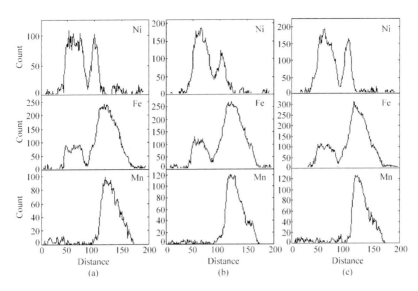

Figure 2.52 Line profiles of Ni, Fe, Mn maps along the direction of film growth. (b) is profiles at GB1. (a) and (c) are the profiles within the grains at the left and right of GB1, respectively

2.9.3 Mapping Valence States of Transition Metals

If the energy-selected electron image can be formed using the fine features at the near edge region, one can map the spatial distribution of the valence states. Here we use Co as an example to illustrate the procedure of mapping valence states. In EF-TEM, an energy window of 12 eV in width is required to isolate the L_3 from L_2 white lines (Fig. 2.42). A five-window technique is introduced: two images are acquired at the energy losses prior to the L ionization edges, and they are to be used to subtract the background for the characteristic L edge signals; two images are acquired from the L_3 and L_2 white lines, respectively, and the fifth image is recorded using the electrons right after the L_2 line that will be used to subtract the continuous background underneath the L_3 and L_2 lines. Then an L_3/L_2 ratio image will be obtained, which reflects the distribution of valence states across the specimen. It must be pointed out that the thickness effect has been removed in the L_3/L_2 image.

A partially oxidized CoO specimen that contains the CoO and Co_3O_4 grain structure was chosen for this study (Wang, et al., 1999a). The CoO and Co_3O_4 phases are separated by clear boundaries and it is an ideal specimen

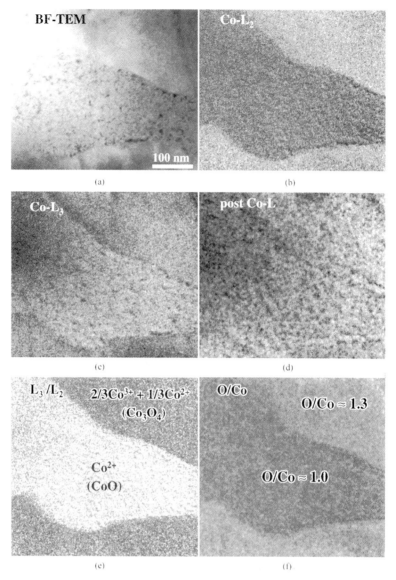

Figure 2.53 (a) Bright-field TEM image of a CoO-Co_3O_4 specimen. Corresponding energy-filtered images recorded from the signals of (b) the Co-L_3 edge, (c) the Co-L_2 edge, and (d) the post-Co-L_2 line. (e) The processed L_3/L_2 image, displaying the distribution of the valence states. (f) The atomic concentration ratio image of O/Co. The continuous background contributed from the single atom scattering has been removed from the displayed Co-L_3 and Co-L_2 images. The O/Co image is normalized in reference to the standard composition of CoO for the low portion of the image in order to eliminate the strong influence on the ionization cross section from the white lines. Each raw image was acquired with an energy window width of $\Delta = 12$ eV except for O-K at $\Delta = 24$ eV

for testing the optimum resolution. Figure 2.53 shows a group of energy-filtered TEM images from a triple point in the CoO-Co_3O_4 specimen. The energy-filtered images using the L_3 and L_2 lines and the post-L_2 line region (Figs. 2.53(b)—(d)) show distinctly the difference in contrast distribution due to a difference in the relative white line intensities. From these three images L_3/L_2 is calculated after subtracting the contribution from the continuous energy-loss region that is due to single atom scattering, and the image clearly displays the distribution of cobalt oxides having different valence states (Fig. 2.53(e)), where the diffraction contrast disappears. The region with the lower oxidation state (Co^{2+}) shows stronger contrast, and the ones with high oxidation states show darker contrast (see the L_3/L_2 ratio displayed in Fig. 2.43(a)). The O/Co image was calculated from the images recorded from the O-K edge and the $L_3 + L_2$ white lines for an energy window width of $\Delta = 24$ eV. Although the energy-filtered O-K edge image exhibits some diffraction contrast and the thickness effect, the O/Co compositional ratio image greatly reduces the effect. The high-intensity region in the O/Co image indicates the relative high local concentration in oxygen (e.g., higher Co oxidation states), the low-intensity region contains relatively less oxygen (e.g., lower Co valence state), entirely consistent with the information provided by the L_3/L_2 image. A spatial resolution of ~ 2 nm has been achieved. This is remarkable in comparison to any existing techniques.

2.10 Energy Dispersive X-ray Microanalysis (EDS)

X-rays emitted from atoms represent the characteristics of the elements, and their intensity distribution represents the thickness-projected atom densities in the specimen. This is the energy dispersive X-ray spectroscopy (EDS), which has played an important role in microanalysis, particularly for heavier elements (Goldstein, et al., 1992). EDS is a key tool for identifying the chemical composition of a specimen. A modern TEM is capable of producing a fine electron probe of smaller than 2 nm, allowing direct identification of the local composition of an individual nanocrystal. Shown in Fig. 2.54 is an EDS spectra acquired from a sample of nanocrystalline Ga-Sr-S. The peak intensity is directly related to the chemical composition of the corresponding element. Quantitative analysis gives Sr : Ga : S = 14 : 26 : 60, e.g., $SrGa_2S_4$. Detailed procedures can be found in Goldstein et al. (1992).

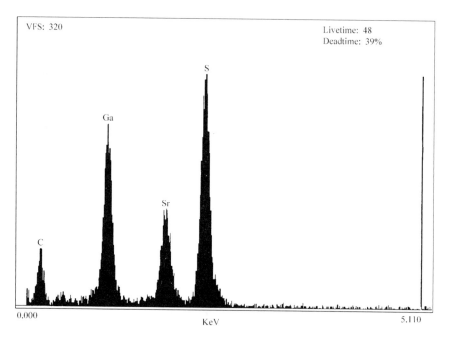

Figure 2.54 EDS spectrum acquired from a specimen of Ga-Sr-S

2.11 Summary

Transmission electron microscopy and associated techniques are powerful tools for characterization of nanophase materials. This chapter mainly introduced the high-resolution imaging in TEM and some newly developed techniques, such as holography and electron energy filtering. High spatial resolution analysis is vitally important for solving many of the practical problems of nanomaterials. Spectroscopy analysis of the solid state effects and the valence states mapping are new directions of quantitative microscopy.

in situ TEM is anticipated to be important for characterizing and measuring the properties of individual nanoparticles, from which the structure-property relationship can be clearly registered to a specific nanoparticle/structure. An emphasis has been placed on new developments in TEM for in situ nanomeasurements of the mechanical and electrical properties of individual nanostructures, aiming to correlate the measured properties with the nanostructure. This is a new direction in TEM and it is important for characterizing nanophase materials. Therefore, TEM is truly a versatile tool not only for structure analysis both crystallographically and chemically, but also a powerful approach for nanomeasurements.

References

Ahmadi, T. S., Z. L. Wang, T. C. Green, A. Henglein and M. A. El-Sayed. Science, 28, 1924 (1996)
Allpress, J. G. and J. V. Sanders. Surf. Sci., 7, 1 (1967)
Ascencio J. A., C. Gutiorez-Wing, M. E. Espinosa, M. Marin, S. Tehuacanero, C. Zorrilla and M. José-Yacaman. Surf. Sci. 396, 349 (1998)
Baibich, M. N., J. M. Broto, A. Fert, F. Nguyen Van Dau, F. Petroff, P. Etienne, G. Creuzet, A. Friederich, and J. Chazelas. Phys. Rev. Lett., 61, 2472 (1988)
Buffat, P.-A., M. Floli, R. Spycher, P. Stadelmann and J.-P. Borel. Faradat Discuss., 92, 173 (1991)
Chapman, J. N., A. B. Johnson, L. J. Heyderman, I. R. , S. McVitie, W. A. P. Nicholson, and B. Bormans. IEEE Trans. Magn., 30, 4479 (1994)
Chapman, J. N., I. R. McFadyen and S. McVitie. IEEE Trans. Magn., 26, 1506 (1990)
Cowley, J. M. and A. F. Moodie. Acta Cryst., 10, 609 (1957)
Cowley, J. M., Diffraction Physics. 3rd revised edn. (Elsevier Science B. V., 1995)
Cowley, J. M., Ultramicroscopy. 41, 335 (1992)
Datye, A. K., D. S. Kalakkad, E. Voekl and L. F. Allard. in Electron Holography. A. Tonomura, L. F. Allard, G. Pozzi, D. C. Joy and Y. A. Ono eds., Elsevier, 199 (1995)
de Heer, W. A., Chatelain, A., Ugarte, D., Science. 270, 1179 (1995)
Dresselhaus, M. S., G. Dresselhaus, and P. C. Eklund, (Eds.). Science of Fullerenes and Carbon Nanotubes. Academic Press, San Diego (1996)
Ebbesen, T. W. Phys. Today, 49, 26 (1996)

Acknowledgment Z. L. Wang is grateful to his collaborators, J. S. Yin, W. A. de Heer, P. Poncharal, T. Ahamed, M. A. El-Sayed, R. L. Whetten, T. Green, M. Mohamadi and J. Petroski. Research was partially sponsored by NSF grant DMR-9733160. Thanks to the Georgia Tech Electron Microscopy Center for providing the facility. YL wishes to acknowledge his collaborators D. J. Sellmyer, S. Bandyopadhyay, S. H. Liou, Z. S. Shan and J. P. Liu. Research at UNL is partially supported by AFOSR F49620-98-1-0098 and CMRA.

Egerton, R. F.. Electron Energy-Loss Spectroscopy in the Electron Microscope. 2nd edn. (Plenum Pub. Co., New York, 1996)
Fan, S.S., Chapline, M.G. et al.. Science. **283**, 512 (1999)
Fortner, J.A. and E.C. Buck. Appl. Phys. Lett.. **68**, 3817 (1996)
Frank, S., P. Poncharal, Z.L. Wang, and W.A. de Heer. Science. **280**, 1744 (1998)
Frost, B.G., L.F. Allard, E. Volkl, and D.C. Joy. in Electron Holography. eds., A. Tonomura, L.F. Allard, G. Pozzi, D.C. Joy and Y.A. Ono, Elsevier Science B.V., (1995)
Gabor, D.. Proc. Roy. Soc.. London. A, **197**, 454 (1949)
Gao, M. and X.F. Duan. Materials Transactions. **39**, 883 (1998)
Goldstein, J.I., D.E. Newbury, P. Echlin, D.C. Joy, A.D. Romig, Jr., C.E. Lyman, C. Fiori and E. Lifshin. Scanning Electron Microscopy and X-ray Microanalysis. Plenum, New York (1992)
Hirayama, T., J. Chen, Q. Ru, K. Ishizuka, T. Tanji and A. Tonomura. in Electron Holography. A. Tonomura, L.F. Allard, G. Pozzi, D.C. Joy and Y.A. Ono eds., Elsevier, (1995)
Hirsch, P.B., A. Howie, R.B. Nicholson, D.W. Pashley and M.J. Whelan. Electron Microscopy of Thin Crystals. Roberts E. Krieger Publishing Company, New York (1977)
Hofer, F., P. Warbichler and W. Grogger. Ultramicroscopy. **59**, 15 (1995)
Iijima, S.. Nature. **354**, 56 (1991)
Ino, S.. J. Phys. Soc. Japan. **21**, 346 (1966)
Ino, S.. J. Phys. Soc. Japan. **27**, 941 (1969)
Kang, Z.C., Z.L. Wang. Philos. Maga.. B, **73**, 905 (1996)
Kroto, H.W. and K. McKay. Nature. **331**, 328 (1988)
Kroto, Heath, O'Brien, Curl. and R.E. Smalley. Nature. **318**, 162 (1985)
Kurata H., and C. Colliex. Phys. Rev.. B **48**, 2102 (1993)
Li, W.Z., Xie, S.S., Qian et al.. Science. **274**, 1701 (1996)
Lichte, H.. Adv. Optical and Electron Microsc. **12**, 25 (1991)
Liu, P., Y. Liu, R. Skomski and D.J. Sellmyer. IEEE Trans. Mag.. in press (1999)
Liu, Y., Z.S. Shan, and D.J. Sellmyer. IEEE Trans. on Magn.. Vol. **32**, 3614 (1996)
Liu, Y., D.J. Sellmyer, B.W. Robertson, Z.S. Shan, and S.H. Liou. IEEE Trans. On Magn.. **31**, 2740 (1995)
Liu, Y., R.A. Thomas, S.S. Malhotra, Z.S. Shan, S.H. Liou and D.J. Sellmyer. J. Appl. Phys.. **83**, 6244 (1998)
Lloyd, S.J., G.A. Botton, and W.M. Stobbs. J. Microsc. **180**, 288 (1995)
Mankos, M., J.M. Cowley and M.R. Scheinfein. Mater. Res. Soc. Bulletin, XX (October) 45 (1995)
Mansot, J.L., P. Leone, P. Euzen, and P. Palvadeau. Microsc. Microanal.

Microstruct. 5, 79 (1994)

Marks, L. D., Rep. Prog. Phys., 57, 603 (1994)

Meirovich, L., Element of Vibration Analysis. McGraw-Hill, New York (1986)

Miyamoto, Y., T. Yoshitani, S. Nakagawa and M. Naoe. IEEE Trans. Magn., 32, 4719 (1996)

Mohamed, M., Z.L. Wang and M.A. El-Sayed. J. Phys. Chem., B, 103, 10255 (1999)

Nakatani, R., T. Dei, Y. Sugita. J. Appl. Phys., 73, 6375 (1993)

Naoe, M., Y. Miyamoto and S. Nakagawa. J. Appl. Phys., 75, 6525 (1994)

Pan., Z.W., Xie, S.S., Chang, B.H. et al., Very long carbon nanotubes, Nature. 394, 631 (1998)

Pearson, D.H., C.C. Ahn, and B. Fultz. Phys. Rev., B 47, 8471 (1993)

Petroski, J.M., Z.L. Wang, T.C. Green, M.A. El-Sayed., J. Phys. Chem., B 102, 3316 (1998)

Pettit, K., S. Gider, S.S.P. Parkin, and M.B. Salamon. Phys. Rev., B 56, 7819 (1997)

Poncharal, P., Z.L. Wang, D. Ugarte, and W.A. de Heer. Science. 283, 1516 (1999)

Reimer, L. (ed.). Energy Filtering Transmission Electron Microscopy, Springer Series in Optical Science. Springer Verlag, (1995)

Ren, Z.F., Huang, Z.P., Xu, J.H., Wang, P.B., Siegal, M.P., Provencio, P.N., Synthesis of large arrays of well-aligned carbon nanotubes on glass, Science. 282, 1105 (1998)

Rinzler, A.G., Hafner, J.H., Nikolaev, P. et al., Science. 269, 1550 (1995)

Salvetat J.P., A.J., Kulik, J.M. Bonard, et al., Adv. Mater. 11, 161 (1999)

Schattschneider, P. and B. Jouffrey, B., in Energy Filtering Transmission Electron Microscopy, ed. L. Reimer, Springer Series in Optical Science. Springer Verlag 71, 151 (1995)

Smith, D.J. and L.D. Marks. Philos. Mag., A 44, 735 (1981)

Stöckli, T., Z.L. Wang, J.-M. Bonard, P. Stadelmann and A. Chatelain. Philos. Mag., B 79, 1531 (1999)

Stöckli, T., J.M. Bonard, A. Chatelain, Z.L. Wang and P. Stadelmann. Phys. Rev. B 57, 15599 (1998)

Tanji, T., M. Maeda, N. Ishigure, N. Aoyama, K. Yamamoto, T. Hirayama. Physical Review Lett., 83, 1038 (1999)

Tonomura, A., Electron Holography. Springer-Verlag, New York, (1993)

Tonomura, A., L.F. Allard, G. Pozzi, D.C. Joy and Y.A. Ono eds., Electron Holography. Elsevier Science B.V. (1995)

Treacy, M.M., T.W. Ebbesen, J.M. Gibson. Nature. 38, 678 (1996)

Wang, Z. L. (ed.). Characterization of Nanophase Materials. Wiley-VCH (1999)

Wang, Z. L., Adv. Materials. **10**, 13 (1998)

Wang, Z. L. and Z. C. Kang. Functional and Smart Materials—Structural Evolution and Structure Analysis. Plenum Press, New York (1998b)

Wang, Z. L., Elastic and Inelastic Scattering in Electron Diffraction and Imaging. Plenum Pub. Co., New York, Chapts. 2 and 3 (1995)

Wang, Z. L., J. Bentley and N. D. Evans. J. Phys. Chem., B (1999a)

Wang, Z. L., M. Mohamed, S. Link and M. A. El-Sayed. Surf. Sci., **440**, L809 (1999b)

Wang, Z. L., J. Petroski, T. Green and M. A. El-Sayed. J. Phys. Chem., B **102**, 6145 (1998)

Wang, Z. L., J. S. Yin, J. Z. Zhang, and W. D. Mo. J. Phys. Chem., B **101**, 6793 (1997a)

Wang, Z. L., J. S. Yin, Y. D. Jiang, and J. Zhang. Appl. Phys. Lett., **70**, 3362 (1997b)

Wang, Z. L., Micron. **27**, 265 (1996)

Wang, Z. L., Micron. **28**, 505 (1997)

Wang, Z. L., P. Poncharal, and W. A. de Heer. Proc. Intern. Union of Pure and Applied Chemistry. **72**, 209 (2000)

Wang, Z. L., T. S. Ahmadi and M. A. El-Sayed. Surf. Sci., **380**, 302 (1996)

Whetten, R. L., J. T. Khoury, M. M. Alvarez, S. Murthy, I. Vezmar, Z. L. Wang, C. C. Cleveland, W. D. Luedtke, U. Landman. Adv. Materials. **8**, 428 (1996)

Wong, E., P. Sheehan, C. Lieber. Science. **277**, 1971 (1997)

Yang, C. Y., J. Cryst. Growth. **47**, 274 (1979a)

Yang, C. Y., J. Cryst. Growth. **47**, 283 (1979b)

Yao, M. and D. J. Smith. J. Microsc., **175**, 252 (1994)

Yin, J. S. and Z. L. Wang. Phys. Rev. Lett., **79**, 2570 (1997)

Yu, D. P., C. S. Lee, I. Bello, X. S. Sun, Y. H. Tang, G. W. Zhou, Z. G. Bai, Z. Zhang, and S. Q. Feng. Sol. Stat. Comm., **105**, 403 (1998)

Yu, D. P., Z. G. Bai, J. J. Wang, Y. H. Zou, W. Qian, J. S. Fu, H. Z. Zhang, Y. Ding, G. C. Xiong and S. Q. Feng. Physical Review. B. **59** (4), 2498 (1999)

Yu, Y.-Y., S.-S. Chang, C.-L. Lee and C. R. Wang. J. Phys. Chem., B **101**, 6661 (1997)

Yuan, J., E. Gu, M. Gester, J. A. C. Bland, and L. M. Brown. J. Appl. Phys., **75**, 6501 (1994)

Zhang, S. F. and P. M. Levy. Phys. Rev. Lett., **77**, 916 (1996)

3 Scanning Electron Microscopy

Alexander J. Shapiro

3.1 Introduction

Scanning electron microscopy (SEM) has become one of the most versatile and useful methods for direct imaging, characterization, and studying of solid surfaces. Just a simple partial enumeration of possible modes of the scanning electron microscope operation shows its capabilities:

Secondary, backscattered, and absorb electron imaging and (SEI, BEI, and AEI, correspondingly), magnetic contrast (Type Iinteraction with secondary electrons, type IIinteraction with backscattered electrons, type IIIpolarization of secondary electrons (SEMPA, scanning electron microscopy with polarization analysis)), X-ray microanalyzer (energy-dispersive spectroscopy (EDS) and wavelength-dispersive spectroscopy (WDS)), X-ray mapping, voltage contrast, electron beam induced current (EBIC) or charge collection microscopy (CCM), cathodoluminescence (CL), scanning deep level transient spectroscopy (SDLTS), electron channeling (ECP), electron backscattered patterns (EBSP) and electron backscattered diffraction (EBSD) with orientation imaging microscopy (OIM), scanning electron-acoustic (thermal wave) microscopy, cryo- or high-temperature microscopy, etc.

Such diversity of applications requires scanning electron microscopes specifically designed for a range of tasks: high-resolution SEM (the surface and topography of solids can be imaged at resolutions approaching 0.3 nm at 30 keV or 2.5 nm at 1 keV without complex and the time-consuming sample preparation required for TEM observation), X-ray microanalyzer (for chemical analysis of small volumes of materials), variable pressure SEM (allows us to investigate specimens in their natural state or under natural environmental conditions, without the need for conventional preparation techniques), and multi-purpose SEM (an instrument for broad variety of applications), for example.

In this chapter we will describe only modes of the SEM operation, which are related to investigations of nanostructured and nanophase materials.

3.2 Basic Principals of Scanning Electron Microscopy

A scanning electron microscope consists of a column, electronics and a number of detectors. A schematic sectional view of a SEM (JEOL JSM-840) column is presented in Fig. 3.1.

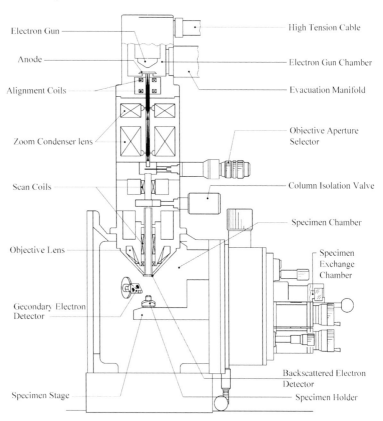

Figure 3.1 Schematic sectional view of a conventional type SEM (JEOL JSM-840)

Primary monochromatic electrons are generated and accelerated down the evacuated column by an electron beam source. The beam energy can be usually selected over a range from a 200 V to 50 keV. Demagnified by the column with the aid of two or more electromagnetic lenses and constricted by the condenser apertures, the electrons are formed into a probe in the sample plane with a spot size from a fraction of a nanometer to micrometers. The electron beam current can be varied from picoamperes to amperes, depending on the choice of aperture and lens strength. Electron bombardment of the sample generates numerous signals: secondary electrons, backscattered electrons, characteristic X-rays, Auger electrons, and photons (cathodoluminescence). The deflection coils raster the electron probe across the sample and, in a synchronized mode, another electron beam runs over a monitor modulating the brightness with the detector output. The ratio of the linear size of the scanned monitor to the linear size of the scanned area on the sample is the image magnification.

3.2.1 Main Parameters of Electron Optics

The relationship between the main parameters of the electron optics are defined by the brightness of the electron source equation:

$$\beta = \frac{4I_b}{\pi^2 d^2 \alpha^2} \tag{3.1}$$

where d is the beam diameter, I_b is the beam current on the sample surface, and α is the beam divergence semi-angle.

This equation shows that the parameters are not independent and proper selection of the above parameters (that means selecting initially the proper type of the scanning electron microscope and detectors) is essential for obtaining the desired results in the selected mode, such as high resolution, microanalysis, high depth-of-field, etc.

Electrons can be emitted from a conductive material either by heating it to the point where the electrons have sufficient energy to overcome the work function barrier of conductor (thermionic sources) or by applying an electric field, which is sufficiently strong that they tunnel through the barrier (field emission sources). The virtual source size (determines the amount of demagnetization required to obtain desired beam diameter), its brightness (the brighter is the source the higher the current density in the electron beam), and the energy spread (a narrow energy spread minimizes chromatic aberrations) of the emitted electrons are the three key parameters of the electron sources.

Let's compare the electron emitters utilized in scanning electron microscopes:

(1) thermionic tungsten (W) hairpin cathodes.
It is a directly heated cathode. Due to evaporation, since the high working temperature is near the melting point, the lifetime of the W cathodes is limited to 60—80 h. The emission current density j_c is the very large. This emitter is very well suited for relatively low magnification imaging and X-ray microanalysis especially with WDS (wavelength-dispersive spectrometer).

(2) thermionic lanthanum or cerium hexaboride (LaB_6, CeB_6) rod cathodes.

These indirect heated emitters are up to 10 times brighter than the W cathode and are the most common electron sources in modern SEMs. But they are quite sensitive to damage during operation and require vacuum in the gun chamber two orders of magnitude better than for W filaments. CeB_6 has longer lifetimes due to a lower evaporation rate than LaB_6.

(3) Schottky field-emission ZrO/W(100) and cold field-emission W(310) point-source cathodes are superior to thermionic sources in terms of source size, brightness, and lifetime. Both are up to 1000 times smaller and up to 100 times brighter than thermionic electron sources. Due to very small energy spread of the cold field-emission source it has improved performance for low-voltage operation. On the other side, the Schottky emitter provides greater emission stability and higher beam current necessary for analytical applications (But the vacuum requirements for field-emission guns are much higher: 1×10^{-10} Pa and 1×10^{-8} Pa for cold-field and Schottky emitters respectively (see Table 3.1).

The main parameters of the sources are presented in Table 3.1.

Table 3.1 Electron source characteristics (after LEO Instrument)

Electron source characteristics

Emitter Type	thermionic	thermionic	cold FE	Schottky FE
Cathode material	W	LaB_6-CeB_6	W(310)	ZrO/W (100)
Operating temperature (K)	2700—2800	1900—2000	300	1800
Cathode radius (nm)	60000	10000	≤100	≤1000
Effective source radius (nm)	15000	5000	2.5(a)	15(a)
Emission current density (A/cm^2)	3	30	17000	5300
Total emission current (μA)	200	80	5	200
Normalized brightness (A/cm^2.sr.kV)	1×10^4	1×10^5	2×10^7	1×10^7
Maximum probe current (nA)	1000	1000	0.2	10
Energy spread at the cathode (eV)	0.59	0.40	0.26	0.31
Energy spread at the gun exit (eV)	1.5—2.5	1.3—2.5	0.3—0.7	0.35—0.7
Beam noise (%)	1	1	5—10	1
Emission current drift (%/h)	0.1	0.2	5	<0.5
Operating vacuum (Pa)	$\leq1\times10^{-5}$	$\leq1\times10^{-6}$	$\leq1\times10^{-10}$	$\leq1\times10^{-8}$
Cathode regeneration	not required	not required	every 6 to 8 h	not required
Sensitivity to external influence	minimal	minimal	high	low

3.2.2 The Minimum Attainable Beam Diameter

The electron beam diameter is the ultimate limiting factor of the scanning electron microscope resolution, assuming that all other factors remain the same.

During demagnetization of the electron source, aberrations (spherical aberration, chromatic aberration, aperture diffraction) of the magnetic lenses and astigmatism increase the electron beam diameter. The effective value can be calculated (Goldstein, et al., 1992).

$$d^2 = d_0^2 + \left(\frac{1}{2}C_s\alpha^2\right) + \left(\frac{\lambda}{\alpha}\right)^2 + \left(C_c\alpha\frac{\Delta E}{E}\right)^2 \qquad (3.2)$$

C_s and C_c are spherical and chromatic aberration coefficients of the lens respectively. α is the semi-convergence angle of the lens. λ is the electron wavelength $= 1.22/E^{1/2}$ nm where E is the energy of the electron beam in eV, and ΔE is the energy spread of the electron beam.

By ignoring the chromatic aberration term at medium energies (Jia and Joy, 1992), one can derive the current i_b of the electron beam as a function of the electron beam diameter:

$$i_b = 1.88\beta \frac{d^{8/3}}{C_s^{2/3}} \qquad (3.3)$$

where β is the brightness of the electron source (amps/sm^2 sterad).

Thus, in order to maintain the necessary minimum electron beam current for obtaining the minimum (threshold) signal contrast in the final image, one can see that a high-resolution scanning electron microscope should have a high brightness electron source and a final lens of very low spherical aberration.

At low electron beam energies, the performance of a SEM degrades substantially because chromatic aberration becomes the dominant factor in determining beam diameter and electron density distribution in the probe. A gaussian probe at 30 keV becomes 2 times broader at 3 keV and is superimposed on the broad wings (Joy, 1987).

The half-angular range of electron paths within the beam is called the beam divergence. It is an important factor in determining the relative influence of the lens aberrations on the size and shape of the focused probe. The beam divergence defines the depth of focus of the SEM and also greatly effects the electron channeling contrast mechanism.

3.3 Contrast Formation and Interpretation

When the primary electrons collide with atoms of a solid surface in the specimen the electrons take part in various interaction. The elementary interaction processes, elastic and inelastic scattering, can initiate modified and secondary radiation: backscattered electrons, electrons with characteristic energy loss, secondary electrons, Auger electrons, characteristic X-rays, Bremsstrahlung or continuum X-rays, photon excitation.

The secondary electrons, backscattered electron, and characteristic X-rays are the most widely utilized signals in a scanning electron microscope.

Elastic collision of the primary electrons with the sample atoms produces backscattered electrons. They undergo a single (the primary electron is scattered "backward" 180 degrees) or multiple scattering event before they leave the sample surface. Most of the backscattered electrons retain 80% of their original energy, but all of the escaped electrons with energies above 50 eV are called backscattered electrons (BSE).

Characteristic X-rays can be formed by the inner shell of the specimen atom ionization process. When the primary electron impact sufficient energy on the atom it can remove core electrons from it. A characteristic X-ray is occasionally emitted when the atom relaxes to a lower-energy state by transition of an electron from a given outer shell to the vacancy in the inner shell. The X-ray is called "characteristic" because its energy equals the energy difference between the two levels involved in the transition and this difference is characteristic of the element.

The Monte Carlo simulation allows visualizing of the primary beam interaction volumes. Figure 3.2 shows the results of the Monte Carlo simulation of the electron beam interaction volume in copper for beam energy of 15 keV. The interaction volume is not defined by the electron beam diameter (regardless of how small it is); its spread (or the range) is almost a micron in dimension. The dimensions of the interaction volume can be estimated by a relatively general equation derived by Kanaya and Okayama, which considers the combine effect of elastic scattering and energy loss due to inelastic scattering (Kanaya and Okayama, 1972).

$$R_{KO} = \frac{0.0276 A E_0^{1.67}}{Z^{0.89} \rho} \quad (3.4)$$

where R_{KO} is the Kanaya and Okayama electron range, A is the atomic weight, E_0 is the primary electron energy, Z is the atomic number and ρ is the density.

Significant fraction (more than 30%) of the beam electrons escapes the specimen surface after multiple elastic scattering as backscattered electrons.

The backscattered fraction is quantified by the backscattered coefficient η.

$$\eta = \frac{n_{BSE}}{n_B} = \frac{I_{BSE}}{I_B} \quad (3.5)$$

where n_B is the number of primary beam electrons incident on the sample and n_{BSE} is the number of backscattered electrons.

For the electron beam energies higher than 3 keV, the backscattering electron signal provides an important SEM contrast mechanisms; compositional contrast, or atomic number contrast. The change in production rate of the backscattered electron with atomic number causes higher atomic

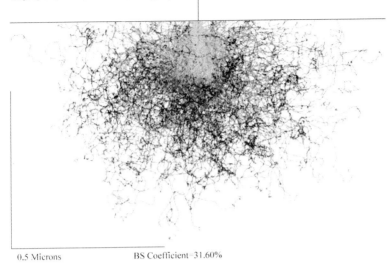

Eo[kV]=15, Z=29, W=63.546, Hho[g/cc]=8.93, Tilt[deg]=0, Traj.No.=500

0.5 Microns   BS Coefficient=31.60%

Figure 3.2 Primary beam interaction in copper for beam energy of 15 keV at 0 degrees tilt as calculated by Monte Carlo single scattering electron trajectory simulation

number elements to appear brighter than lower atomic number elements.

The backscattered coefficient η is generally considered monotonic with atomic number. But if higher precision is required, it can be calculated by the equation derived by Reuter (1972):

$$\eta = -0.0254 + 0.016Z - 1.86 \times 10^{-4} Z^2 + 8.3 \times 10^{-7} Z^3 \quad (3.6)$$

Another important observation: over the range from 10 keV to 50 keV, the backscattered coefficient is practically independent from primary electron beam energy (Heinrich, 1981).

A formula to explain electron backscattering as a function of energy below 10 keV is derived by R. Gaudin, et al. (1995).

But the backscattered electron coefficient is strongly dependent from the angle between the incident beam and the local surface. Primary beam interaction in copper for beam energy of 15 keV at 70 degrees tilt as calculated by Monte Carlo single scattering electron trajectory simulation is presented in Fig. 3.3. The backscattered coefficient in this case is 63.4% compared to 31.6% at 0 degrees tilt.

When observing rough surfaces, be aware that, since the backscattered coefficient is sensitive to both topographical and compositional contrasts, the image will contain a superposition of both effects.

A very small fraction (0.01% to 0.1%) of the backscattered electrons BSE_I undergoes a single high-angle scattering event; the other BSE_{II} are subjects to a multiple small angle scattering event. But BSE_Is are important for high-resolution imaging since this event is extremely localized. The impact

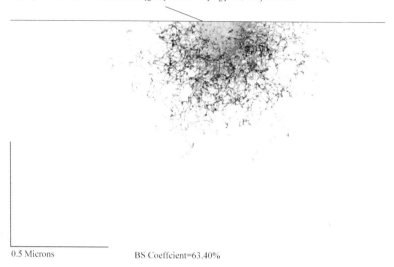

Figure 3.3 Primary beam interaction in copper for beam energy of 15 keV at 70 degrees tilt as calculated by Monte Carlo single scattering electron trajectory simulation

parameter is less than 0.01 Å for Si at 20 keV. At the same time, the probability of the single high-angle event is a function of Z^2, where Z is the atomic number of the element, thus providing strong chemical contrast.

During inelastic scattering events, energy is transferred from the primary electrons to the atoms of the sample. That causes small path change of the primary electron and the ionization of electrons of the specimen atoms in a cascade process. These ionized electrons, designated secondary electrons (SE), then leave the atoms with kinetic energy less than 50 eV (a majority have energies less than 10 eV). The rate of energy loss with distance traveled in the sample by primary beam is of the order of 1 to 10 eV/nm and is energy and material dependant. The secondary electrons, because of their low energy, can escape the sample only from a near-surface layer (~10 nm). There are only two areas from which secondary electrons are able to escape from the sample. First, secondary electrons created in the area where the primary electrons enter the sample (designated as SE_I), and from the area where backscattered electrons exit the sample surface (SE_{II}); the SE_{III} component is generated from spurious interactions in the column and specimen chamber, as presented in Fig. 3.4.

Calculations made by means of Monte Carlo simulations (Joy, 1984) in aluminum show that the SE_I electrons leave the sample surface with a Gaussian intensity profile which has a full width at half maximum intensity (FWHM) of about 2 nm (Joy, 1984).

The SE_{II} component of the signal leaves the sample also with Gaussian intensity profile but a FWHM is of the order of 0.2 to 0.4 of the electron range

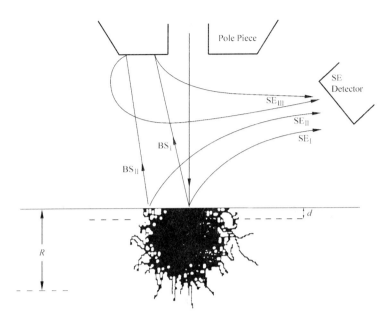

Figure 3.4 The origin of the SE_I and SE_{II} secondary electron signals in the SEM. d is the escape depth for the secondary and R is the incident electron range. The SE_{III} component is generated from spurious interactions in the column and specimen chamber

or fractions of a micrometer. This signal contains information related to the backscattered electrons and sample interaction (Joy, 1991).

The distance which the electron will penetrate will fall rapidly as the incident energy reduced. Figure 3.5 shows Monte Carlo calculations of the interaction volume of electrons in carbon at 1, 3 and 10 keV electron beam energies at the same scale.

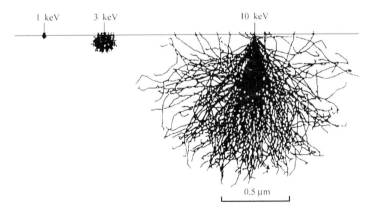

Figure 3.5 The variation of electronic range with initial energy of electron. Data plotted for carbon with assumed density of 1 g/cm³ (after Joy, 1984)

The total intensity distribution of secondary electron production at high and low electron beam energies is shown in Fig. 3. 6.

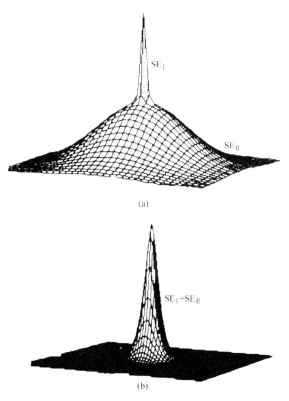

Figure 3.6 A schematic view of the emission profile of secondary electrons from a surface (top) at high electron beam energy (20 keV) and (bottom) at low energies (2 keV) (after Joy, 1991)

At low electron beam energy (~ 2 keV) the interaction volume is the same for both SE_I and SE_{II}. That means that the SE_I and the BSE signals at low electron beam energy contain high-resolution information (Fig. 3. 6).

The TEM observations (Josell, et al., 1998) of the Al/Ti multilayers system (bright-field image (Fig. 3. 7)) and the SEM high-resolution Images (Fig. 3. 8 and Fig. 3. 9, SE and BSE images of a fracture surface) show that the multilayers are composed of columnar grains.

At the same time, the comparison reveals that a modern FEG-SEM has a real advantage over TEM observations, maintaining necessary resolution without difficulties of the TEM specimen preparation, but also shows the topography of the fracture service.

This picture shows that the backscattered electron in the FEG-SEM contains high-resolution information at 5 keV.

"Macroscopic" relief feature which is important in the topographical

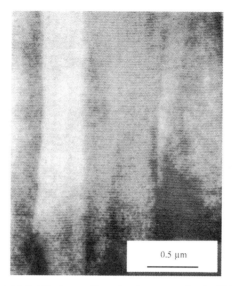

Figure 3.7　TEM (bright field) observation of columnar grains in Al/Ti multilayer system

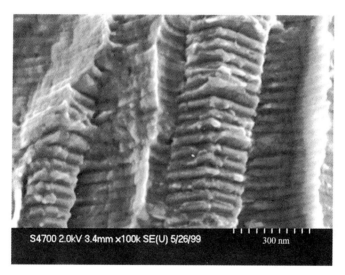

Figure 3.8　Secondary electron Image of a fracture surface of the Al/Ti multilayer system

contrast generation is terrace sidewalls or other sharp edges. At "normal" SEM energies, such edges are overbrigtened by BSE, which diffuse to the sidewall and escape there.

At low incident energies the edges are much sharper and are not overbright compare Fig. 3.10 and Fig. 3.11, Which show SE images of a contact on silicon substrate at 10 keV and 1 keV, respectfully.

Figure 3. 9 BSE image of a fracture surface of the Al/Ti multilayer system

Figure 3. 10 SE images of a contact on silicon substrate at 10 keV

Another advantage of the low-voltage SEM operation is the observation of non-conducting materials. When a non-conductive sample is scanned with an electron beam, an electron charge is accumulated if the sample is receiving more electrons than it is emitting, or emitting more than it is receiving. That effect will result in image distortion and thermal or radiation damage of the sample. One way to treat this problem is coating the sample with a thin film of conductive material. Another is to operate the SEM at electron beam energy when the total electron yield of the SE and BSE is unity. As a result of deviations from unity condition, negatively charged areas will appear bright and positively charge dark in the SEM image. One has to be careful is

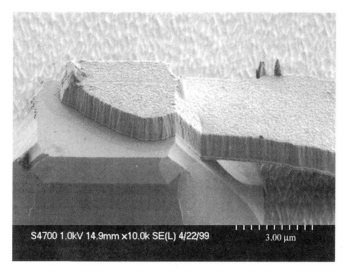

Figure 3.11 SE image of a contact on silicon substrate at 1 keV

selecting the beam energy since the unity condition could be different for different parts of the sample. A procedure for selecting the operation conditions is described by D.C. Joy and C.S. Joy (1996).

3.4 Secondary Electron Detectors

3.4.1 Everhart-Thornley Detector

An electron detector developed by T. E. Everhart and R. F. M. Thornley in 1960 is still the most popular electron detector. Practically all SEM (including the most modern ones) are equipped with the Everhart-Thornley (E-T) detector. The detector is positioned in the SEM chamber, where the objective lens magnetic field value is negligible (Everhart and Thornley, 1960).

The principle of operation is very simple:

(1) Secondary electron mode: since the secondary electron energy is very low ($<$50 eV), a positive bias (80—250 V) is applied to the Faraday cage to allow them to enter the cage. The collection efficiency is almost 100%. A large positive potential (10 to 12 kV) is applied to the face of the scintillator, which is coated with thin metallic film. Interaction of the accelerated electrons with the scintillator produces photons, which by total internal reflection in a light guide conducted to a photomultiplier. Very high gain of the photomultiplier ($10^5 - 10^6$) allows low noise and high bandwidth

amplification. Since this type of detector collects all of the low-energy electrons in the chamber, the total signal consists of the all three types of the secondary electrons (SE_I SE_{II}, and SE_{III}) and thus provides combined secondary and backscattered electron information. The E-T detector provides excellent easily recognizable topographic imagery.

(2) Backscattered electron mode: when a negative bias (-50 V) is applied to the Faraday cage, all of the low-energy electrons (secondary) are rejected and only backscattered electrons are detected. The backscattered electron efficiency of this detector is relatively low and depends on the specimen tilt (the specimen tilt changes the backscattered electron distribution relative to the detector).

In both modes of operation, the E-T detector can be used at TV rate.

A system of two opposite E-T detectors and a BSE/SE converter plate below the polepiece can be implemented for the enhancement or suppression of different types of contrast by recording the A + B or A − B signals from opposite detectors (Kässens and Reimer, 1996).

3.4.2 In-iens Secondary Electron Detector

In a high-resolution microscope, the polepiece of the objective lens induces a strong magnetic field toward a specimen for final beam focusing. The magnetic field rolls up the SE_I and SE_{II} (but not the SE_{III} electrons because the SE_{III} interactions occur outside the magnetic field region) and they are detected by an in-lens detector.

For example, the Hitachi S-4700 FEG-SEM provides two secondary electron detectors placed below the objective lens (lower detector E-T type) and above it (upper detector). The design of the upper-detection system is presented in Fig. 3.12.

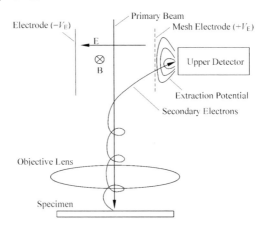

Figure 3.12 Design of the upper-detection (in-lens) system (Hitachi S-4500 FE-SEM)

To minimize the influence of the primary electron beam, an E cross B (E×B) filter which generates crossed electrostatic and magnetic fields (E×E field) is located in front of the detector in order to realize high detection efficiency of secondary electrons. The electrostatic field (E) is generated by a pair of electrodes, and the electrode facing the detector is made of a mesh structure, so that the secondary electrons can pass through the electrode and be detected by the upper detector. A pair of coils generates a magnetic field (B), which is perpendicular to the electrostatic field (E). Secondary electrons coming into the E×E filter are deflected toward the upper detector by the E×E field. Since the strength of the E×E field is adjusted for detecting secondary electrons, no backscattered electrons can pass directly through the mesh electrode because the energy of backscattered electrons is much higher than that of secondary electrons.

The comparison of the images obtained with upper and lower (Figs. 3.13 and 3.14, respectfully) detectors under the same conditions reveals the following:

(1) The magnetic field from the objective lens effectively extracts the secondary electrons from the holes in the sample.

(2) The resolution of the SE image obtained with the upper detector is higher than with the lower detector. This can be explained by the fact that the secondary electrons from the sample surface are captured by the magnetic field of the objective lens and the lower detector is collecting mainly the SE_{III} electrons, which have the backscattered electron resolution.

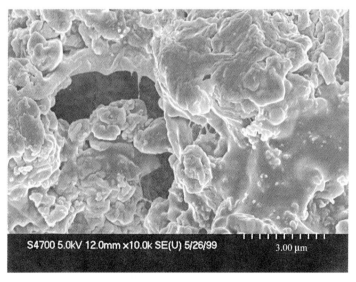

Figure 3.13 SE image obtained with upper detector

Figure 3.14 SEM image obtained with the SE lower detector. The images have been obtained on the Hitachi FE-SEM S-4700 (we thank Paula Cecere for help).

3.5 Dedicated Detectors

Many different types of dedicated detectors (for compositional, topographical, orientational, and magnetic contrasts) have been developed to improve the efficiency and sensitivity of the specific signal collection. A detailed description of the dedicated detectors is given in Goldstein et al. (1992).

3.5.1 Solid-State Diode Detector

The Si pnjunction (usually divided annular type) is a "standard" backscattered electron detector in conventional SEMs and the principals of its operation are well known (Gedcke, et al., 1978). It is sensitive only to high-energy backscattered electrons (not affected by secondary electrons). The divided annular design of the detector allows separation of the compositional, topographic and other types of contrasts by collecting differential or combined signals from different parts of the detector. This "standard" detector has, however, two significant limitations: it can not be operated at TV rate (due to its narrow bandwidth) and it has an energy threshold (due to electron absorption in the surface electrode and silicon dead layer).

3.5.2 Scintillator Backscattered Electron Detector

The "standard" unbiased Everhart-Thornley detector can selectively detect backscattered electrons but with very low efficiency, due to its small collection angles. A large scintillator made of the same material as the light guide can be placed above the sample thus providing collection of backscattered electron over much larger solid angles (Robinson, 1975) and with a much higher collection rate. This type of detector has another advantage over the "standard" solid-state detectors—the backscattered electron images can be observed at TV rates.

3.5.3 BSE-to-SE Conversion Detectors

Placing a biased (-50 V) grid around the sample can separate secondary and backscattered electrons. A positively biased Everhart-Thornley detector is used to collect secondary electrons generated by backscattered electrons (which are not significantly affected by the bias) from interactions with the chamber walls and a converter target (covered with high secondary electron yield material), which is attached to the objective polepiece. This arrangement allows increase of the collection angle.

Another implementation of the BSE-to-SE conversion is utilized in the microchannel plate detector system (Joy, 1995; Helbig, et al., 1987; Autrata and Hejna, 1991). The BSE-to-SE conversion occurs within the detector, which consists of one or more annular arrays of glass capillaries (12—23 mm diameter and 0.5 to 1 mm long). During bombardment of a capillary's entrance by backscattered electrons, secondary electrons are produced which are then accelerated along the capillary (covered on the inside with high secondary electron yield material) by a potential applied to the exit surface of the arrays. This arrangement creates a cascade of secondary electrons with gains approaching 100. In addition, by applying positive or negative bias to the front surface of the detector array, one can select secondary or backscattered electron imaging.

Since the yield of the secondary electrons increases at lower energies of the backscattered electrons, the detector is uniquely suited for backscattered electron detection in the low-voltage electron microscopy.

3.5.4 Multi-detector System

Image contrast formation in the low-voltage SEM is quite different from that in conventional SEM (Joy and Joy, 1996; Reimer and Kässens, 1994) due to the reduced electron beam range and design of a magnetic "snorkel" lens. In addition, the contamination of the surface becomes more visible. The multi-detector systems provide the means for enhancement or suppression of different types of contrasts. Kässens and Reimer (1996) have used two opposite Everhart-Thornley detectors and a BSE/SE converter plate below the polepiece to study contrast effects in low-voltage electron microscopy. They came to a conclusion that the use of a differential between the two detector (SE or BSE) signals for topographical imaging results in a better contrast and shows fewer artifacts. Hejna (1995) implemented two scintillation backscattered detectors with a high voltage applied to scintillators. One detector collects BSE emitted at low takeoff angles, the second at high takeoff angles. He concluded that detection of material contrast is possible with high-takeoff detector, but one should be careful in interpreting images acquired at energies below 1 keV.

3.5.5 Electron Backscattered Diffraction (EBSD)

First studied in 1954 (Alam, et al., 1954) and introduced 20 years later (Venables and Harland, 1973) to the SEM, the technique of electron backscattered diffraction (EBSD) has become a very valuable technique in materials science. It is a diffraction technique for obtaining local crystallographic information from the surface of bulk samples, texture and grain boundary misorientation in multiphase materials, crystal deformation, etc. Information can be obtained on the orientation of crystals with a spatial resolution well below one micron as an image of a sample by crystalline contrast and to characterize the orientation of the grains by electronic interferometry (pseudo-Kikuchi patterns). The diffraction lines obtained on a phosphor screen or photographic plate are characteristic of the grain orientation. These lines are introduced into a computer that calculates the orientation matrix of the observed grain.

The surface quality of the sample is crucial in order to generate the electron diffraction pattern. In most cases, a fracture surface or an electropolished finish are required to obtain acceptable pseudo-Kikuchi patterns.

By combining the EBSP information with X-ray compositional analysis (in some cases when two phases have similar composition but different crystal structures, like TiO_2, X-ray analysis will not be able to make the distinction), one can obtain accurate identification (Goehner and Michael, 1996) of sub-micron sized regions without the specimen preparation difficulties associated with studies by transmission electron microscopy (TEM).

Dedicated computer software is used to facilitate the indexing of the electron backscattered patterns (Wright and Adams, 1992) and on-line phase identification is possible (Michael and Goehner 1993). Detailed descriptions of the electron backscatter diffraction techniques are given by Wilkinson and Hirsch (1997).

3.5.6 Magnetic Contrast

There are three contrast mechanisms which allow direct observation of the magnetic microstructure in the SEM (detailed descriptions of all types of magnetic contrasts are given by Newbury, et al., 1987).

Magnetic contrast typeI (Cellota and Pierce, 1982) is obtained when a leakage magnetic field from the surface of the sample deflects the secondary electrons and thus may create an angular distribution asymmetry relative to the secondary electron detector. This asymmetry leads to a difference in the secondary electron collection efficiency from adjacent magnetic domains and the magnetic contrast. Since the secondary electron yield rises and the lateral range of the backscattered electron is smaller when the primary beam energy is lowered, low-voltage operation of the SEM is preferable. The magnitude of the magnetic contrast typeI can reach 50%. Theoretical explanations of the magnetic contrast typeI have been given, for example, by Wardly (1991), Wells (1986), and Kotera et al. (1995; 1997).

Magnetic contrast typeII is originated due to deflection during elastic scattering of the primary beam electrons by the magnetic forces in the sample (Fathers; 1973). When magnetic domains in the sample are magnetized parallel to the surface, they can be visualized by magnetic contrast typeII. The magnetic forces bend the scattering electrons toward the surface of the sample or in the opposite direction. This action changes the backscattered coefficient accordingly and thus visualizes the magnetization orientation of the domains in the sample. The effect is very small—0.3% (Newbury et al., 1987)—and strongly dependent on the orientation of the domain magnetization relative to the primary electron beam. That is why the sample surface must be well prepared and the parameters of the SEM very thoroughly defined (Newbury, et al., 1974). Two methods have been implemented to quantify magnetic contrast typeII: the Monte Carlo electron trajectory simulation method by Newbury, et al. (1973) and analytical expressions (process of electron scattering and magnetic deflection) by Fathers, et al. (1973) and Wells (1976).

Magnetic contrast typeIII is derived from detection of spin polarization of the secondary electrons by the exchange interaction near the surface of ferromagnetic materials. Unguris, et al. (1982) have determined that the secondary electron magnetic moments are parallel, and as a result their spins antiparallel, to the magnetic vector at the point of the origin in the sample. Development of a sensitive electron spin analyzer, which is small enough to fit into the chamber of a SEM, brought to the spin polarized SEM or the scanning electron microscopy with polarization analysis (SEMPA) technique (Cellota and Pierce, 1982; Unguris, et al. 1985; Scheinfein, 1990). High spatial resolution of a Fe-SEM in combination with electron spin analyzer allows direct imaging of the magnitude and the direction of the magnetization of very small magnetic structures (~ 10—20 nm). Another important advantage (compare to MOKE, the magneto-optic Kerr effect microscopy (Kranz and Hubert, (1963), for example) is that SEMPA observes the magnetization independently from the topography, although both can be measured at the same time by different detectors. If the microscope is equipped with an X-ray analyzer, the compositional maps can be also correlated with the topographic and magnetic information.

3.5.7 X-ray Spectrometers

It is hard to overestimate the contribution that X-ray microanalysis has made in materials science. It is a fast and relatively simple method, which provides quantifiable compositional information on almost all elements at the sample surface. The inelastic scattering of primary electrons in a SEM result of ionizing of inner shells of atoms in the sample, followed by de-excitation of those atoms, which produce characteristic (of the element) X-rays, continues (Bremsstrahlung) X-rays and electrons. By analyzing the intensities of the characteristic X-rays one can determine the chemical composition of the solid surface. The limitation in resolution of the X-ray microanalysis is imposed by electron elastic scattering (Balaic, et al., 1999). By operating the FE-SEM at beam energies below 5 keV, the depth and lateral resolution can be reduced to 100 nm. The detection limit of this technique is also very low 0.01% (Goldstein et al. 1992). Detailed description of the theory and experimental methods and applications can be found in numerous books and overviews (Williams, et al., 1995; Scott, et al., 1995; Reed, 1993; Jones, 1992; Heinrich and Newbury, 1991; Newbury and Joy, 1991, for example).

The detection of photons is usually accomplished by two techniques: energy dispersive spectrometry (EDS) and/or wavelength dispersive spectrometry (WDS).

3.5.7.1 Wavelength-Dispersive Spectrometers

The wavelength dispersive spectrometer (WDS) is based on the principle of diffraction. The characteristic X-ray photons excited by the electron beam are sorted out using a diffracting curved crystal and a gas-proportional counter. They are positioned in such a way to satisfy the Bragg reflection conditions. It is necessary to employ sequentially different crystals and multilayers with different layer spacings in order to be able acquire characteristic lines from all elements. The WDS has very good energy resolution (2—20 eV). Though geometrical collection efficiency is quite poor, it is still possible to get very high count rates (due to fast gas-proportional counter and high beam current), up to 50 000 s^{-1} without degradation in energy resolution (Wollman, et al., 1997). On the other hand, qualitative analysis performed with WDS is relatively slow in comparison with EDS since the entire energy range has to be scanned with all available crystals.

3.5.7.2 Energy-Dispersive Spectrometers

An EDS system usually employs a solid-state detector made of lithium-drifted silicon (Si(Li)), lithium-drifted germanium (Ge(Li)), or high-purity germanium (HPGe) detectors. The photon's energy is measured through photoelectric capture in the diode. The detectors and preamplifiers (FET transistor) must be operated at sufficiently low (liquid nitrogen) temperature to minimize electronic thermal noise. Due to the presence of a Be window (to protect the detector crystal from contamination and possible loss of vacuum) and limited thickness of a detector crystal, X-ray detection is limited to a range from 1 keV to 20 keV with resolution ~130 eV at MnKα. Implementation of an ultra-thin window allows detection of X-rays down to 200 eV. The spectrum is accumulated simultaneously at all energies by multichannel analyzer. Since the energy dispersive spectrometers provide very stable, inexpensive operation, and allow fast quantitative compositional analysis, they became "standard" attachments to the scanning electron microscopes.

Recently a new EDS has been proposed (Lesyna, et al., 1993; Silver, et al., 1996) and developed (Irwin, et al., 1996; Wollman, et al., 1997) ——X-ray microcalorimeter spectrometer. In this system, X-ray energy is converted to heat in an X-ray absorber connected to a thermometer. The measurement of the temperature change defines the absorbed photon energy. The X-ray spectrometer could be based on different cryogenic detectors: transition-edge sensor (TES) microcalorimeter (Wollman, et al., 1997), semiconductor thermistor microcalorimeter (Lesyna, et al., 1993; Silver et al., 1996), semiconductor-insulator-superconductor (SIS) tunnel junctions (Mears, et al., 1996), normal-insulator-superconductor (NIS) microcalorimeter (Nahum and Martinis, 1995), superconducting tunnel junction (STJ) detectors (Frank, et al., 1998), but a TES detector based system (μcalEDS) is the only one which will be soon available as a commercial instrument. This low-energy Al-Ag superconducting microcalorimeter TES detector can detect X-rays from 0 keV to 2 keV with the best energy resolution of 2.0 eV±0.1 eV at 1.5 keV (Wollman, et al., 1999). Since the microcalorimeter EDS X-ray detector combines the excellent energy resolution of a wavelength-dispersive spectrometer with the parallel energy detection of EDS, it allows accurate chemical identification of small particles (0.6 to 0.8 micron) during high-resolution SEM operation. For higher energies, a Mo-Cu superconducting TES microcalorimeter could be more suitable; its resolution is 4.5 eV ± 0.1 eV at 6 keV.

The geometrical collection efficiency can be improved by implementing

X-ray optics which can focus or collimate X-rays. The polycapillary X-ray optic was developed by Kumakhov and Komarov (1990). Thousands of fused glass capillaries are directed toward a focal point and, due to multiple grazing-incidence reflections, deflect X-rays of a wide energy range over a wide angle. Balaic, et al. (1999) Kirkland, et al. (1995) and Angello, et al. (1997) used X-ray optic to improve efficiency for WDS, Carpenter and Taylor (1993) for EDS, and Wollman, et al. (1997) for the microcalorimeter EDS.

The comparison of different X-ray spectrometers is presented in Table 3.2. (partially from Wollman, et al., 1997).

Table 3.2 Comparison of X-ray Spectrometers

Spectrometer type	Energy resolution (eV)	Maximum count rate (s^{-1})	Solid angle (msr)	Collection efficiency (msr)
EDS (large area)	175 (at 6 keV) 145 (at 6 keV)	30 000 5 000	150 150	115 115
EDS (high resolution)	130 (at 6 keV)	3 000	25	19
WDS (several diffracting crystals)	2—20	50 000	8—25	0.8—2.5*
Microcalorimeters: a) Semiconductor	7—8	100	0.07	0.04
b) TES	7 (at 6 keV) 2 (at 1.5 keV)	150 500	0.05	0.03
c) TES with polycapillary optics	7 (at 6 keV) 2 (at 1.5 keV)	150 500	4 8	2 4
d) NIS	18—30	2 500	0.2	0.1
SIS tunnel junction	29 (at 6 keV)	2 500	0.01	0.005

* Because a WDS spectrometer only accepts X-rays of a narrow energy band, its practical collection efficiency is further reduced (up to several orders of magnitude) when scanned over the entire energy range.

3.6 Conclusions

High-resolution scanning electron microscopy with X-ray microanalysis has proven that it can be applied successfully in studying nanophase and nanostructured materials. For example, morphological and compositional information can be obtained simultaneously even for small particles (Osker, et al., 1995); insulators can be studied at low beam energies without coating; crystallographic information can be obtained in FE-SEM from electron backscattered diffraction patterns (EBDP) (Goehner and Michael, 1996) and elastic strain and small lattice rotations were measured using electron back scattered diffraction (Wilkinson and Hirsch, 1997); it is possible by means of SEMPA to visualize the magnetic domain structure and the exchange coupling processes in multilayers (Cellota, et al., 1995); depth information can be obtained in a block of a copolymer system by changing the beam energy (Harrison, et al., 1998); chemical analysis can be performed by means of microcalorimeter EDS on particles as small as 0.08 mm in diameter (Wollman, et al., 1998), and so on.

References

Alam M.N., M. Blackman. and D.W. Pashley. Proc. Roy. Soc.. **221** 224 (1954)
Angello R., J.V. Howard. J.J. McCarthy. Oharra D. Micros. Microanal. **3** 889—890 (1997)
Autrata R., J. Hejna. Scanning. **13**, 275—287 (1991)
Balaic D.X, Z. Barnea, K.A. Nugent, S.W. Wilkins, H. Yamada, S. Masui, R. Garnet. Microbeam Anal 4(Suppl); 632—639 (1999)
Carpenter D.A., and M.A. Taylor. Microbeam Anal 2 (Suppl) S84—S85 (1993)
Cellota R.J., D.T. Pierce. and J. Unguris. MRS Bu; etin Vol. XX, no. 10 30—33 (1995)
Cellota R.J. and D.T. Pierce. in Microbeam Analysis 1982 (San Francisco Press, San Francisco. 469 (1982)
Everhart T.E. and R.F.M. Thornley. J. Sci. Instr.. **37**, 246—248 (1960)
Fathers D.J., J.P. Jakubovics, et al.. Phys. Status Solidi. A **20**, 535 (1973)
Frank M., L.J. Hiller, J.B. le Grand, C.A. Mears, S.E. Labov, M.A. Lindeman, H. Netel, D. Chow, and A.T. Barfknecht. Review of Scientific Instruments. ——69, Issue 1, 25—31 (1998)
Gaudin R., D.C. Joy, P. Hovington. Scanning. Vol. **17**, 63—64 (1995)
Gedcke D.A., J.B. Ayers, and P.B. DeNee. SEM. **1**, 581 (1978)
Goehner R.P. and J.R. Michael. J. of Res. of Nat. Inst. of Stand. Tech.. **101** 301 (1996)
Goldstein J.I., D.E. Newbury, P. Echlin, D.C. Joy, A.D. Romig Jr. C.E. Lyman, C.E. Fiori, and E. Lifshin. Scanning Electron Microscopy and X-Ray Microanalysis. 2nd Edn. plenum, New York, 59 (1992)
Harrison C, M. Park, P. Chaikin, R.A. Register, D.H. Adamson, N. Yao. Macromolecules. **31**: (7) 2185—2189 (1998)
Heinrich K.F.J. and D.E. Newbury. ed. Electron Probe Quantification. Plenum Press, New York and London, (1991)

Heinrich K. F. J.. Electron beam X-Ray Microanalysis. Van Nostrand-Reinhold, New York (1981)
Hejna J.. Scanning. Vol. 17, 387—394 (1995)
Helbig H. F., R. D. Rydgren, and L. Kotorman. Scan Microsc. 1, 1491—1499 (1987)
Irwin K.D, G.C. Hilton, D.A. Wollman, J.M. Martinis. Applied Physics Letters. 69: (13), 1945—1947 (1996)
Jia Y., and D.C. Joy. Acta Microscopica. 1, 20 (1992)
Jones I. P.. Chemical Microanalysis, The Institute of Materials. (1992)
Josell D., D. van Heerden, D. Read, J. Bonevich, and D. Shechtman.. J. Mater. Res.. Vol. 13, No 10 (1998)
Joy D.C. and C.S. Joy, Micron.. 27, No. 3—4, 247—263 (1996)
Joy D.C.. EMAG87, Munchester. 8—9, 175—180 (1987)
Joy D.C.. Hitachi Inst. News. 18, 3 (1995)
Joy D.C.. Monte Carlo studies of high-resolution images. Microbeam Analysis. ed. by Romig A. and Goldstein J.I. San Francisco Press, San Francisco (1984) pp. 81—86
Joy D.C.. Journal of Microscopy. Vol. 161, Pt. 2, February (1991), pp. 343—355
Kanaya K. and S. Okayama. J. Phys.. D.: Appl. Phys.. 5, 43 (1972)
Kässens M. and L. Reimer. Journal of Microsc. Vol. 181. Pt. 3 277—285 (1996)
Kotera M, M. Katoh, H. Suga. Jpn. J. Appl. Phys.. 1 34: (12B) 6903—6906 (1995)
Kotera, K. Yamaguchi, A. Fujita, et al.. Jpn. J. Appl. Phys.. 1 36: (12B) 7726 (1997)
Kranz J. and A.Z. Hubert. Angew. Phys. 15 220 (1963)
Kumakhov M.A. and F.F. Komarov: Phys. Rev.. 191 289—350 (1990)
Lesyna L., D. Di Marzio, S. Gottesman, and J. Kesselman. Low Temp. Phys.. 93 779—784 (1993)
Mears C.A., S.E. Labov, M. Frank, M.A. Lindeman, L.J. Hiller, H. Netel, and A.T. Barfknecht. Nucl. Instrum. Methods Phys. Res.. A 370 53—56 (1996)
Michael J.R. and R.P. Goehner. MSA Bulletin. 23, 168 (1993)
Nahum M. and J.M. Martinis, Appl. Phys. Lett.. 66 3203—3205 (1995)
Newbury D.E. and D.C. Joy. in Analysis of Microelectronic Materials and Devices. Chap. 2.4 (1991)
Newbury D.E. et al.. Appl. Phys Lett.. 23 488 (1973)
Newbury D.E. et al.. Appl. Phys. Lett.. 24 98 (1974)
Newbury D.E. et al.. Advance Scanning Electron Microscopy and Microanlysis. Plenum Press, New York and London pp. 147 (1987)
Osker B., R. Wurster, and H. Seiler. Scanning Microsc. Vol. 9, No. 1 63—73 (1995)
Reed S. J. B. Electron Microprobe Analysis. Cambridge University Press,

(1993)
Reimer, L. and M. Kässens. Proc. 13th Int. Congr. On Electron Microsc. 1 73 (1994)
Robinson V. N. E.. SEM 51 (1975)
Scheinfein M. R., J. Unguris, M. H. Kelley, D. T. Pierce, R. J. Celotta. Review Of Scientific Instruments. **61**: (10) 2501—2526, Part 1 (Oct. 1990)
Scott V. D., G. Love, and S. J. B. Reed. Quantitative Electron-Probe Microanalysis. Sec. Ed. Ellis Horwood, London (1995)
Silver E, M. LeGros, N. Madden, J. Beeman, E. Haller. X-ray Spectrometry. **25**: (3) 115—122 (1996)
Unguris J. et al.. J. Microsc. **139**, RP1 (1985)
Unguris J. et al.. Phys. Rev. Lett.. 49, 72 (1982)
Venables J. A.. and C. J. Harland. Phil. Mag. 27 1193 (1973)
Wardly G. A.. J. Appl. Phys.. 42, 376 (1991)
Wells O. C.. In Use of Monte Carlo Calculations in Electron Probe Microanalysis and Scanning Electron Microscopy. ed. by K. F. J. Heinrich, D. E. Newbury, and H. Yakovitz, NBS publ. **460**, Washington, DC. 139 (1976)
Wells O. C.. J. Microsc. **144**, RP1 (1986)
Wilkinson A. J. and P. B. Hirsch, Micron 48 No. 4, 279—308 (1997)
Williams D. B., J. I. Goldstein, and D. E. Newbury. X-ray Spectrometry in Electron Instruments. Plenum Press, New Yoork and London (1995)
Wollman D. A., S. W. Nam, D. E. Newbury, G. C. Hilton, I. K. D., N. F. Bergren, S. Deiker, D. A. Rudman, and J. M. Martinis. submitted to The proceedings of LTD. 8 (1999)
Wollman D. A., G. C. Hilton, K. D. Irwin, L. L. Dulcie, N. F. Bergren, D. E. Newbury, K. S. Woo, Y. Liu Benjamin, C. Diebold Alain, and J. M. Martinis. Characterization and Metrology for ULSI Technology. In: 1998 International Conference, ed. by D. G. Seiler, A. C. Diebold, W. M. Bullis, T. J. Shaffner, R. McDonald, and E. J. Walters, 799—804 (1998)
Wollman, D. A, K. D. Irwin, G. C. Hilton, L. L. Dulcie, D. E. Newbury, J. M. Martinis. Journal of Microscopy-Oxford. **188**: Part 3 196—223 (1997)
Wright S. J. and B. L. Adams. Metall. Trans.. A **23**, 759 (1992)

4 Scanning Probe Microscopy

Chunli Bai and Chen Wang

4.1 Overview

A lot of surface analysis techniques have been developed in the past half century, playing important roles in the structural studies on metal and semiconductor surfaces. However, each technique is limited in some respect. Probably the most notable advances in the past nearly two decades are related to the invention and growing of the family of scanning probe microscopy (SPM), and associated surging interests. There have been numerous publications covering the exciting developments in this field. This chapter is dedicated to a few representative directions of the research activities, hope to draw the attention of the readers to the promising potentials that the technology has brought forth. It is not intended to review such a vast scope in this chapter, and interested readers are encouraged to search the relevant literature.

Scanning tunneling (STM) has shown that it is possible to control and scan a tip over a conducting surface with angstrom precision. This same generic principle of STM has been applied to many other novel scanning probe microscopes. The developments in the areas of atomic force microscopy (AFM), lateral force microscopy (LFM), magnetic force microscopy (MFM), ballistic electron emission microscopy (BEEM), scanning ion conductance microscopy (SICM), photon scanning tunneling microscopy (PSTM), and near-field scanning optical microscopy (NSOM) have been highlighted by the results from each area which illustrate the potential of these techniques to provide new information about the physical properties of surfaces on an atomic or nanometer scale.

STM and related techniques have entered many disciplines in physics, chemistry, biology, metrology and materials science. SPMs allow not only the study of surface structures, but also modification of surfaces and interfaces from one micrometer down to the atomic scale.

This chapter is organized to present the principles and main applications of three representative members of the SPM family, including STM, AFM and BEEM. These techniques have their distinctive and unique characteristics for investigating both conductive and insulating surfaces, and metal-semiconductor interfaces. The basic principles for each technique are closely related, and their core application fields complement each other very nicely, providing a range of powerful microscopic analytical approaches.

4.2 Scanning Tunneling Microscopy

4.2.1 Introduction

The principle of STM makes it a representative one in the family of SPM in many aspects, and the most extensively investigated one. The main difference between the STM and other microscope is that there is no need for lenses and special light or electron sources. Instead the bound electrons already existing in the sample under investigation serve as the exclusive source of radiation (Binnig and Rohrer, 1985; Bai, 1995; Stroscio and Kaiser, 1993; Guntherodt and Wiesendanger, 1994).

Both STM and scanning tunneling spectroscopy (STS) based analysis techniques are capable of observing the surface structure of conductors and semiconductors with high resolution in real time under different experimental conditions such as in vacuum, in air, and solutions, and at low temperatures. The information about the surface electronic structure can also be obtained by using STS.

When an atomically sharp tip is scanned across a plane of atoms in the constant-current imaging mode yielding an image with sub-angstrom corrugations, it is not known exactly from experiments what role distance might play in equation (4.1). Moreover, since tunneling involves states at the Fermi level, which may themselves have a complex spatial structure, we must expect that the electronic structure of the surface and tip may enter into the equation in a complex way. Equation (4.1), based on analogy with the one-dimensional tunneling problem, is rather simple and therefore the full three-dimensional tunneling problem as it relates to STM must be considered. These theoretical concepts have been carefully addressed by many groups (Tersoff and Hamann, 1985; Lang, 1986; Chen, 1990).

In addition to delineating the atomic topography of a surface STM has made it technically possible to probe directly the electronic structures of materials at an atomic level by spatially resolved tunneling spectroscopy, which is also a vital part of STM studies.

Conventional STM is based on the control of the tunneling current I through the potential barrier between the surface to be investigated and the probing metal tip. If a small bias voltage is applied between the sample surface and the tip (in the best case, an atomically sharp tip), a tunneling current will flow between the tip and sample when the gap between them is reduced to a few atomic diameters. It takes advantage of the strong dependence of the tunneling probability of electrons on the electrode separation. The tunneling current (at a low bias voltage V_T and low temperature) behaves as (Tersoff and Hamann, 1985):

$$I = 18 \frac{V_T}{10^4 \Omega} \frac{K}{d} A_{eff} e^{-2Kd} \qquad (4.1)$$

where $2K(\text{Å}^{-1}) = 1.025 \Phi^{1/2}(\text{eV})$, Φ is the average work function, assumed equal to the mean barrier height between the two electrodes. $A_{eff} = \pi \times (1/2 L_{eff})^2$ is the effective area determining the lateral resolution $L_{eff} \approx 2 \times [(R_t + d)/K]^{1/2}$ which applies when the separation d becomes smaller than the radius R_t of the tip. For typical metals ($\Phi \approx 5$ eV) the predicted change in current I by one order of magnitude for the change $\Delta d \approx 1$ Å has been verified. If the current is kept constant to within, e.g., 2%, then the gap d remains constant to within 0.01 Å. This fact represents the basis for interpreting the image as simply a contour of constant height above the surface.

The STM can be operated in either the constant-current mode or the constant-height mode, as shown in Fig. 4.1 (Binnig and Rohrer, 1987). In the basic constant-current mode of operation, the tip is scanned across the surface at constant tunneling current, which is maintained at a preset value by continuously adjusting the vertical tip position with the feedback voltage V_z. In the case of an electronically homogeneous surface, constant current essentially means constant so the topographic height of surface features of a sample can be measured by raster scanning the tip in x-y plane over the surface and deriving the height of the surface from V_z. The height of the tip $z(x,y)$ as a function of position is read and processed by a computer and displayed on a screen or a plotter.

Alternatively, in the constant-height mode a tip can be scanned rapidly across the surface at nearly constant height and constant voltage V_z while the tunneling current is monitored, as shown in Fig. 4.1(b). In this case the feedback network is slowed down to keep the average tunneling current constant or turned off completely. The rapid variations in the tunneling current due to the tip passing over surface features are recorded and plotted as a function of scan position.

Each mode has its own advantages. The basic constant-current mode was originally employed and can be used to track surfaces which are not atomically flat. The height of surface features can be derived from V_z and the sensitivity of the piezoelectric driver element. On the other hand, a disadvantage of this mode is that the finite response time of the feedback network and of the piezoelectric driver set relatively low limits for the scan speed. The constant-height mode allows for much faster imaging of atomically flat surfaces since the feedback loop and the piezoelectric driver do not have to repined to the

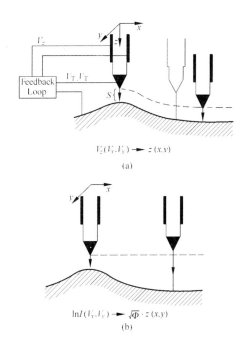

Figure 4.1 Schematic view of two modes of operation in STM. S is the gap between the tip and the sample. I and V_T are the tunneling current and bias voltage, respectively, and V_z is the feedback voltage controlling the tip height along the z direction. (a) constant-current mode and (b) constant-height mode (Binnig and Rohrer, 1987)

surface features passing under the tip. Fast imaging is important since it may enable researchers to study dynamic processes on surfaces and may reduce data-collection time. Fast imaging also minimizes the image distortion due to piezoelectric creep, hysteresis and thermal drifts. In contrast to the constant-current mode, however, deriving the topographic height information from the recorded variations of the tunneling current in the constant-height mode is not easy because a separate determination of $\Phi^{1/2}$ is required to calibrate z. In both modes, the tunneling voltage and/or the z position can be modulated to obtain, in addition, information about local spectroscopy and/ or the spatially resolved local tunneling barrier height, respectively.

For a more detailed description of other modes of operation, such as various tracking modes and differential microscopy, the reader is referred to the available literature (Bai, 1995; Stroscio and Kaiser, 1993).

4.2.2 STM Studies on Metals

STM and STS have established their unique and complementary function to the classical ones to observe the surface structures of conductors and semiconductors with high resolution in real time under different experimental conditions such as in vacuum, air, and solutions, and at low temperatures. The information about the surface electronic structure can also be obtained by using STS.

In theory, the metal surface structure can be inferred from the crystal structures, while in practice reconstruction of metal surfaces for the purpose of lowering surface energy makes surfaces more complicated especially for those induced by chemisorption on the metal surface. Many surface analysis techniques (including X-ray diffraction, helium diffraction, low energy electron diffraction, high energy ion scattering, transmission electron microscopy and field ion microscopy) have been employed in the study of metal surface structure during the past few years, but the absolute surface structures can hardly be determined, especially the reconstruction of a metal surface by gas chemisorption. Alternatively, STM has its unique advantages over the study of metal surfaces since it can observe the metal surface structures with or without periodic structure and can also be used to study the microscopic mechanism of nucleation and growth of surface reconstruction induced by gas chemisorption by comparing different STM images of the surface with various gas coverage. Theoretically, energy bands of s, p, d, f and their hybridization all exist in metal, but their contributions to tunneling current will be quite different, leading to various contrasts for geometric structure of the metal surfaces (Lang, 1986; Chen, 1990).

STM images reflect the contour of the constant surface state density of the surface. For a clean metal surface, the surface contour usually represents just the shape of the surface potential barrier which closely responded to the atomic positions on the surface; thus STM images indicate directly the geometric structure of the metal surface. Prevalent metal surface reconstruction provides a good test ground for STM practices.

There exist various reconstructions on the surface of Pt, Ir and Au. In particular, the (1×2) reconstruction on the (110) surface has attracted much attention, although many models have been proposed and received certain experimental support. STM applied to the (110) surface of Au (Binnig, et al., 1983a; 1983b) observed clearly parallel hills on the surface usually running several hundred A along the [110] direction. Most of the hills are separated by 8 A thus forming the (1×2) reconstructed ribbons, which are further separated by steps and (1×3) channels. This causes the [001] direction to be more disordered than the [110] direction. High-resolution images can also reveal an increased density of (1×3) channels and a transition from (1×4) channel to two (1×2) channels. According to the depth and symmetry of the (1×3) channel, a convincing missing-row model was proposed that various reconstruction and surface disorders were driven by the formation of Au(111) facets consisting of monatomic height steps (111). Facets with two free rows generated the (1×2) reconstruction of the missing-row type. Three-row facets produce the (1×3) reconstruction and the combination of two- and three-row facets can give other local reconstructions and cause surface disorder.

The STM studies (Woll, et al., 1988) of the Au(111) surface of clean and annealed single crystals have determined the flat surfaces. No individual Au atoms were observed except monatomic periodic steps with periodicity 6 A oriented along the [112] direction in some areas and multiple atomic steps in other regions, which depends on the local slope of the surface. The STM study (Hallmark, et al., 1987) of the vicinal Au(111) surface revealed a terrace eight atom wide as well as monatomic and multiple atomic height step. The STM study proves that Au(111) surface is not as stable as expected and undergoes reconstruction. Moreover, the surface reconstruction is not the simple contraction of the top layer atoms as predicted by the model. In the STM study (Kuk, et al., 1987) of the Au(111) film prepared by epitaxial evaporate deposition, the atomically resolved STM image is obtained both in UHV and in air.

The STM study (Wintterlin, et al., 1988) of the Al(111) surface in vacuum also gives an atomically resolved image as shown in Fig. 4.2. Most of the Al(111) surface is found to consist of the atomically flat terrace several hundred A wide separated by monatomic height steps.

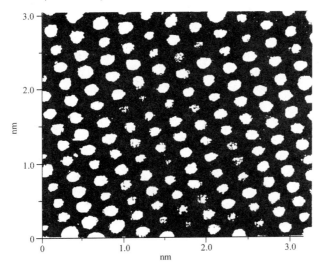

Figure 4.2 Topography of the Al(111) surface obtained in vacuum. Scan range is 34 A × 34 A (Wintterlin, et al., 1988)

Recent progress on manipulating individual atoms on metal surfaces suggested a novel method to construct metallic clusters using substrate metal atoms (Meyer, et al., 1997). It is clear that STM is capable of resolving individual atoms of the metal surface with close-packed structure in some cases. However, we can usually observe the main characteristics of the metal surface by STM in real time, such as steps, terraces and flatness. Compared with the case of semiconductor, the drop of resolution in the STM study on metal surface is possibly due to its large atom density and free electron density as well. In addition to the instrument properties, the sharpness of the tip as well as the sample properties are all important factors.

4.2.3 STM Studies on Semiconducting Surfaces

4.2.3.1 Si(111)-7×7 and Na/Si(111) Surface

Clean surface of elemental semiconductors, such as silicon and germanium, are easy to prepare technically. They provide a very rich field for analyzing various surface reconstruction phenomena.

Properties of the semiconductor surface are very sensitive to surface preparation procedures. For example, the form of the reconstruction of the cleaved Si(111) surface (namely 2×1 pattern) differs from that of the flash annealed surface (7×7 pattern). Notably, a clean Si(111) surface has been known for over 30 years to reconstruct into 7×7 structure upon heating to approximately 1173 K. Electron diffraction observations indicated that a superlattice exists in the plane of the surface with a unit cell that is seven times larger laterally in each direction than a simply terminated surface would be. In spite of many hints from a plethora of surface techniques, the details of this structure remained unsolved for 25 years until the STM observation of this 7×7 structure in real space (Binnig, et al., 1983). Since then, this surface has been considered the standard sample for checking the performance of the STM.

Similar STM images and LEED patterns can be obtained for either n-type or p-type Si(111) surface after annealing in vacuum at 1173 K. STM images of Si(111) 7×7 surface are shown in Fig. 4.3, together with a dimer-adatom-stacking-fault(DAS) model proposed by Takayanagi et al. (Takayanagi, et al., 1985). The 7×7 unit cell has a rhombic configuration and contains 12 adatoms at the topmost layer and 4 vacancies at four corners, which are usually referred to as corner holes. There are two triangular subunits separated by the shorter diagonal of the rhombus and the right one has a stacking fault. The STM topographic image of Fig. 4.3 closely follows the expected geometric contours of this surface, as calculated using atomic charge superposition methods which ignore electronic-structure contributions to the STM image. Within the cell there are 12 distinct protrusions, and at each corner a deep depression.

The electronic structure of Si(111) 7×7, like the geometric structure, is also an intriguing topic. Photoemission and inverse photoemission spectroscopies have provided energy information of the occupied and unoccupied states, respectively, but unfortunately it is the average of at least 10 atoms. The CITS (current imaging tunneling spectroscopy) method (Hamers, et al., 1986) allows real space imaging of the spatial distribution of

Figure 4.3 7×7 reconstruction pattern of Si(111) surface which can be accounted for by the DAS model

surface electronic states. Thus one can identify these state with specific features of the structure of the 7 × 7 surface. Real space images of these states can be subsequently obtained by taking the difference between current images just above and below the observed onsets in conductance. The method is demonstrated as very helpful in identifying different bond states.

The closely related surface reconstruction of Si(111) 5×5 and 9×9 have also been examined by the STM. Extending the interpretation of the 7 × 7 tunneling image to those of the 5×5 and 9×9, it is seen that the DAS model can be scaled down to the 5×5 and up to the 9×9 structure, accounting for the protrusions as adatoms on T4 sites on the mesh halves, and the excess of filled-state signal between corner and middle protrusions as rest-atom dangling bonds.

Owing to the relatively simple electronic properties of the alkali metals, their adsorption on the metal and semiconductor surfaces has been one of the most interesting topics, from both an experimental and a theoretical point of view . Some of the well-known phenomena accompanying alkali adsorption are the promotion of chemical reactions or the reduction of the work function. For instance, it is known that the oxygen uptake rate of the Si (111) 7×7 surface decreased by as much as 40% upon structural transformation of the clean 7×7 to a Na-induced 3 × 1 surface. STM studies of this surface provide detailed information regarding the structural transformation with atomic resolution. From the STM image obtained after depositing Na which showed a good 3 × 1 LEED pattern, it is readily seen that the 7×7 structure is completely replaced by domains of the 3×1 structure formed by a single layer of Na atoms. These domains are rotated by 120° from each other. A 3 × 1 surface image with atomic resolution is depicted in Fig. 4.4. Each 3 × 1 unit cell contains two Na atoms so that the saturation coverage is 2/3, which agrees with the saturation coverage deduced from the RHEED study on the alkali-metal-induced 3 × 1 structure on the Si(111) surface.

Figure 4.4 Na adsorption caused conversion of the reconstruction pattern from 7 × 7 to 3 × 1 (Jeon, et al., 1992)

To understand the origin of the island formation and the mechanism of the 7 × 7 to the 3 × 1 structural conversion, the images at various stages of Na deposition have been obtained (Jeon, et al., 1992). At the early stage of the conversion, the 3 × 1 regions are found on the lower terraces adjacent to the steps. The 7 × 7 periodicity in the uncovered regions could be observed until the entire surface converted to the 3 × 1 structure, indicating the importance of the dimer wall breaking up for the conversion to the 3 × 1 structure. Further Na deposition resulted in a reversal of the contrast of the 7 × 7 and 3 × 1 LEED pattern intensities until finally the 7 × 7 pattern disappeared completely. The 7 × 7 structure, in addition to the 3 × 1 structure, is the only periodic structure observed during Na deposition. This is an important difference from the surface obtained after annealing the Na-saturated 3 × 1 surface, in which case the Na layer could be gradually peeled off and various intermediate structures ($n × n$, $n = 2, 5, 7, 9$) are observed in the regions where the Na layer is removed.

Based on the STM results, a general picture of the 7 × 7 to 3 × 1 conversion can be conjectured as follows: the Si adatoms back bonds breakup upon Na deposition and annealing. The adatoms are confined to the unit cells because of the dimer walls. Near the steps, however, the unit cell structure is imperfect, and breakup of the dimer walls and adatom diffusion occur rather easily. Therefore, the 3 × 1 conversion starts at the steps; Si adatoms diffuse into the corner holes of the 7 × 7 structure or to the steps, concomitant with the restacking of the rest-atom layer, also under stress. A new substrate structure, i. e., the unreconstructed Si (111) is then stabilized by the 3 × 1 adlayer. Once the 3 × 1 structure is formed, it expands by converting the adjacent 7 × 7 unit cells. If the annealing temperature during Na deposition is lower or higher than the optimum temperature range, a complete conversion to the 3 × 1 surface does not take place.

Tunneling $I-V$ curves measured from the Na adsorbed 3 × 1 regions show that the Na monolayer-covered surface is insulating, but the tunneling gap closed when a Na double layer was formed. When oxygen is adsorbed onto the Na monolayer-covered surface, the surface morphology does not change very much. However, a small amount of oxygen adsorption on the Na double-layer-covered surface caused dark patches in the STM image, although the LEED pattern remained unchanged. This result is of significance in understanding the metallization of the alkali metal overlayer and the mechanism of negative electron affinity.

4.2.3.2 GaAs(110) and Na/GaAs(110) Surfaces

Both bulk and surface properties of gallium arsenide are most attractive to many researchers because of its potential applications in various fields. Studies of the clean GaAs surfaces are mainly focused on (110) surface since it is easy to cleave along this direction. Of course, studies on other surfaces are also performed under practical conditions.

The GaAs(110) surface has a chain like structure similar to the Si(111)-2 × 1 surface. The chain consists of alternating Ga and As atoms. The two chemical elements in the unit cell offer the possibility of chemically differentiating between the two atoms; an important technique which would be valuable to many inhomogeneous systems.

Feenstra et al. (Feenstra, et al., 1987) have obtained the first voltage-dependent STM images of the vacuum cleaved GaAs(110) surface with atomic resolution. Images show either only Ga atoms or only As atoms depending on the bias voltage. Not only did they identify the two kinds of chemically different atoms in real space, but also determined precisely the positions of As and Ga atoms on the surface. It is theoretically predicated that the STM images of the occupied state (at negative bias voltage) should reveal the positions of As atoms, and the images of the unoccupied states (at positive bias voltage should reveal that of Ga atoms. Figure 4.5 shows an STM image acquired at the sample bias voltages of -2.0 V. It was further found that the difference in the lateral positions of state-density maxims between occupied and unoccupied states is 0.21 nm on average, which is larger than the geometric separation of ~ 0.1 nm (Feenstra, et al., 1987). This large separation is a manifestation of the surface buckling showing in the electronic structure of the surface. It turns out that the buckling of the surface atoms causes an increase in charge transfer between the Ga and As, along with repulsion of the surface state bands. The separation between the occupied and unoccupied state can then be used as a measure of the surface buckling when compared with theoretical calculations. Feenstra et al. (Feenstra, et al., 1987) thus concluded that the buckling angle the amount of rotation out of the (110) plane is larger than 23°, with a most probable value from 23° to 35°.

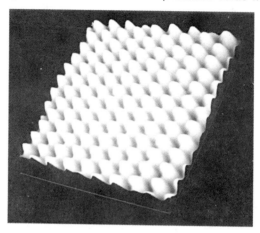

Figure 4.5 STM image of clean GaAs surface, recorded at sample bias of −2 V and tunneling current of 0.02 nA

The geometry and the coverage-dependent growth structures formed by Na on GaAs (110) surface have been investigated by STM at room temperature (Bai, et al., 1993). When the coverage is low, Na adatoms reside on the bridge site encompassing one Ga and two As surface atoms, as illustrated in Fig. 4.6. It is interesting that the Na adatoms appear to form low-density linear chains with the nearest Na-Na distance of 8 Å along the [110] direction and exhibit no long-range order in the [001] direction. The Na chains are separated by more than three As rows and are short in length (Fig. 4.6). The formation of the ordered one-dimensional chain at low coverage suggests

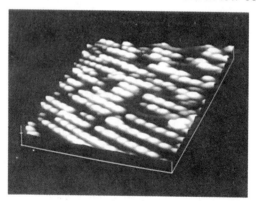

Figure 4.6 Chain-like arrangement of Na adatoms on GaAs (110) surface (Bai, et al., 1993)

that Na atoms can move easily on the GaAs (110) surface at room temperature, and that repulsive interactions along the [001] direction can be significant at low coverage. The fact that the short chains form readily indicates that the minimum nucleation size is small, no more than a few atoms.

When the Na coverage is increased from ≈ 0.07 ML to ≈ 0.09 ML (1 ML = 2 Na per substrate unit cell), the chains become slightly disordered. Some of them pack closer together to form domains of local 2 × 2 structure. With increased Na coverage, the surface is found to be dominated by clusters at about 0.11 ML and completely covered by bigger clusters at about 0.13 ML. The saturation coverage of Na is thus estimated to be roughly 0.1 ML. The $I-V$ curves recorded over the various Na-covered surfaces show no evidence of metallic characteristics.

It is interesting to compare the Na/GaAs(110) with the Cs/GaAs(110), in which Cs atoms form a zigzag chain at low coverage and, at high coverage, a high-density C 4 × 4 overlayer consisting of a five-atom polygon (Whiteman, et al., 1991). In contrast, neither the zigzag chain nor a high-density two-dimensional ordered structure is observed from Na/GaAs (110). As mentioned, the coverage is close to saturation.

There are many important factors to be considered concerning adsorptions on semiconducting surfaces, such as changes in the electronic energy through the change of bond lengths and angles, and the dipole-dipole interaction through the charge transfer. The overlayer can grow quite differently even for the same atom if the adsorbate-substrate interaction is dominant over the adsorbate-adsorbate interaction. Therefore, different nucleation processes may occur, such as the results of Sm, Ce Yb and Mg on GaAs(110) (Li, et al., 1991a; 1991b).

4.2.4 Organic Molecules Studied by STM

An important application of STM is to investigate the fine structures of various organic and biological molecules in their surface adsorbed state. Many interesting observations have been obtained in the past decade, and stimulated vast interests to better understand the electron tunneling principles in association with the properties of molecules. High resolution STM image of molecules (aromatic molecules, metal phthalocynines or MPc, etc.) (Gimzewski, et al, 1987, Lippel, et al., 1989, Chiang, 1997, Strohmaier, et al., 1998) adsorbed on metals were demonstrated in the past decade, together with the successful studies of a number of self-assembled molecular layers, such as liquid crystals (Foster, et al., 1988; Smith, et al., 1989; 1990) and long-chain alkane molecules (Rabe, et al., 1991; Cyr, et al., 1996), etc., on inert surfaces such as graphite and MoS2. Another example is the high resolution imaging of organic adsorbates on electrode surface under electrolyte using electro-chemical STM(ECSTM)(Wan, et al., 1995; 1997). These results have not only provided a direct venue to look into the structural characteristics of molecules, but also invoked deeper interest in the electronic properties of the molecules, such as the front orbital distributions and molecular polarizabilities. Furthermore, interactions of adsorbates with substrates are also an important aspect. In general, the assembled structures of adsorbed organic molecules are mainly governed by the electrostatic multipole-multipole interaction, steric repulsion, intermolecular and surface forces.

So far, much attention has been given to understanding the contrast mechanisms associated with various molecules. STM images of nCBs(4-n-alkyl-4'-cyanobiphenyls, where n is the number of carbons in the alkyl group, on graphite show that the aromatic groups or cyclohexanes enhance the tunneling efficiency and appear as bright contrast (Foster, et al., 1988; Smith, et al., 1989; 1990). Contrast differences were also prevalent in the derivative of linear alkanes (Rabe, et al., 1991; Cyr, et al., 1996) and planar molecules (Stromaier, et al., 1998). Figure 4.7 is an STM image revealing large uniform regions of molecular arrays of copper(II) octaalkoxyl-substituted phthalocyanine molecules (denoted as CuPcOC8). The example shows that linear alkanes could considerably increase the adsorption barrier, and therefore improve the adsorption and immobilization of phthalocyanines on inert surfaces such as graphite.

Figure 4.7 STM image of self-assembled mono-molecular layer of CuPcOC8 on graphite surface. Scanning area is 14.6 nm × 14.6 nm. The tunneling condition is 398 pA, −648 mV

Several theoretical models have been proposed and generated considerable interest. Spong et al. (Spong, et al., 1989) suggested that local work function modifications by the adsorbates is the origin for observed contrast. By taking into consideration the polarizability of adsorbed molecules, one has

$$\Phi = \phi - e\mu/\varepsilon_0 \qquad (4.2)$$

where Φ is the barrier height, ϕ is the barrier height for clean surface, μ is the dipole moment of the adsorbates, and ε_0 is the permitivity of free space. Thus the dipole moment of the organic molecules could reduce the barrier height of a substrate such as graphite. As a result, the work function of the substrate surface will be periodically varied by the adsorbates and becomes a function of position, such as in the case of linear alkane molecules (Fig. 4.8).

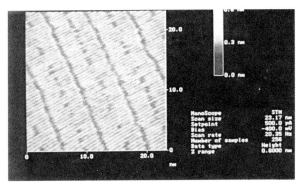

Figure 4.8 Linear alkane molecules($C_{34}H_{70}$) resolved by STM. Scan area 20 nm × 20 nm. Tunneling condition is -400 mV, 300 pA

On the other hand, resonant tunneling mechanism is also proposed in terms of the positions of front orbitals (Foster, et al., 1988; Smith, et al., 1989; 1990), lowest unoccupied molecular orbitals (LUMO) and highest occupied molecular orbitals (HOMO). This model would provide a direct linkage between experimentally observed fine structures with the intrinsic orbitals. In many cases, the front orbitals are separated from the Fermi level of the substrate by more than an electron volt. Therefore, in order to account for the imaging at small tunneling bias voltages, one should consider the factors that could decrease the energy difference between HOMO or LUMO of molecules and the substrate's Fermi level, such as the accompanying pressure within STM gap, interactions between molecules and the substrate, and the applied electric field. It is worth noticing that solid evidence is still needed to fully understand the contrast differences for organic and biological molecules. The results should be very beneficial for further studies aimed at recognizing individual functional groups within molecules.

4.3 Atomic Force Microscopy

4.3.1 Introduction

With regard to the operational principles of STM described in the previous section, it is important to notice that the "tunneling" phenomenon utilized by STM requires that this instrument be used only for conductors or semiconductors. For a non-conducting material, its surface must be covered with a thin conductive layer. However, the existence of conductive film often lowers the resolution and thus limits the usefulness of the STM. It was chiefly this limitation that prompted the development of the atomic force microscope (AFM) in 1986 (Binnig, et al., 1986; Sadrid, 1990), which can be applied to image conductors and non-conductors in air, liquids or vacuum.

The concept of force microscopy has led to widespread interest in almost all branches in surface studies. As a result, several highly effective microscopies have appeared, such as lateral force microscopy or friction force microscopy with the hope for new chemical mapping method of surfaces (Noy, et al., 1995; Sinniah, et al., 1996). The principles behind the schemes require detailed knowledge of the interaction, both long and short range, especially the interactions between the probe with various functional end-groups of the surface. This area of research still remains challenging yet very promising.

The operational principle of AFM is explained in Fig. 4.9. The cantilever, which is extremely sensitive to weak force, is fixed at one end; the other end has a sharp tip which gently contacts the surface of a sample. When the sample is being scanned in x-y direction, because of the ultra small repulsive force between the tip atoms and the surface atoms of the sample, the cantilever will move vertically with respect to the surface of the sample, corresponding to the contours of the interaction force between the tip and surface atoms of the sample. The topographic images can be obtained either by recording the deflection of the cantilever at each point (variable deflection mode) or by keeping the force constant using an integral feedback loop and recording the z-movement of the sample (constant-force mode). The STM yields images related to the surface electronic densities near the Fermi level. The AFM tip may be kept in direct contact with the surface (contact mode) or it may be vibrated above the surface (non-contact mode). For high-resolution imaging and most routine topographic profiling, the repulsive force (10^{-9} — 10^{-8} N) or contact mode is usually used. In the non-contact mode, the van der Waals force, the magnetic force, or the electronic force is detected.

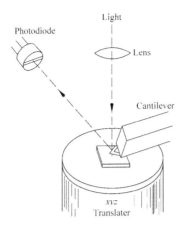

Figure 4.9 Schematic of AFM operation principles where a laser beam deflection scheme is used to detect the movement of the cantilever (Binnig, et al., 1986)

4.3.2 The Force Sensor

An AFM can be designed to look very similar to an STM. The primary difference is that it is convenient to move the sample rather than the delicate, and sometimes bulky, force sensor (composed of tip and cantilever). Much of the technology for implementing the AFM is now well developed. Techniques for vibration isolation, scanning, sample approach, feedback control and image processing are taken with little modification from the STM. Construction of the force sensing cantilever stylus and the measurements of the deflection of the cantilever still need careful consideration. The force sensor is a crucial component for AFM, determining its sensitivity and resolution. While the force sensor senses the force across a sample, the role of the cantilever is to communicate this information to the outside world. When the AFM is operated in the contact mode, in order to register a measurable deflection with small forces, the cantilever must flex with a relatively low force constant. The data acquisition rate in the AFM is limited by the mechanical resonant frequency of the cantilever. To achieve an imaging bandwidth comparable to that obtainable in the STM, AFM cantilevers should have resonant frequencies >10 kHz. Fast imaging rates are not just a matter of convenience, because the effects of thermal drifts are more pronounced with slow scanning speeds. If the scanning rate is too high or the cantilever resonance is too low, the inertia of the cantilever will cause the stylus tip from tracking steep downward slopes. The combined requirements of a low spring constant and a high resonant frequency can be met by reducing the mass of the cantilever stylus assembly.

High lateral stiffness in the cantilever is desirable to reduce the effects of lateral forces in the AFM. Frictional forces can cause appreciable lateral bending of the cantilever, leading to associated image artifacts. Investigations have indicated that choosing a "V" or "X" shape of the lever can yield substantial lateral stiffness. An example of cantilever is given in Fig. 4.10.

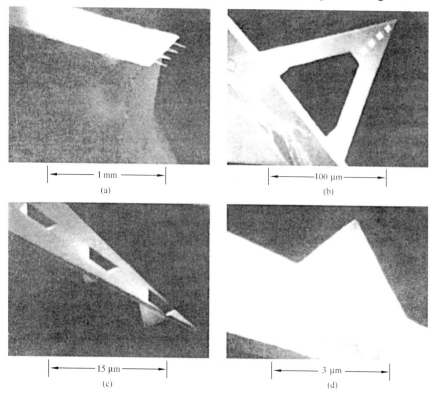

Figure 4.10 Illustration of an AFM cantilever probe (Albrecht, et al., 1990)

For high-resolution topographical imaging, the cantilever stylus used in the AFM should satisfy the following criteria: ① a low force constant, ② a high resonant frequency, ③ a high mechanical Q, ④ a high lateral stiffness, ⑤ short lever length, ⑥ incorporation of a mirror or electrode for deflection sensing, and ⑦ a sharp protruding tip.

For imaging atomically flat samples one can simply use the end of a cantilever to act as an effective local tip. For imaging rougher samples, however, one desires a sharp, protruding tip of known shape so that the interaction between the sample and the cantilever can be characterized more precisely.

Fabrication processes have been developed to produce SiO_2 cantilevers with integrated conical tips or tetrahedral tips (Albrecht, et al., 1990). The low mass of these cantilevers allows them to have high resonant frequencies (10—100 kHz) with force constants (typically 0.0006—2 N/m) which are small enough to detect forces of less than 10^{-8} N. Images which show atomic periodicity of several types of crystalline samples, including graphite, MoS_2, WTe_2, $TaSe_2$, and boron nitride (BN), were obtained by using the cantilevers described above.

4.3.3 Illustration of AFM Applications

As AFM can image both conductors and nonconductors, and can image samples in air, liquids or vacuum, it is finding a wide range of applications. The first published atomic periodicity images were layered compounds: graphite, molybdenum disulfide, boron nitride. Figure 4.10 shows the AFM image of the amino acid dl-leucine in which the white dots represent topographic peaks of methyl groups at the end of individual dl-leucine molecules (Fig. 4.11) (Hansma, et al., 1988). The positions of the methyl groups in this molecular crystal agree with the positions predicted from X-ray diffraction analysis of the sample, demonstrating that the surface is a simple termination of the bulk. AFM also succeeded in resolving individual organic molecules at nitronyl nitroxide single-crystal surfaces (Fig. 4.12) (Ruan, et al., 1991).

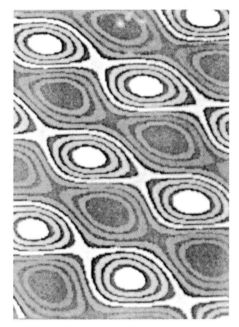

Figure 4.11　AFM image of dl-lencine molecular crystal (Hansma, et al., 1988)

Besides these atomically flat surfaces, the AFM has made it possible to image organic molecules on substrates. Surfaces in air are typically covered

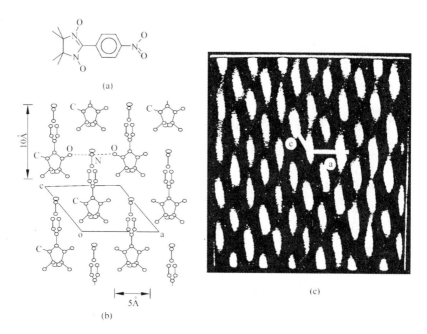

Figure 4.12 (a) Molecular structure and (b) the lattice coordination of NTDIOO crystal; (c) is the AFM imaging of the a-c plane (Ruan, et al., 1991)

with a layer of liquid contaminants, and when the probe comes near a sample there is a capillary force that pulls the tip towards the sample (with a typical value of 10^{-7} N). Operating the AFM within water could reduce this undesirable and often destructive force and allow better control of the force between the tip and the sample.

Biological applications are one of the major aspects for AFM studies. AFM has been used to image various cell surfaces, such as red blood cells (Fig. 4.13) (Zhang, et al., 1995), crystallized lipid bilayers, DNA (Li, et al., 1999) and purple membranes. In Fig. 4.13, an example is given on the high resolution of the surface of a human red blood cell. By sectioning the surface into 29 imaging zones, AFM can be used to study each individual zone respectively, and putting together the images one could obtain a complete mapping of the whole surface morphology (Zhang, et al., 1995).

Although the AFM has been utilized to image some biological samples, a major challenge for researchers is to develop new sample preparation techniques that will allow smaller details to be imaged. The sample must be rigid enough not to be damaged or distorted by the applied force. As mentioned earlier, imaging with the sample in liquid eliminates meniscus forces that can pull the stylus destructively into the sample, and thus makes it possible to use smaller forces, usually 10^{-9} N or less.

An application of STM and AFM of particular interest is the in-situ study in

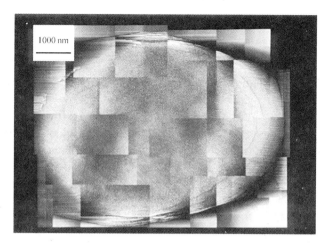

Figure 4.13 Sectioning profiling of the surface of a human red blood cell by using AFM (Zhang, et al., 1995)

electrochemistry. Special requirements have to be satisfied in the design of an STM for an in situ electrochemical investigation because the electrolyte is highly conducting, which makes the detection of tunneling current difficult. With the advances in AFM, especially the non-conducting AFM tips and optical-beam-deflection techniques, AFM has been used successfully to study the electrode surfaces under potential control in a fluid electrolyte.

AFM has also been used successfully to image Au(111) electrode surface showing atomic periodicity while the electrode was under potential control in a fluid electrolyte (Manne, et al., 1991). AFM images in electrolyte taken at $+0.7$ V prior to Cu deposition showed large areas exhibiting Au(111) structure. By sweeping the potential the Cu can be stripped down to an under-potential-deposited monolayer and finally returned to a bare Au(111) surface.

The images revealed that the under-potential-deposited monolayer has a correlation with different electrolytes. Specifically, for a perchloric acid electrolyte the Cu atoms are in a close-packed lattice with a spacing of 0.29 nm. For a sulfate electrolyte they are in a more open lattice with a spacing of 0.49 nm. As the deposited Cu layer grew thicker, the Cu atoms converged to a (111)-oriented layer with a lattice spacing of 0.26 nm for both electrolytes. Images were also obtained of an atomically resolved Cu monolayer in one region and an atomically resolved Au substrate in another in which a 30° rotation of the Cu monolayer lattice from the Au lattice is clearly visible. The observation of a complete adsorption and desorption cycle for a metal onto another metal surface with such high resolution demonstrates that the AFM is a valuable technique in studying other electrochemical processes in situ.

4.3.4 Force Spectrum Analysis

The core of the AFM studies is the understanding of probe-surface interactions. Though available knowledge provides general ideas of this interaction (Israelachivili, 1985), the microscopic behavior is far from being clarified. The key factors in determining these forces, such as molecular arrangements, various functional end-groups, adsorption properties, have all been pursued extensively. Adhesions associated with small molecules such as water and organic species are important not only to understanding of the adsorbed films, but also to the enhancement of resolution power of SPM, since the hydration force is considered as an essential factor in improving the resolution of SPM in ambient and solvent conditions.

By recording the interacting force between probe and surface while changing their separations, one can obtain rich information about the nature of the probe-surface interaction. This so-called force spectrum has been proven to be an effective method to measure probe-sample interactions and intermolecular forces (Lee, et al., 1994; Moy, et al., 1994; Florin, et al., 1994). By analyzing the characteristics of the force curves, it is possible to retrieve information concerning the nature of the interacting force, van der Waals, electrostatic, elastic and so on.

As shown in Fig. 4.14, the adhesions measured by AFM display rather different behaviors on surfaces of mica, CaF_2 (cleavage plane(111)) and KCl (cleavage plane (100)) (Tang, et al., 1999). The three surfaces have different degrees of inertness in ambient condition. The adhesion force is literally determined by the density of surface free energy, yet there are a number of uncertainties such as the effects of adsorbed surface film thickness, contact geometry and area, and so on. By measuring the characteristic force spectrum at controlled humidity, one could find that the measured adhesion is mainly determined by the gas-liquid interface. The examples of CaF_2 and KCl represent two surfaces of very different deliquescence conditions. The solubilities for CaF_2 and KCl are 1.5 mg at 18℃ and 35.98 g per 100 g H_2O at 25℃, respectively. At sufficient humidity, the surface of KCl will dissolve and the effect is manifested in the increase of adhesion. This effect could be used to monitor the surface quality and relevant variations.

There has been growing interest in using the force spectrum method to probe the characteristic interactions between biological species, as well as a large variety of organic molecules and polymers, etc., and the revealed information could be far-reaching. Examples can be found in the exploration on the binding strength of cell adhesion. Proteoglycans from marine sponges produced a value of 400 pico-newtons, leading to the belief that a single pair of molecules could hold a weight as high as 1600 cells (Dammer, et al., 1995). Using the measured force and binding energies, the effective rupture

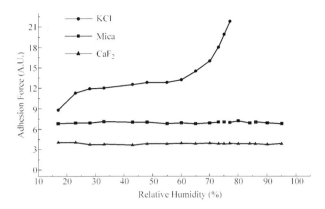

Figure 4.14 Force spectrum measurement of three substrate surfaces (mica, KCl, CaF$_2$) showing the effects of humidity on surface adhesions (Tang, et al., 1999)

length of avidin-biotin could be estimated (Moy, et al., 1994). The technique has also been applied to investigate the bonding strength bases of nucleic acids (Lee, et al., 1994) and ligand-receptor pairs (Florin, et al., 1994). Force measurements have also been used to study the local elastic properties such as polysaccharidries chains (Fig. 4.15) (Rief, et al., 1997; Overney, et al., 1996).

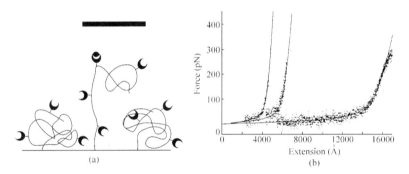

Figure 4.15 Force spectrum of elongation of single molecule of polysaccharides (Rief, et al., 1997)

There is no doubt that surface energy is not the sole factor in adhesion measurements. Other factors such as contact geometry and deformations are also crucial in making the technique a quantitative one. Hertzian model for spherical and cylindrical contacts provides a good simplification (Johnson, et al., 1971; Derjaguin, et al., 1975), which also points to the necessity to have a regularly shaped probe. This method is successfully used to study the adhesion mapping of various films, including self-assembled films, polymeric films (Tian, et al., 1998), and other surface adsorbates, and is also an important factor in surface frictional analysis (Noy, et al., 1995; Sinniah, et al., 1996). The precision force measurements have also prompted explorations of novel sensory systems that could detect interactions of single molecules, which is of particular interest to biomaterial studies.

4.3.5 Lateral Force Microscopy

In an AFM operating in the contact mode, the interaction force is manifested as a displacement of a cantilever (with or without a tip), where the magnitude of the force is the product of the displacement and the spring constant of the cantilever. The lateral force microscope (LFM) is an extension of the AFM which allows the measurement of two forces simultaneously through the bending and twisting of the cantilever in the imaging process (Meyer and Amer 1990; Hipps, et al., 1992). It measures topography just like the AFM, but the additional sensor is used to measure frictional forces between the sample and the tip from the same scan.

It is instructive to examine briefly the nature of the cantilever response to frictional and surface forces. In the latter case, the direction of the acting force is normal to the surface of the sample and results in vertical bending (z-axis direction) of the free end of the anchored cantilever. By contrast, in the lateral or frictional regime, where the cantilever is presumed in contact with the surface, upon scanning, the cantilever undergoes a torsion (twist) motion about its long axis (in the x-y plane). Both motions are orthogonal to each other. This orthogonality is what enables that simultaneous, yet independent, acquisition of topographic images and frictional data.

To measure the cantilever displacement, the direction of a laser beam reflected off the backside of the cantilever is monitored with a position-sensitive detector (PSD). In the case of surface topography, the bending of the cantilever is recorded with a segmented PSD, typically a bicell, which consists of two photoactive (e.g., Si) segments (anodes) that are separated by 10 m and have a common cathode. Additionally, lateral forces induce a torsion of the cantilever which, in turn, causes the reflected laser beam to undergo a change in a direction perpendicular to that due to surface corrugation. Thus, with a simple combination of two orthogonal bicells, i.e., a quadrant PSD, one is able to measure the deflection of the cantilever independently yet simultaneously in two orthogonal directions (Meyer and Amer, 1990). This capability is unique to the optical beam deflection method which, due to its very nature, measures the orientation of the cantilever and not only its displacement.

It should be noted that the torsion force constant depends critically on the tip length. Additionally, the tip should be centered at the free end of the cantilever. Both factors can be readily controlled with the use of microfabricated cantilevers and tips.

Figure 4.16 shows the AFM and LFM images of Langmuir-Blodgett film

4-penty 1-4′-methoxy-phenylbenzoate on the surface of a glass substrate(Li, et al., 1999). These two images were obtained simultaneously in the same area. Figure 4.16(a) is the image recorded by AFM in contact mode where it shows the height difference, whereas in Fig. 4.16(b) the contrast is mainly due to the difference in friction coefficients for two materials. It should be noted that since the friction is proportional to load, therefore the contrast should increase with increasing load (note that adhesion should also be counted as part of the applied load).

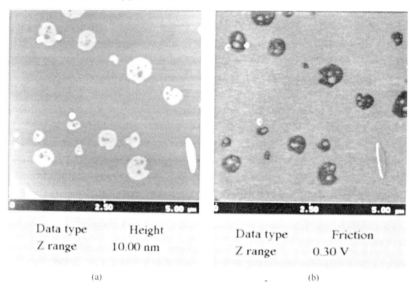

Figure 4.16 (a) Contact mode AFM and (b) LFM images of the domains of self-assembled molecular domain

4.3.6 Force Microscope Operating in Non-contact Mode

As mentioned, the force microscope can be operated in contact mode or non-contact mode. The AFM described in the last section and the LFM mentioned above are operated in the contact mode to measure the repulsive forces between the tip and the sample. A force microscope operated with relatively large tip-sample separation can be used to measure longer range interactions between tip and sample across areas that are orders of magnitude larger than that of a single atom. Three basic forces can be mapped across a sample in the non-contact mode by detecting the deflection of the cantilever under the influence of the desired force. These consist of electrostatic, magnetic, and van der Waals forces, in the order of the complexity of the interaction. The first deals with monopoles, the second with dipoles, while the third requires a quantum mechanical treatment. These forces and their derivatives can be as small as 10^{-12} N and 10^{-14} N/m, respectively, which are much smaller than those encountered with AFM techniques. Much effort has been made to investigate the magnetic force in the force microscope. This kind of microscopy has developed into an independent field and is called magnetic force microscopy (MFM).

For distances of a few angstroms to a few tens of angstroms, van der Waals forces are significant. They can be used to measure topography with a resolution of a few nanometers. For instance, a scanning force microscope was developed (Wickramasinghe, 1990) that works with attractive rather than repulsive forces and scans the tip 3 nm to 29 nm away from the surface. It allows one to measure attractive forces down to 10^{-11} N that are one thousand times smaller than the forces typically encountered with contact mode AFM. The attractive-force gradients encountered by the tip as it approaches the sample surface change the spring constant of the cantilever and hence its resonance frequency. The vibrating transducer is driven at a fixed frequency, and this results in a change in the vibration amplitude of the cantilever. This change is detected by a sensitive laser probe and is utilized in a feedback loop to stabilize the sample to average tip spacing, as the sample is raster scanned in x and to record an image. The image recorded in this way represents contours of constant-force gradient across the sample.

In this way the force microscope can detect a surface relief of as little as a few nanometers, and because it senses topography from a distance, it can inspect features inside the deep, narrow clefts. It promises to be valuable for mapping rough surfaces and for non-contact examinations of finished microcircuits. The non-contact force microscope has also been used to monitor the quality of silicon surfaces with atomic resolution (Giessibi, 1995; Sugawara, et al., 1995).

4.3.7 Force Microscope Operating in Tapping Mode

Applications of the non-contact mode operation are limited due to the nature of the interacting forces. The tip must be vibrated close to the sample surface with low energy because van der Waals forces are relatively weak. However, moving the vibrating tip closer to the surface increases the chances of getting the tip stuck in the water layer which covers the surface of all samples exposed to the atmosphere. Tapping the tip-to-sample separation, which is typically 5 and 10 nanometers, defines the lateral resolution of a force microscope operated in the non-contact mode. Due primarily to these limitations, the non-contact force microscope has found only limited applications.

To overcome the limitations of non-contact techniques, Digital Instruments has developed a new technique for operating the force microscope, the so-called tapping mode. This technique vibrates the cantilever with a larger amplitude than the non-contact mode operation. In the tapping mode, high aspect ratio tips with small radii of curvature are used and the vibrating tip contacts the sample surface many times per data point. The cantilever oscillation is damped when the tip contacts the water layer and the sample surface, but the larger vibration amplitude gives the cantilever sufficient energy to overcome the surface tension of the adsorbed water layer. The force imparted onto the sample by the cantilever can be very small because small shifts in the vibration amplitude can be detected. The tapping mode allows delicate samples to be imaged with normal forces on the order of fractions of a nano-newton and shear forces that are essentially zero. The applied force is significantly lower than the force applied by the contact AFM. The lateral resolution in the tapping mode compared favorably to the contact AFM because high vibration frequencies allow the tip to contact the sample surface many times before it translates laterally by one tip diameter. Therefore, the tip shape defines the lateral resolution in the tapping mode just as it does with the contact AFM. In practice, this mode can be used to get a topographic image of any sample regardless of its conductivity or mechanical composition.

There is a fast growing interest in using tapping force microscopy to study the surface properties, especially soft materials. The approach, though in essence is a modified contact mode AFM, has minimized the effect due to friction or other lateral forces. The key factors in the measurements are the amplitude and phase variations as compared with free resonant vibrations. Much effort is seen to establish the quantitative description of the variations (Spatz, et al., 1995; Burnham, et al., 1997). It is believed that the description is not a trivial exercise as it involves both interface and bulk properties of the sample material.

Figure 4.17 is a tapping mode AFM topographic image of microphase separations of poly(styrene-butadiene-styrene) (SBS) triblock copolymer film (Bai, et al., 1999). The polymer SBS (S/B=3/7) has a molecular weight of $M_w = 81600$. The SBS film was prepared by dissolving SBS in butanone solution with a concentration of 8.1 mg/L and then rapidly distributing it onto freshly cleaved mica and immediately drying it under an infrared lamp prior to the AFM observations. The characteristic morphologies consisting of hills and valleys were obtained. The hills are likely to grow from long worm-like to mesh-like microphase domains. Such topographic difference will be caused by the difference of the free surface energies between polystyrene and polybutadiene. By considering the Tanaka, et al.'s model for explaining the hill formation of the material with higher surface energy (Tanaka, et al., 1996), the hill structure here can be interpreted as the polystyrene, which has higher surface energy than polybutadiene.

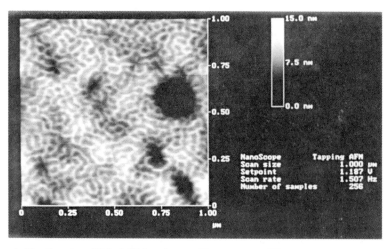

Figure 4.17 Tapping AFM of SBS polymer. The contrast difference is a direct reflection of phase separation (Bai, et al., 1999)

4.3.8 Magnetic Force Microscopy

The magnetic force microscope (MFM) is essentially a kind of force microscope operating in the non-contact mode except that the tungsten or silicon tip is replaced by a nickel or iron tip which is magnetized along its length (Martin, et al., 1988; Mamin, et al., 1988). Tips coated with ferromagnetic thin films were also investigated successfully by several groups and have become commercially available recently. Magnetic thin film tips have the substantial advantage of a significantly reduced tip stray field as compared to bulk-wire tips. Another advantage is that their magnetic properties can be controlled by choosing an appropriate coating material. Thus, it is possible to measure selected components of the sample field by coating tips with high coercivity films and suitably magnetizing them in an external field.

The magnetic forces measured in the MFM are purely magnetostatic. They arise from the magnetic dipoles in the tip interacting with dipoles in the sample. When the cantilever, which is oscillated at its resonant frequency, is brought near, a magnetic sample and the tip encounter a magnetic force gradient, the effective spring constant and, hence, the resonance frequency is shifted. By driving the cantilever above or below the resonant frequency, the oscillation amplitude varies as the resonance shifts. An image of magnetic field gradients is obtained by recording the oscillation amplitude as the tip is scanned over the sample.

Much of the reason for this excitement is the fact that MFM is the only magnetic imaging technique that can provide high resolution (10—100 nm) with essentially no special sample preparation. This microscope enables us to look at the structure of materials such as magnetic heads that determine the definition, uniformity and strength of the magnetic disks and other media, giving insight into both head performance and quality of the storage medium. Good quality images can be taken even when the magnetic material is covered with a thin overcoating, an important feature when imaging many technological important samples. A resolution of 10 nm has been shown on rapidly quenched FeNdB thin films by MFM (Grutter, et al., 1990).

An example can be seen in the observation of the static changes of domain configuration in $(YGdBi)_3(GaFe)_5O_{12}$ garnet using MFM in an external magnetic field (Tian, et al., 1997). Sequential variations of the magnetic domain configuration in the garnet are observed and could be associated with the strength of the applied field. It has been demonstrated that the undisturbed magnetic domain structure could be obtained with a soft magnetic tip. The stray field emanating from a hard tip could magnetize the garnet and alter the widths of the domain region, an effect which can be minimized by choosing appropriate tip-sample separation.

When an external magnetic field H_{ext} was applied to the garnet, one could readily observe sequential changes in the magnetic domains in the garnet using a hard magnetic tip (since hard tips have the advantage of most likely preserving their magnetization throughout experiments). As shown in the series of images in Fig. 4.18, it is evident that the observed domain

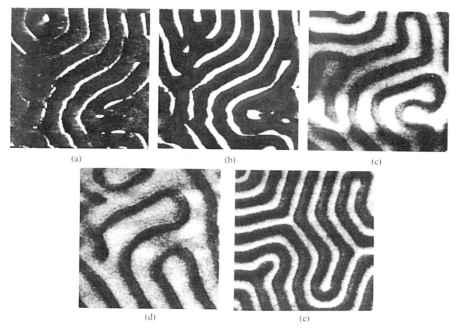

Figure 4.18 The variation of magnetic domains in magnetic garnet under the influence of external field. Scan area is 80 μm x 80 μm, tip-sample separation is 200 nm. From (a) to (e), H_{ext} = 0, 21, 43, 61 and 0 Gauss, respectively (Tian, et al., 1997)

configuration has been strongly affected by applied field. With increasing H_{ext}, the widths of the bright domains increase gradually and the widths of the dark domains decrease slightly, while the domain configuration also changes appreciably (Fig. 4.18(c)—(d)). When H_{ext} is raised to 75 G (image not shown), there is no discernible image contrast in MFM image, which indicates the whole garnet has been saturated completely.

Since the tip-sample separation is kept constant during MFM scanning, the impact of the hard tip on the garnet is fixed and tends to reduce the widths of the bright domains. During the static magnetization process, the width of the bright domains is always increasing with H_{ext} increasing, indicating that H_{ext} is dominating. On the other hand, the magnetization density of the hard tip could be reduced, which only affects the overall image contrast.

Since the first MFM image was obtained in the late 1980s, the field has grown rapidly. MFM is already a powerful tool for magnetic imaging, with many scientific and industrial applications. Future advances in MFM will include improved tips, matching of tip-sample characteristics, and determination of field strengths and domain structure from force gradient images, etc.

4.4 Ballistic-Electron-Emission Microscopy

The discovery and application of semiconductor materials have resulted in the necessity for a complete understanding of the fundamental characteristics of semiconductor surfaces and interfaces, such as the influence of thin film deposition on semiconductor surfaces, the transport properties of interfaces, carrier mobilities, quantum-well depths, etc. Subsurface interface electronic properties are not directly accessible to conventional surface analytical techniques. Although conventional Schottky-Barrier (SB) characterization methods, including photoemission, photoresponse, current-voltage, and other techniques, can be used for indirect investigation of the properties of interfaces, they are limited by their lack of spatial resolution for probing the variation of SB properties over the interface plane. Ballistic-electron-emission Microscopy (BEEM) is the first method for direct spectroscopic investigation and imaging of subsurface or interface electronic properties with high spatial resolution.

4.4.1 The Principle of BEEM

BEEM, developed by Kaiser and Bell (Kaiser and Bell, 1988), utilizes STM in a three-electrode configuration (Fig. 4.19). The sample for BEEM investigation consists of at least two layers separated by an interface of interest, in general, is a metal/semiconductor SB heterojunction. The STM tunneling tip is positioned near the surface of the heterojunction to emit ballistic electrons into a metal/semiconductor structure via vacuum tunneling. These low-energy electrons have typical attenuation lengths of about 10 nm in the metal base, and some of the injected electrons may propagate through a surface layer to the subsurface metal/semiconductor SB interface before scattering. For a base-tip tunnel bias less than the base-collector barrier height V_b, there will be no ballistic-electron current into the collector since the injected ballistic electrons have insufficient energy to surmount the energy barrier. But, as the base-tip bias V is increased above V_b (Fig. 4. 19(b)), a dramatic increase in base-collector current I_c occurs. The spatial variation in I_c reveals the differences in electronic structure at the different areas of the interface. Moreover, the I_c-V spectrum provides a direct probe of interface electronic structure, including the important SB height, defect structure at the interface, quantum-mechanical reflection of electrons at the interface, and ballistic-electron transport properties of the base film.

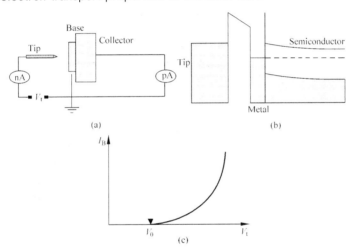

Figure 4. 19 Schematics of BEEM. (a) experimental setup; (b) energy diagram; (c) schematic BEEM spectrum with a threshold which corresponds to the Schottky barrier height of metal-semiconductor interface

The collector current (I_c) can be described by

$$I_c = RI_t \frac{\int_0^\infty D(E_x) \int_{E_{min}}^{E_{max}} f(E) dE_t dE_x}{\int_0^\infty D(E_x) \int_0^\infty [f(E) - f(E + eV)] dE_t dE_x} \quad (4.3)$$

where I_t is the tunneling current, $D(E_x)$ is the transmission probability for an electron to tunnel through the vacuum barrier, $f(E)$ is the Fermi function, and

R is a measure of attenuation due to scattering in the base layer. R is taken to be an energy-independent constant, since ballistic-electron attenuation lengths in metals are nearly independent of energy for $E - E_F$ of less than 2 eV. E_{max} and E_{min} can be written as:

$$E_{max} = [m_t/(m - m_t)] \cdot [E_x - E_f + (V - V_b)]$$
$$E_{min} = E_F - e(V - V_b)$$

m_t is the electron's effective mass parallel to the interface within the semiconductor, and m is the free-electron mass.

The equation can be fitted to the experimental spectra by adjustment of V_b and R. Most importantly, as a consequence of the parabolic conduction band minimum and K_t conservation, Equ. (4.3) predicts that the I_c-V spectrum behaves as $(V - V_b)^2$ for the voltages just above the threshold value V_b, independent of the fitting parameters.

There is a critical angle for electron propagation in the base outside of which electrons may not be collected. The critical angle is a dominant effect in transmission and total internal reflection in a base which depends on the incident energy and interface band structure. In addition, it focuses the electrons which provides high spatial resolution of interface properties. As mentioned above, since only electrons with small transverse momenta in the base may be collected, scattering in the base serves mainly to reduce the number of electrons collected rather than to reduce resolution. For example, for Au on Si or GaAs with $m_t/m = 0.067$, this critical angle is less than 6°, so only electrons within 6° of normal incidence may be collected. For a tunneling voltage just above threshold (i.e., $V - V_b$ is about 0.3 V), a lateral resolution of the order of 1 nm is achieved for a 10 nm thick base layer. The resolution obtained by the BEEM experiment is in agreement with this treatment. This demonstrates that critical-angle reflection defines the spectrum shape and spatial resolution of the BEEM spectrum, and is a dominant effect in interface carrier transport.

4.4.2 BEEM Experiments

There is not much difference between the apparatus used for BEEM and STM experiments. In order to keep the surface clean and to isolate the vibration from outside, for some samples it is necessary to perform the BEEM experiment in a vacuum or an inert gas. The criteria to be satisfied in the design of the BEEM are the same as those for the UHV STM. However, in order to measure the collector current I_c, a high-sensitivity (gain $\sim 10^{11}$ V/A), low-impedance (10 Ω) current amplifier should be added to the electronics. The typical collector current is less than 100 pA and is weaker than tunneling current by a factor of at least 10. This leads to the stricter requirements for the stability, reproducibility, and high signal-to-noise ration (SNR) of the BEEM system. The software for data acquisition should have the capability of acquiring the STM topographic and the corresponding BEEM images simultaneously. The I_c signals should be averaged many times at each surface location to improve the spectral SNR.

The complexity of schottky-barrier (SB) formation phenomena, including the role of interface-defect formation, electrode interdiffusion, and chemical reaction, is expected to induce inhomogeneity into interface structure and electronic properties. For example, the experimental results obtained by other techniques demonstrate that the properties of the Au/GaAs SB interface are strongly affected by interface-defect formation, pronounced interdiffusion and alloy-formation phenomena between the Au and GaAs electrodes. In contrast to Au/GaAs, the Au/Si SB shows simple, reproducible SB characteristics.

The topography and the corresponding BEEM images of Au/Si (100) and Au/GaAs (100) SB heterojunctions have been obtained. The samples were prepared by evaporation of 1.6 nm diameter Au disk electrodes, 10 nm thick, on chemically etched n-type Si (100) and chemically etched n-type GaAs (100) wafers. The STM image shows smooth topography at the Au electrode surface for the Au/Si heterostructure. The BEEM image of an Au/Si SB interface also shows homogeneous electronic properties. In contrast to the Au/Si heterostructure, a large degree of interfacial heterogeneity was observed for the Au/GaAs system, with domains of high ballistic-electron transmittance about 2 nm to 20 nm in size. BEEM images of the chemical etched Au/GaAs (100) interfaces, prepared whether with or without air exposure before Au deposition show the same heterogeneous interface-defect structure. This persistence of the interface heterogeneity indicates that the defect structure is not simply the result of substrate surface contamination.

BEEM can also be used for the investigation of interface band structures. GaAs has a direct conduction-band minimum at the zone center and two higher indirect minima at the L and X points of the Brillouin zone, while Si has only a single conduction-band minimum along the [100] direction. For example, the Au/GaAs spectra display a threshold region which, in contrast to the Au/Si spectra, is seen by comparing typical spectra of Au/GaAs and Au/Si heterostructures (Bell and Kaiser, 1988).

The BEEM spectroscopy for the Au/GaAs system, obtained by fitting the theory according to Equ. (4.3) with three threshold values, is indicated in Fig. 4.20. For comparison. For the case of three thresholds, the agreement between theory and experiment is excellent, yielding threshold values of 0.89, 1.18 and 1.36 eV (Bell and Kaiser, 1988). The first threshold value is in agreement with the commonly accepted value for the Schottky-barrier energy for Au/GaAs, namely 0.9 eV. The differences between the upper thresholds and the lower one, 0.29 and 0.47 eV, agree well with the expected relative energies of the three lowest conduction-band minima in GaAs: 0.29 and 0.48 eV for the separation between the direct minimum and the satellite minima at the L and X points, respectively. As a sensitive test of the agreement between experiment and theory, the experimental and theoretical derivative spectra, dI_c/dV versus V, are compared with Fig. 4.20(b). The thresholds, marked by arrows, appear as steps in the derivative spectra and the theory agrees well with the experimental BEEM spectrum. The changes in slope at the thresholds show relative magnitudes which are in agreement with the ordering of the different effective masses of the three minima.

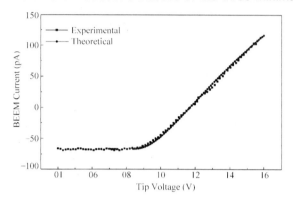

Figure 4.20 A typical BEEM I_c-V spectrum obtained on Au/n-Si(111) system (dotted line). Theoretical simulation (solid line) reveals the Schottky barrier height as 0.78 eV

It is worth noticing that BEEM techniques can be used for observing not only the multiple thresholds in the conduction-band structure, but it can also be applied to measuring many I_c-V spectra on different locations of the heterojunction, i.e., spatial variations in the relative energies of different thresholds can be detected, providing a direct measure of the local variation of the conduction-band structure. Variation in local interface band structure may result from variation in interfacial strain or diffusion-induced non-stoichiometry. For semiconductor/semiconductor strained-layer interfaces, the local variation in band structure plays an important role in determining the interface electronic properties. The ballistic-electron spectroscopy methods are applicable to investigation of such and many other systems.

BEEM has also been employed to study Au/Si(111), Au/CdTe, NiS_2/Si and $CoSi_2$/Si interface structures, providing new information about these interfaces.

4.4.3 Ballistic-Hole Spectroscopy of Interfaces

BEEM methods enable direct spectroscopy of subsurface interface electronic structure on a nanometer spatial scale. The energy-resolved spectroscopy methods based on BEEM have been limited to investigation of electron transport. However, the properties of hole transport in semiconductors and meals, through interfaces, and through tunnel barriers is central to the understanding and development of quantum-well, superlattice and other important systems.

Hecht, et al. (Hecht, et al., 1990) developed a ballistic-hole spectroscopy of interface transport and interface electronic structure based on BEEM. This new method has enabled direct measurement of subsurface valence-band SB height and valence-band structure. Comparison of BEEM ballistic electron and hole spectroscopes of subsurface interface conduction and valence bands is shown in Fig. 4.21. In the case of BEEM the ballistic-electron distribution created by tunneling is collected after transport through a multilayer structure (Fig. 4.21(a)). In contrast, ballistic-hole spectroscopy (Fig. 4.21(b)) is based on tunneling of electrons from the interface structure to the tunneling tip with the tip bias positive with respect to the structure (tunneling bias, V, negative). For negative V, where tunneling probes occupied states in the base, the emission of an electron from the base electrode of the interface structure creates a hole in the conduction band of the base material. Since hole-attenuation lengths are as large as several hundred angstroms, the hole may propagate ballistically through the base electrode and to the subsurface interface. Transmission through the interface is allowed if the hole energy (measured with respect to the base conduction-band

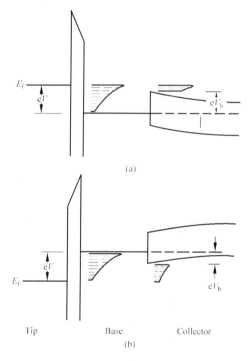

Figure 4.21 Energy diagrams for BEEM electron and hole spectroscopies. (a) Ballistic electron spectroscopy of a metal/n-type semiconductor system. (b) Ballistic hole spectroscopy of a metal/p-type semiconductor system. eV_b is the Schottky-barrier height (Hecht, et al., 1990)

minimum) is less than the threshold defined by the valence-band maximum ($E_F - eV_b$). Since the ballistic-hole energy is directly determined by the applied bias, an accurate hole spectroscopy of interface transport and interface valence-band structure is possible. Since the incident hole energy is simply controlled by the bias voltage, direct spectroscopy of interface valence-band structure can be performed.

Ballistic-hole spectroscopy can be performed by measuring the hole current transmitted through the interface and reaching the collector as a function of the tunnel bias V applied between the tip and base. Ballistic-hole and ballistic-electron spectra for the Au/Si(100) system are displayed in Fig. 4.21. The combined spectra show that a region of zero observed collector positive tunneling bias (tunneling tip negative with respect to the base) directly indicates the position of the conduction-band minimum. As expected, for negative tunneling bias, the observed collector current is opposite in sign to that observed for positive V. The current provides a spectroscopy of ballistic-hole interface transport. Further, the ballistic-hole spectrum threshold directly yields the barrier height formed by the valence-band maximum at the subsurface Schottky barrier.

The detailed characteristics of ballistic-hole creation and transport are compared to those for ballistic electrons in the experimental derivative spectra (Fig. 4.22(a)). The marked difference in spectral shape above threshold between the hole and electron derivative spectra reveals a fundamental asymmetry between the collected distributions for electron and hole injection. For ballistic-electron spectroscopy at a positive tunneling bias, the collected electrons originate from the top of the tunneling distribution where the distribution is maximum. The number of electrons created per unit energy remains nearly constant with the bias. However, for ballistic-hole spectroscopy at negative tunneling bias the ballistic-hole distribution originates from the bottom of the tunneling distribution where the distribution is at a minimum. Therefore, the number of holes created per unit energy decreases with increasing bias. This fundamental asymmetry between BEEM electron and hole spectroscopy is directly revealed in the experimental derivative spectra for an Au/Si(100) SB interface (Fig. 4.22(b)). The ballistic-hole derivative spectrum displays an abrupt maximum and a sharp decay resulting from the decay in the ballistic-hole distribution with tunneling bias above threshold. In contrast, a smooth increase is observed in the ballistic-electron derivative spectrum. It is particularly significant that the accurate understanding of this contracting behavior presents a sensitive and important test of interface transport spectroscopy and theory.

BEEM has seen considerable progress and many BEES studies have been carried out which have resulted in much improved structural resolution for interfacial features(Sirrinhaus, et al., 1995, Rubin, et al., 1996).

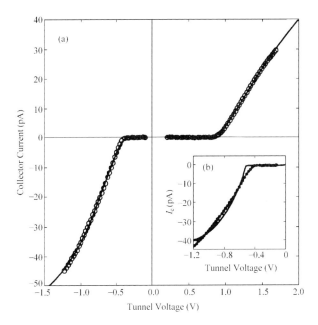

Figure 4.22 (a) Experimental ballistic hole (negative bias) and ballistic electron (positive bias) spectra measured at 77 K for Au/p-type Si(100) system. (b) Comparison of ballistic hole spectrum with theoretical evaluation (Hecht, et al., 1990)

4.5 Applications of STM and BEEM in Surface and Interface Modifications

Above we described the applications of SPM on surface topographies and electronic structures. In fact, being a surface analysis technique to study the surface properties, SPM also can be a tool to modify or etch many surfaces on nanometer scale. It has already been used in direct surface indentation, lithography, electron beam induced deposition and etching, single-atom manipulation, and so on. This is an important application aspect of STM and will be introduced in this section.

Fabrication of artificial features on materials surfaces using SPM is a very attractive and yet intriguing subject actively pursued in the past several years. Extensive investigations were specially dedicated to the revealing of the physical and chemical nature of the process. The effort leads to a promising research direction that might serve to lay the foundation for the next generation of microlithography technology. The progress may also be far reaching which could be potentially beneficial to not only material sciences, but also other branches of physical sciences, as well as chemistry and biology.

Research efforts in this field have been very fruitful and yielded remarkable progress both experimentally and theoretically (Stroscio and Eigler, 1991; Eigler and Schweizer, 1990). As the controls of experimental conditions and observation of fabricated features were gradually improving, STM began to show aspects that could be related to various physical and chemical origins. So far, the surface fabrication or modification processes using STM were the most thoroughly studied one, especially for several model systems. Among them, HOPG and other lamellar compounds form an important part.

4.5.1 Surface Nanofabrication with STM

For the situation of STM, the tip is so close to the targets as to make the electron current highly spatially confined. It is the key to the extremely high spatial resolution of STM images. As with lens-focused electron beams, the high spatial confined electron beam can also cause structure defects, phase changes, chemical reactions and the shifts of adsorbates, or induce chemical deposition and etching in small regions on the sample surfaces, which makes STM be used in nanofabrication area. The small distance (~ 1 nm) between tip and sample causes electrons to tunnel to (or from) a region on the sample that is approximately a few nanometers in diameter, with an even smaller major distribution area. Thus the surface fabrication produced by STM must be performed on the nanometer scale; i.e., STM can nanofabricate as exemplified by Fig. 4.23. Furthermore, that it is also possible to manipulate single atoms adsorbed on surface with STM.

Figure 4.23 STM fabricated "CAS" on graphite surface. The line width is around 20 nm

It is clear at this stage that the fabrication process should be jointly determined by the strong external electric field during the operation and the properties of the materials. The mechanisms behind the fabrication involve various aspects of the interactions between atoms/molecules and low-energy electrons under a strong electric field. The injected low-energy electrons are considered a unique approach to provide energy for chemical reactions to proceed at nanometer scale. The fact that there are a variety of fabrication outcomes for different materials under comparable external conditions provides positive support to this view. For example, similar voltage pulses could generate significantly different features on HOPG (Albrecht, et al., 1989), 1T-TaS$_2$ and MoS$_2$. One may notice these materials possess metallic, semi-metallic and semiconducting properties respectively. It is therefore necessary to take into account the physical properties of the involved material. The consequence of this attention may help to develop novel approaches to study local properties. On the other hand, various effects related to strong electric field have been noticed by a number of researchers. Field emission of electrons, evaporation of ions, and field induced contacts were several prevalent mechanisms to account for these variations (Tsong and Chang, 1995; Chang, et al., 1994).

Since surface modification is essentially performed within a nanometer sized tunnel junction, the relevant physical quantities would be field strength and current density, and the junction properties include geometry of the potential barrier. The final outcome of the modification should be jointly affected by these factors. However, under certain circumstances, the result may be related to one of the above mentioned factors dominantly, as will be discussed in the following presentation, thus providing a possibility of looking into the different aspects of the external factors.

It has been demonstrated that under ambient conditions, nanometer scaled features have certain correlation with the characteristics of the specific tunnel junction. Both craters and mounds could be generated with the same junction in a relatively controllable fashion (Fig. 4.24) (Wang, et al., 1997). It was found that certain correlations could be established with the general behavior for a typical tunnel junction under a strong electric field. Namely, the mound-like features could be attributed to the deposition of tip materials due to the high current density that accompanies the electrical breakdown of the junction, and the crater-like features are mainly due to the damage of the field emitted electrons. The dimensions of the generated craters were also found to display certain variations in accordance with junction thickness or tip-sample separation.

These results could help to clarify the impacts of the factors of the tunnel junction on the fabrication process and to gain better control of the process. It is demonstrated that, under comparable conditions, the depth of the asgenerated craters has monotonic correlation with the pulse duration, while the apparent surface diameters do not show significant changes (Wu, et al., 1999). The characteristic "V" shaped cross-sectional profile of the craters is consistent with the previously reported results. This is considered as an indication of the region affected by the injected low-energy electrons through anisotropic diffusions at the vicinity of the injection path (Wang, et al., 1996). Owing to the inevitable scattering of the electrons by the target atoms, energy transfer would occur which could precipitate subsequent events.

This is believed to be the direct evidence that the electron induced gasification reaction of the carbon atoms did involve during fabrication rather than field evaporation mechanism. The removal rate of the carbon atoms is estimated to be on the order of 10^5/s (Fig. 4.25) (Wu, et al., 1999). In

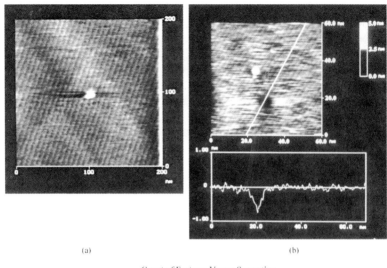

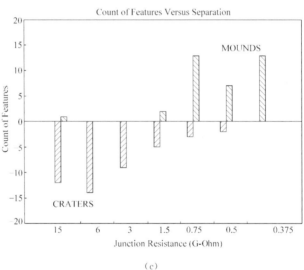

Figure 4.24 (a, b) Illustration of crater and mound-like features generated by STM on graphite surface. (c) The associated correlation with the static resistance of the tunnel junction (Wang, et al., 1997)

addition, the characteristics of the fabrication are shown to be related to the specific reaction type, i.e.,

$$C(solid) + H_2O + 1.82 \text{ eV} \rightarrow CO(gas) + H_2(gas)$$
$$C(solid) + 2H_2O + 1.85 \text{ eV} \rightarrow CO_2(gas) + 2H_2(gas)$$

and

$$C(solid) + O_2(gas) \rightarrow 2CO(gas) + 2.29 \text{ eV}$$
$$C(solid) + O_2(gas) \rightarrow 2CO_2(gas) + 4.08 \text{ eV}$$

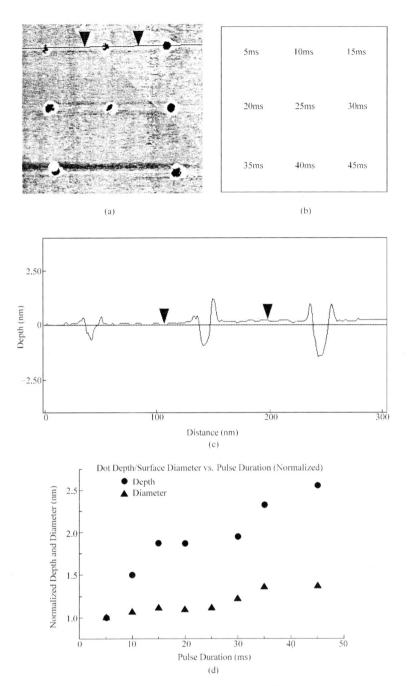

Figure 4.25 Time correlated fabrication results by STM. (a) and (b) show the fabrication results and the corresponding pulse duration; (c) is the cross-sectional profile of the first three craters; (d) dependence of the depth and width of the craters on pulse duration (Wu, et al., 1999)

Evidence has been accumulated that nanofabrication processes by STM are indeed quite sensitive to the operating environment.

Lens-focused electron beams, ion beams and X-rays can also be used in nanofabrication. Although the STM seems unlikely to become competitive in some areas of nanofabrication, such as wafer-scale resist patterning, it has its own characteristics. First, STM can work either in tunneling mode or field emission mode. When working in the last mode, an applied voltage of lower than 10 V can produce enough electric field to make electrons emitted from the tip pass over the barrier because the distance between tip and sample is still very small. These emitted electrons with certain current and energy will not diverge because of the small moving distance, which results in nanometer beam diameter on the substrate surface. These electrons can induce bond breaking and chemical reactions on nanometer scale. Secondly, b moving the tip to the sample to contact it, the tip of STM can also produce local contact forces and electrostatic forces in a small region of sample surface to create indentations directly. Lastly, at present, STM is the only instrument that can provide nanometer sized beam of very low-energy electrons(0—20 eV). The importance of low energy is obvious when it is considered that many processes, such as migration, bond breaking, chemical reactions that would be interesting to control have activation energies $<$ 10 eV per atom which require the controlling beam to have low energy.

4.5.2 Single Atom Manipulation

In addition to being useful for direct writing, lithography, electron-beam induced deposition and etching, STM can also be used to move adsorbates such as metal particles, atom clusters, or a single atom from one place to another, i.e., manipulate these small particles. This capability would be useful for building devices out of small particles, perhaps from particles of different materials. It could also be useful for studying interactions between particles and between particles and substrates. Eventually it may become possible to build on dismantled molecules atom by atom.

The simplest adsorbate on a surface is single atoms. It is possible to move these atoms according to a predetermined path using STM (Stroscio and Eigler, 1991; Eigler and Schweizer, 1990). It is known that the tip of an STM always exerts finite force on an adsorbate atom. This force contains both van der Waals and electrostatic contributions. By adjusting the position and the voltage of the tip, the magnitude and direction of this force may be tuned. This, taken together with the fact that it generally requires less force to move an atom along a surface than to pull it away from the surface, makes it possible to see these parameters such that the STM tip can pull an atom across a surface while the atom remains bound to the surface.

Eigler and Schweizer (Eigler and Schweizer, 1990) successfully transferred repeatedly Xe atoms back and forth between the tip and the substrate surface. The electrical conductance between the tip and the substrate depends on the position of the Xe atom, which results in a switching device with a low-conductance state when the Xe atom is on the substrate and a high-conductance state when the Xe atom is on the tip. This atomic switch is a bistable element, components such as this are vital in developing microcircuits.

Although the chemical identity and the structure of the outermost atoms of the tip were not known, it was for any given tip and bias voltage that there was a threshold height. Only when the tip-sample distance was lower than the threshold, i.e., the tunnel current was high enough, were the adsorbed xenon atoms moved. Simple investigations showed that neither the magnitude nor the sign of the applied voltage had significant effect on the threshold tip-sample distance. This suggested that the dominant force between the tip and the xenon atom is due to the van der Waals interaction.

The generally accepted theory of Tersoff and Hamann (Tersoff and Hamann, 1985) demonstrated that for small bias and constant current an STM image corresponds to a map of constant local density of states at the Fermi level. The extent to which an adsorbate will be "visible" to the STM depends on how it locally changes this state density. Lang(Lang, 1986, 1987) has shown that for single-atom adsorbates on metal surfaces, the crucial parameters in determining the apparent height of the atom in a low-bias STM image are the s-state and p-state densities due to the adsorbate at the Fermi level. Now Xe, like other rare gas atoms, when adsorbed on a metal surface makes virtually no contribution to the state density at the Fermi level. The STM studies indicated that the Xe atom appears as a nearly cylindrically symmetric 1.53 ± 0.02 Å high protrusion from the Ni(110) surface. Similar images for Xe adsorbed on the Pt(111) surface at 4 K have also been obtained (Weiss and Eigler, 1992).

To perform atomic scale modifications, one relies on the fact that the STM tip and the sample can interact through a variety of different mechanisms. For example, ionization followed by field evaporation has been suggested(Lyo and Avouris, 1991) for the reversible transfer of Si atoms between the tip and Si surface by using STM. Several mechanisms have been proposed to account for the reported manipulation of atoms, including possible field evaporation of negative ions, and electromigration of atomic adsorbates(Haberland, et al., 1989; Ralls, et al., 1989). Calculations for Xe atoms adsorbed on metals by using this model indicate that there should be substantial vibrational heating for tunneling current $>$ 100 pA. The observed sideways motion at smaller tip-sample separation may be due to the increased van der Waals attraction to the tip as the tip is brought closer to the surface.

In all, a number of kinds of atoms and molecules have been successfully manipulated, such as characters formed by xenon atoms and CO molecules, nanometer atom rings of iron atoms. Recent achievements on moving Cu atoms from different surface sites shows the possibility of constructing metallic clusters from individual atoms (Bartels, et al., 1997).

4.5.3 Interfacial Modification with BEEM

Analogous to STM fabrication on a surface, storing information at a subsurface region covered by a protective metal film has its unique advantage in applications and provides an approach to probe interfacial properties. Fernandez, et al. (Fernandez, et al., 1990) first reported their observation that modification of subsurface electronic properties can be induced by ballistic electrons in some specific Au/n-Si system when the energy of injected electrons is several eVs higher than the Schottky barrier. Their typical modification result was an area of a few hundred angstroms in diameter with decreased BEEM current, surrounded by a ring of enhanced contrast. In addition, protrusions were sometimes observed in the topographic image, indicating slight structural changes in Au film. Since the modification results were strongly dependent on the composition of interfacial impurity layer, the authors attributed the observed variations in BEEM current to be related to the changes at the Au/n-Si interface. As a plausible explanation, they proposed that the ballistic electron current at high bias enhances the interdiffusion of Au and Si at the interface. The created Au-Si intermix layer at the interface thus decreased the ballistic transmission. However, the later discovery of locally enhanced ballistic transmission after a voltage pulse in some cases (Hallen, et al., 1991) has implied that the previously assumed Au-Si alloyment mechanism appears insufficient. BEES measurements performed inside and outside the modified region have further exhibited that there exists a huge discrepancy in their natures (Qiu, et al., 1998). On the other hand, the dynamic studies of modification process have revealed a well-defined linear relationship between the area of the modified interface region and the applied voltage (Qiu, et al., 1998) (Fig. 4.26), as well as the duration of the applied pulse, indicating the modification is very likely to be associated with the accumulation of injected electrons.

Although the underlying mechanism is not thoroughly understood, it is more likely that certain chemical processes might be involved. Further investigations of the compositions of the modified regions should be crucial to understand the principles of the interfacial fabrications.

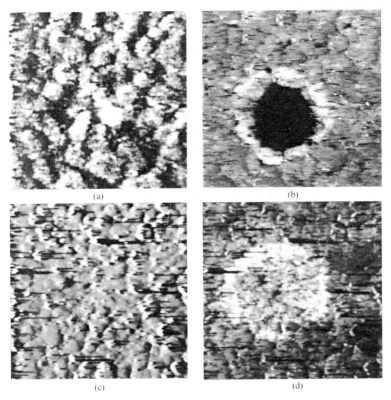

Figure 4.26 (a—d) Two kinds of BEEM fabrication results showing either reduced or enhanced electron transmission properties in the fabricated region. Scan areas for (a) and (b), (c) and (d) are 80 nm × 80 nm, 120 nm × 20 nm, respectively. The imaging condition is −1.30 V, 2.50 nA for (a) and (c). (Qiu, et al., 1998)

4.6 Concluding Remarks

The successful achievements of G. Binnig, H. Rohrer and their colleagues at the IBM Zurich Laboratory initiated a surge of research and engineering activity. This brought about rapid advances in STM technology and led to the development of many other novel scanning probe microscopes, such as atomic force microscope (AFM), lateral force microscope (LFM), magnetic force microscope (MFM), ballistic-electron-emission microscope (BEEM), scanning ion-conductance microscope (SICM), near-field optical microscope (NSOM), scanning thermal microscope, and scanning tunneling potentiometry (STP). These microscopes take advantage of the remarkable ability to control the spatial position of the tip relative to the sample, and provide new information about the physical properties of surfaces on an atomic or nanometer scale. This chapter is devoted to brief descriptions of the operational principles and some applications of these new types of microscopes.

In its brief history, STM has developed into an invaluable and powerful surface and interface analysis technique. However, STM has certain limitations and its operation at the atomic resolution is far from routine. A better knowledge of the role played by the microscopic structure of the probing tip is needed. The size, nature and chemical identity of the tip influence not only the resolution and shape of a STM scan but also the electronic structure to be measured. Although the well-defined tip at the end with one atom can be achieved, a tip with known geometry is not available by a convenient technique, such as electrochemical etching or grounding. The apex of such a tip is also limited by the mechanical stability of the tip during the STM scanning process.

The appearance of STM and STS ushers in a new era for the studies of the geometrical and electronic structure of metal and semiconductor surfaces, opening new ways for the investigation of the relations between the surface structure and the electronic structure. At present, nevertheless, STM is used only in the study of the solid surface with a known structure, being difficult in applying STM to the solid surface with an unknown structure. The prospect for the future of STM and related SPMs is sure to be as exciting as that in the past decade.

References

Albrecht, T.R., S. Akamine, T.E. Carver, C.F. Quate. J. Vac. Sci. Technol.. **A8**, 3386 (1990)
Albrecht, T.R., M.M. Dovek, M.D. Kirk, C.A. Lang, C.F. Quate, D.P.E. Smith. Appl. Phys. Lett.. **55**, 1727 (1989)
Bai, C., T. Hashizume, D. Jeon, T. Sakurai. J. Vac. Sci. Technol.. **A11**, 525 (1993)
Bai, C.L.. Scanning Tunneling Microscopy and its Application. Springer, Heidelberg (1995)
Bai, C.L., J.W. Li, Z. Lin, J. Tang, C. Wang. Surface and Interface Analysis. in press (1999)
Bartels, L., G. Meyer, R.H. Rieder. Phys. Rev. Lett.. **79**, 697 (1997)
Bell, L.D., W.J. Kaiser. Phys. Rev. Lett.. **61**, 2368 (1988)
Binnig, G., C.F. Quate, C. Gerber. Phys. Rev. Lett.. **56**, 930 (1986)
Binnig, G., H. Rohrer, C. Gerber, E. Weibel. Phys. Rev. Lett.. **50**, 120 (1983a)
Binnig, G., H. Rohrer, C. Gerber, E. Weidel. Surf. Sci.. **131**, L379

(1983b)
Binnig, G., H. Rohrer. Scientific American. 253(2), 50 (1985)
Binnig, G., H. Rohrer. Angew. Chem. Int'l. Ed. Eng., 26, 606 (1987)
Burnham, N. A., O. P. Behrend, F. Oulevey, G. Gremand, P-J Gallo, D. Gourdon, E. Dupas A. J. Kulik, H. M. Pollock, G. A. D. Briggs. Nanotechnology. 8, 67 (1997)
Chang, C. S., W. B. Su, T. T. Tsong. Phys. Rev. Lett., 72, 574 (1994)
Chen, C. J.. Phys. Rev., Lett., 65, 448 (1990a)
Chen, C. J.. Phys. Rev., **B42**, 8841 (1990b)
Chiang, S.. Chem. Rev., 97, 1083 (1997)
Cyr, D. M., B. Venkataraman, G. W. Flynn, A. Black, G. M. Whitesides. J. Phys. Chem., 100, 13747 (1996)
Dammer, U., O. Popescu, P. Wagner, D. Anselmetti, H. J. Guntherodt, G. N. Misevic. Science. 267, 5201 (1995)
Derjaguin, B. V., V. M. Muller, Y. P. Toporov. J. Colloid and Interface Sci., 53, 314 (1975)
Eigler, D. M., E. K. Schweizer. Nature. 344, 524 (1990)
Feenstra, R. M., J. A. Stroscio, J. Tersoff, A. P. Fein. Phys. Rev. Lett., 58, 1192 (1987)
Fernandez, A., H. D. Hallen, T. Huang, R. R. Burnham, J. Silcox. Appl. Phys. Lett., 57, 2826 (1990)
Florin, E. L., V. T. Moy, H. E. Gaub. Science. 264, 415 (1994)
Foster, J. S., J. E. Frommer. Nature. 333, 542 (1988)
Giessibi, F. J.. Science. 267, 68 (1995)
Gimzewski, J. K., E. Syoll, R. R. Schlittler. Surf. Sci., 181, 267 (1987)
Grutter, P., A. Wadas, E. Meyer, H. Heinzelmann, H. R. Hiber, H. J. Guntherodt. J. Vac. Sci. Technol., **A8**, 406 (1990)
Guntherodt, H. J., R. Wiesendanger, eds., Scanning Tunneling Microscopy. I, 2nd edn. Springer (1994)
Haberland, H., T. Kolar, T. Reiners. Phys. Rev. Lett., 63, 1219 (1989)
Hallen, H. D., T. Huang, A. Fernandez, J. Silcox, R. A. Buhrman. Phys. Rev. Lett., 69, 2931 (1991)
Hallmark, V. M., S. Chiang, J. F. Rabolt, J. D. Swaler, R. J. Wilson. Phys. Rev. Lett., 59, 2879 (1987)
Hamers, R. J., R. M. Tromp, J. E. Demuth. Phys. Rev. Lett., 56, 1972 (1986)
Hansma, P. K., V. B. Elings, O. Marti, C. E. Braker. Science. 242, 209 (1988)
Hecht, M. H., L. D. Bell, W. J. Kaiser, L. C. Davis. Phys. Rev., **B42**, 7663 (1990)
Hipp, H., H. Bielefelt, J. Colchero, O. Marti, J. Mlynek. Ultromicroscopy. 42—44 1498 (1992)
Israelachivili, J. N.. Intermolecular and Surface Forces. Academic Press (1985)
Jeon, D., T. Hashizume, X. Wang, C. Bai, K. Motai, T. Sakurai. Jpn. J.

Appl. Phys., **31**, L501 (1992)

Johnson, K.L., K. Kendall, A.D. Roberts. Proc. R. Soc. Lond., **A 324**, 301 (1971)

Kaiser, W.J., L.D. Bell. Phys. Rev. Lett., **60**, 1406 (1988)

Kuk, Y., P.J. Silverman, H.Q. Nguyen. Phys. Rev. Lett., **54**, 1452 (1987)

Lang, N.D., Phys. Rev. Lett., **56**, 1164 (1986)

Lang, N.D., Phys. Rev. Lett., **58**, 45 (1987)

Lee, G.U., L.A. Chrisey, R.J. Colton. Science. **266**, 771 (1994)

Li, J.W., C. Wang, G.Y. Shang, Q.M. Xu, Z. Lin, J.J. Guan, C.L. Bai. Langmuir. **15**, 244 (1999)

Li, Y.Z., J.C. Patrin, M. Chander, J.H. Weaver. Phys. Rev., **B 44**, 12903 (1991a)

Li, Y.Z., J.C. Patrin, Y. Chen, J.H. Weaver. Phys. Rev., **B 44**, 8843 (1991b)

Lippel, P.H., R.J. Wilson, M.D. Miller, C. Woell, S. Chiang. Phys. Rev. Lett., **62**, 659 (1989)

Lyo, I.W., Ph. Avouris. Science. **253**, 173 (1991)

Mamin, H.J., D. Rugar, J.E. Stern, B.D. Terris, S.E. Lambert. Appl. Phys. Lett., **53**, 1563 (1988)

Manne, S., P.K. Hansma, J. Massie, V.B. Elings, A.A. Gewirth. Science. **251**, 183 (1991)

Martin, Y., D. Rugar, H.K. Wickramasinghe. Appl. Phys. Lett. **52**, 244 (1988)

Meyer, G., N.M. Amer. Appl. Phys. Lett., **57**, 2089 (1990)

Meyer, G., L. Bartels S. Zophel, E. Henze, K.H. Rieder. Phys. Rev. Lett., **78**, 1512 (1997)

Moy, V.T., E.L. Florin, H.E. Gaub. Science. **265**, 257 (1994)

Moy, V.T., E.L. Florin, H.E. Gaub. Science. **266**, 5183 (1994)

Noy, A., C.D. Friesbie, L.F. Rozsnyai, M.S. Wrighton, C.M. Lieber. J. Am. Chem. Soc., **117**, 7943 (1995)

Overney, R.M., D.P. Leta, C.F. Pictroski, M.H. Rafailovich, Y. Liu, J. Quinn, J. Sokolov, A. Eisenberg, G. Overney. Phys. Rev. Lett., **76**, 1272 (1996)

Qiu, X.H., G.Y. Shang, C. Wang and C.L. Bai. Appl. Phys., **A 66**, s91 (1998)

Rabe, J.P., S. Buchholz. Science. **253**, 424 (1991)

Ralls, K.S., D.C. Ralph, R.A. Burnham. Phys. Rev., **B 40**, 11561 (1989)

Rief, M., F. Oesterhelt, B. Heymann, H.E. Gaub. Science. **275**, 1295 (1997)

Ruan, L., C. Bai, H. Wang, Z. Hu, M. Wan. J. Vac. Sci. Technol. B **9**, 1134 (1991)

Rubin, M.E., G. Medeiros-Ribeiro, J.J. O'Shea, M.A. Chin, E.Y. Lee, P.M. Petroff, V. Narayanamurti. Phys. Rev. Lett., **77**, 5268 (1996)

Sadrid, D., Scanning Force Microscopy. Oxford Press, Oxford (1990)

Sinniah, S. K., A. B. Steel, C. J. Miller, J. E. Reutt-Robey. J. Am. Chem. Soc., 118, 8925 (1996)
Sirrinhaus, H., E. Y. Lee, H. von Kanel. Phys. Rev. Lett., 74, 3999 (1995)
Smith, D. P. E., H. Horber, Ch. Gerber, G. Binnig. Science, 245, 43 (1989)
Smith, D. P. E., J. K. H. Horber, G. Binnig, H. Nejoh. Nature, 344, 641 (1990)
Spatz, J. P., S. Seiko, M. Moller, R. G. Winkler, P. Reineker, O. Marti. Nanotechnology, 6, 40 (1995)
Spong, J. K., H. A. Mizes, L. J. LaComb Jr, M. M. Dovek, J. E. Frommer, J. S. Foster. Nature, 338, 137 (1989)
Strohmaier, R., J. Petersen, B. Gompf, W. Eisenmerger. Surf. Sci., 418, 91 (1998)
Stroscio, J. A., D. M. Eigler. Science, 254, 1319 (1991)
Stroscio, J. A., W. J. Kaiser, eds, Scanning Tunneling Microscopy. Academic Press (1993)
Sugawara, Y., M. Ohta, H. Ueyama, S. Morita. Science, 270, 5242 (1995)
Tanaka, K., A. Takahara, T. Kajiyama. Macromolecules, 29, 3232 (1996)
Tang, J., C. Wang, M. Z. Liu, M. Su, C. L. Bai. Chin. Sci. Bull., in press (1999)
Tersoff, J., D. R. Hamann. Phys. Rev., B 31, 805 (1985)
Tian, T., C. Wang, G. Y. Shang, N. X. Wang, C. L. Bai. J. Magn. Magn. Mat., 171, 135 (1997)
Tian, F., C. Wang, Z. Lin, J. W. Li, C. L. Bai. Appl. Phys., A 66, s591 (1998)
Tsong, T. T., C. S. Chang. Jpn. J. Appl. Phys., 34, 3309 (1995)
Wan, L. J., S. L. Yau, K. Itaya. J. Phys. Chem., 99, 9507 (1995)
Wan, L. J, K. Itaya. Langmuir, 13, 7173 (1997)
Wang, C., C. L. Bai, X. D. Li, G. Y. Shang, I. Lee, X. W. Wang, X. H. Qiu, T. Fang. Appl. Phys. Lett., 69, 348 (1986)
Wang, C., X. D. Li, G. Y. Shang, X. H. Qiu, C. L. Bai. J. Appl. Phys., 81, 1227 (1997)
Weiss, P. S., D. M. Eigler. Phys. Rev. Lett., 69, 2240 (1992)
Whiteman, L. J., J. A. Stroscio, R. A. Dragoset, R. J. Celotta. Phys. Rev. Lett., 66, 1338 (1991)
Wintterlin, J., J. Wiechers, T. Gritsch, H. Hofer, R. J. Behm. J. Microscopy, 152, 423 (1988)
Woll, C., S. Chiang, R. J. Wilson, P. H. Lippl. Phys. Rev., B 99, 7988 (1988)
Wu, J., C. Wang, G. Y. Shang, X. H. Qiu, C. L. Bai. J. Appl. Phys., in press (1999)
Zhang, P. C., C. L. Bai, Y. M. Huang, H. Zhao, Y. Fang, N. X. Wang, Q. Li. Scanning Microscopy, 9(4), 981 (1995)

5 Optical Spectroscopy

Tiejin Li

5.1 Introduction

Research on nanostructured materials is an active and rapidly developing interdisciplinary field involving chemistry, physics, and biology and is related to the technological development of materials science, information science and microelectronics. During the past few years, studies in these fields such as the synthesis of nanocrystals materials or quantum dots (QDs) to construct organized assemblies on nanoscale dimensions, and the microscopic mechanisms of surface and interfacial electron transfers have made great progress. In the areas of technological interest, all will benefit significantly from multidisciplinary research. Indeed, nanostructured organized assemblies with optoelectronic functionality will play a key role in advanced materials in the next century.

From the viewpoint of physics and chemistry, the QDs building blocks with photoactive or electroactive units are considered first. In other words, nanostructured species are characterized by the nanoscale ordered arrangement of their components, such as metal or semiconductor clusters and nanocrystals, as well as supramolecular chemistry by the nature of the intermolecular noncovalent interactions that hold these components together on nanoscales to form spontaneously supramolecular chemical organized assemblies.

What is supramolecular chemistry? In his Nobel lecture, Lehn (1988, 1990) put it succinctly: "Supramolecular chemistry is the chemistry of the intermolecular bond, covering the structural and functional entity formed by association of two or more chemical species." The types of interactions may include metal ion coordination, electrostatic interactions, hydrogen bonding, and van der Waals forces, which provide new comprehensive methods for chemical assembly. The unifying power and interdisciplinary nature of supramolecular chemistry have attracted practitioners from various domains of scientific research and has led to the progressive incorporation of different areas of nanostructured physics, nanochemistry, nanobiology and nanotechnology. In addition, research in supramolecular nanostructured chemistry has progressively eliminated barriers between different disciplines, ensuring that scientific thoughts can be easily constructed. The area of research should be kept open and welcoming so that all those who come from different areas of science and technology may communicate in a common language.

Supramolecular nanostructured assemblies of inorganic QDs materials and organic functional molecules with positive cooperativity based on molecular recognition processes and molecular assembly may be built into self-organizing nanostructured molecular architectures displaying intriguing linear and nonlinear photophysical properties. An assembly subunit can serve as a modular of an information carrier to control the cooperative interactions and coherent interactions of photons, electrons, ions, protons, excitons, polarons, and magnetons and their transport processes. However, the linear and nonlinear spectroscopic technique would provide a powerful tool for the identification and application of nanostructured assembly materials.

The novel features of nanostructured assembly systems displaying higher forms of supramolecular behavior are characterized by self-organization, regulation, cooperativity, responsibility, and replication. No matter what inorganic QDs, hybrid organic-inorganic molecular materials are used, nanostructured functional materials constructed by molecular recognition and directed self-assembly reveal new horizons. From microscopic molecules to nanostructured assemblies, scientific ideas have been introduced in an attempt to encompass the broad field concerned with the development of nanochemistry and nanostructured physics. Some of the latest results have indicated the effects of diminishing dimension and quantum confinement, linear and nonlinear optics, exciton and its fine structure, relaxation processes, coherent effects, and so on. Furthermore, electron transport of surface and interface in supramolecular nanostructured systems is of fundamental significance.

It is the purpose of this chapter to review the optical and spectroscopic properties of some QDs and their supramolecular nanostructured assemblies.

5.2 Nanoclusters and Nanocrystals

Nanoclusters and nanocrystals can be considered as finite arrays of atoms and molecules. They represent a new class of materials with hybrid molecular-solid state properties. There is clear evidence that nanoclusters and nanocrystals can exhibit structure and properties quite distinct from those of bulk systems. These materials might be termed quantum dots (QDs); an electron or exciton can be confined in zero-dimensional space. Quantum states in the QDs are size dependent, leading to new photophysical and spectroscopic properties. For instance, optical transitions can be tuned simply by changing the size of the

QDs. Alivisatos(1996), Brus and Bawendi (Norris, et al., 1997), described current research on semiconductor nanocrystals. They indicated that the affinity to join the dots into complex assemblies creates many opportunities. Recent dramatic advances in the preparation of optoelectronic materials open a doorway. It is essential to change optical properties by changing the nanosizes of materials that are not newly synthesized compounds. It can be clearly suggested that electronic energy structures change, and surface and interfacial properties of quantum dots will affect light-induced electron transport processes. Some experimental results will be explained as follows.

5.2.1 Absorption and Photoluminescence Spectroscopic Evidence for Quantum Confinement

As the size of the bulk materials becomes smaller and approaches the Bohr radius of the bulk exciton, quantum confinement effects will become apparent. The electronic excitations shift to higher energy, and the oscillator strength will be concentrated in just a few transitions, so changing the density of electronic states and leading to blue shifts of the absorption band edge (Alivisatos, et al., 1996a). That is not sensitive for routine absorption spectra of the QDs for surface states and absorption of outside molecules. The luminescence of the QDs is, by contrast, quite sensitive to surface conditions associated with surface localized states. For strong confinement, an analytic approximation obtained by Brus for the lowest excited state for small R is:

$$E(R) = E_g + \frac{\hbar^2 \pi^2}{2 \widetilde{R}^2}\left(\frac{1}{m_e^*} + \frac{1}{m_h^*}\right) - \frac{1.8e^2}{\varepsilon_2 \widetilde{R}} + \text{small terms} \quad (5.1)$$

where $\hbar$ is planck constant, m_e and m_h are the effective masses of the electron and hole, respectively, ε is the dielectric constant, and R is the average sphere radius of the cluster (Yoffe, 1993). The second term represents the blue shift of the optical absorption band edge due to the quantum confinement of carriers, and the third term represents the red shift induced by the Coulomb interactions between electrons and holes. In fact, these two terms are the coupling effects in the QDs. For naked QDs, the second term is dominant, which has been verified by many articles. But the effects of the third term have not been experimental determined.

The optical properties of two prototypical semiconductor QD systems have been described by Brus and Bawendi et al. (Norris, et al., 1997). Owing to the sample redundant residual inhomogeneities, the absorption and photoluminescence (PL) band could be broadened and diffused. For improvement of absorption information the most common technique has been transient differential absorption spectroscopy (TDA), also called pump-probe or hole-burning spectroscopy. More recently, they have begun to utilize a simpler optical technique, photoluminescence excitation (PLE) spectroscopy.

When PLE is combined with fluorescence line narrowing (FLN) spectroscopy, both absorption and emission information representative of a "single dot" are revealed. A nanosecond Nd:YAG/dye laser system was used as excitation sources of FLN. The laser power was attenuated [$\sim 20(\mu J/mm^2)/pulse$] to ensure that the detected luminescence was linear with respect to the excitation intensity. The luminescence was dispersed through a monochromator and detected by a time-gated optical multichannel analyzer (OMA) to obtain spectra.

CdSe crystal is generally a direct gap wurtzite structured semiconductor. Its QDs is strongly confined systems. Qualitatively, all of these effects can be readily observed in the absorption spectra of Fig. 5.1. Both PLE and FLN techniques are demonstrated in Fig. 5.2 along with absorption and luminescence results for 19 Å effective radius CdSe QDs sample. A significantly narrowed and structured spectral band is revealed by PLE and FLN. For example, a vibration [longitudinal optical (LO) phonon] progression is clearly higher resolved in the PLE spectrum, while the absorption and the PLE spectra show the two lower exciton features. This FLN spectrum can be used to extract a model "single dot" emission lineshape. PLE spectra of some samples revealed much more exciton structure. PLE results for a 28 Å radius CdSe sample are shown along with its absorption and full luminescence spectra in Fig. 5.3. Bawendi et al., Norris, et al. (1997) had reported a series of PLE spectra of high quality in Fig. 5.4. The QDs are arranged (top to bottom) in the order of increasing radius from ~15 Å to ~43 Å. Quantum confinement

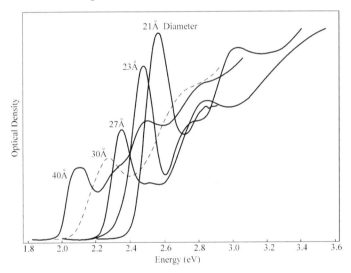

Figure 5.1 Optical absorption vs. size for CdS QDs shows the shift to higher energy in smaller sizes, as well as the development of discrete structure in the spectra and the concentration of oscillator strength into just a few transition (adapted from Alivisatos, et al., 1996)

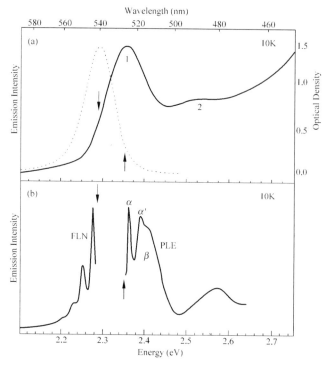

Figure 5.2 (a) Absorption (solid line) and full luminescence (dotted line) spectra for ∼ 19 Å CdSe QDs. (b) FLN and PLE spectra for the sample. An LO-phonon progression is observed in FLN. Both narrow (α, α') and broad (β) absorption features are resolved in PLE. The downward (upward) arrows denote the excitation (emission) position used for FLN (PLE) (adapted form Norris, et al., 1997)

clearly blue shifts the transitions (>0.5 eV) with decreasing size. The actual excitation/emission position is indicated with arrows in the full luminescence spectra, shown in Fig. 5.5.

The surface effect of the QDs can influence the luminescent radiative lifetime and transition probability. Bhargava et al. (1994) reported that the coated nanocrystals of Mn-doped ZnS with sizes varying from 3.5 nm to 7.5 nm could yield high luminescent efficiency of 18%. Furthermore, a luminescent decay is at least 5 orders of magnitude faster than the corresponding Mn^{2+} transition in the bulk crystals at the same time. The separation of the ZnS:Mn QDs is maintained by coating with the surfactant methacrylic acid. It could be assumed that this enhancement in efficiency results from the electronic passivation of surface with a close-packed molecular layer and is related to the photopolymerization of the surfactant. In the ZnS:Mn QDs systems the Mn^{2+} ion d-electron states act as efficient luminescent centers while interacting strongly with the s-p electronic states of the host nanocrystal into which external electronic excitation is normally

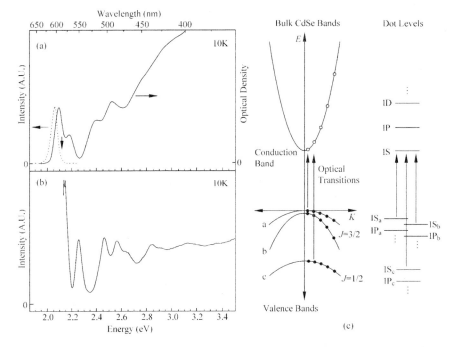

Figure 5.3 (a) Absorption (solid line) and full luminescence (dotted line) spectra for ~ 28 Å CdSe QDs. In luminescence the sample was excited at 2.655 eV (467.0 nm). The downward arrow marks the emission position used in PLE. (b) PLE scan for the same sample. (c) A simplistic approach to describing the quantum dot spectrum (adapted from Norris. et al., 1997)

directed. The band-to-band excitation in ZnS is used to excite the Mn^{2+} emission. The subsequent transfer of electron and hole into the electronic levels of the Mn^{2+} ion leads to the characteristic emission of Mn^{2+} ($4\,T\,1—6\,A\,1$ transition) in ZnS. The perturbation by the host crystal fields renders the otherwise spin forbidden d-d transition partially electric dipole allowed (through the mixing of opposite spin states). PL and PLE spectra for the QDs of ZnS:Mn to compared to bulk ZnS:Mn powder is shown in Fig. 5.6.

The results of our experiment on TiO_2 nanoclusters (smaller than 5 nm) coated with a molecular layer of stearic acid (ST) (i.e., the negative hydrophilic carboxylic groups of stearic acid bind to the surface of TiO_2 QDs) are somewhat different. This dipole layer induced an attractive potential to the electron inside TiO_2 QDs; i.e., this layer introduced a red shift tendency of the absorption band edge. The band structures of TiO_2 depend mainly on the crystal field effect, from which no pronounced change of band structure could be introduced by the interface dipole layer. However, the influence of the dipole layer on the binding energy of the exciton, that is, the Coulomb interaction between electrons and holes in the QDs, is considerable. The dipole layer might become a trap center for the exciton, which could enhance

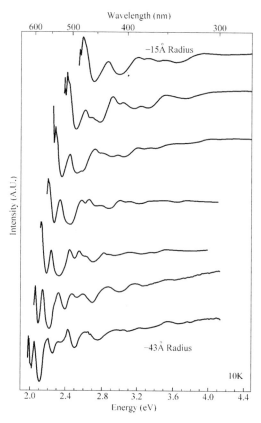

Figure 5.4 Normalized PLE scans for seven different sizes CdSe QDs samples. Size increases from top to bottom and ranges from ~15 Å to ~43 Å in radius (Adapted from Norris, et al., 1997)

the exciton binding energy more significantly and induce a red shift of the absorption band edge of the QDs. Furthermore, the interface deformation potential might also become another kind of center for the self-trapped exciton, producing a red shift of the absorption band edge in Fig. 5.7.

The exciton trapping effects of the surface modification could increase not only the binding energy of the exciton but also its lifetime, which are related to the explanation of the peculiar photoluminescence of surface-coated TiO_2 QDs. For bulk TiO_2, even in the single-crystal state, it is difficult to observe photoluminescence at room temperature, but a photoluminescence band could be observed at 77 K, which is assigned to the emission of impurity-bound excitons or self-trapped excitons, and the same phenomenon occurs for the naked TiO_2 QDs in hydrosol. On the contrary, photoluminescence of TiO_2 QDs coated with ST could be observed at room temperature. As seen from Fig. 5.8, these differences are expected. The self-trapped exciton could not form directly by an optical transition, and the efficiencies for self-trapped

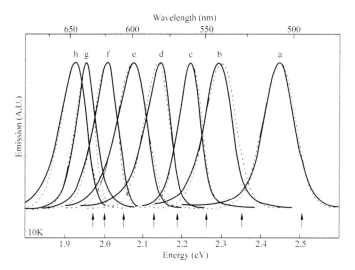

Figure 5.5 Full luminescence spectra for eight different size (A~15 Å, B~19 Å, C~21 Å, D~24 Å, E~27 Å, F~33 Å, G~44 Å and H~50 Å) CdSe QDs (solid lines). Arrows indicate FLN excitation positions and PLE emission positions eV). Dotted lines show the calculated results (adapted from Norris, et al., 1997)

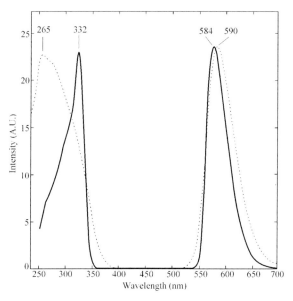

Figure 5.6 The PLE and PL spectra of bulk (solid lines) and QDs (dotted lines) ZnS: Mn. Note that the position of the PLE peak reflects the band edge of the host (adapted from Bhargara, et al., 1994)

exciton generation in bulk TiO_2 and naked TiO_2 QDs are small. As a result, there is a potential barrier to the self-trapped exciton from the free exciton,

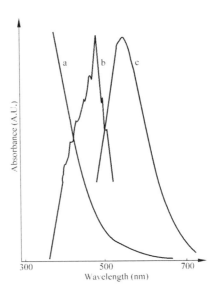

Figure 5.7 Absorption (a), excitation (b), and photoluminescence spectra (c) of coated TiO$_2$ QDs (adapted from Zou, et al., 1991)

and the nonradiative electron-phonon coupling is strong, so emission due to the self-trapped exciton is not seen at room temperature, but at 77 K sufficient long-lived self-trapped exciton could be emitted. However, the ST modification of TiO$_2$ QDs enhanced the interfacial dipole effect, which modified the surface electronic structure on QDs, canceling the barrier for self-trapped exciton absorption. Transitions to the energy level of self-trapped exciton could occur under this condition. Moreover, along with the giant oscillator strength effects of the localized exciton, the sufficient long-lived self-trapped

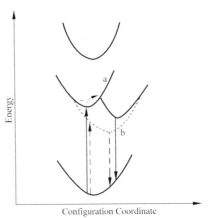

Figure 5.8 The energy level diagram of bulk TiO$_2$. (a) The transition for naked QDs. (b) The energy level modified (dotted lines) and the transition (dashed lines) for coated TiO$_2$ QDs (adapted from Zou, et al., 1991)

excitons are excited under light illumination to contribute to the photoluminescence at room temperature.

For investigation of the factors that affect the red shift of the optical absorption band edge, a series of TiO_2 QDs coated with a layer of different surfactants (such as sodium stearate, dodecylbenzenesulfonate sodium (DBS), oleic acid) were assembled. For the TiO_2 QDs coated with surfactant, we found that the red shift of the optical absorption band edge was a universal phenomenon, and the smaller the cluster size the larger the red shift (Li, et al., 1992). In the case of band-to-band transitions of semiconductor quantum dots, coating with organic molecules that form a close-packed monolayer of coordinating ligand on the QDs surface is very effective for surface electronic passivation of the QDs. Bawendi et al. (Murray, et al., 1995) showed that the quantum efficiency of photoluminescence of CdSe QDs coated with trioctylphosphine oxide and selenide at room temperature was more than 10%, and CdSe particles with uniform diameters (~20 Å) have been arranged by self-organization into three-dimensional quantum dot superlattices.

5.2.2 Raman and FTIR Studies on the QDs and Its Supramolecular Assemblies

Inelastic light scattering (Raman spectroscopy) and FTIR spectroscopy are an ideal tools to study the surface structured properties of the QDs. With surface-enhanced Raman scattering (SERS), it is possible in many cases to obtain an enhancement of $>10^4$ in the Raman scattering signal from surface molecules directly on the surface of nanostructured materials. Recent advance in SERS has allowed studying single molecules adsorbed on single nanoparticles at room temperature. For single rhodamine 6G (R6G) molecules adsorbed on the selected silver nanoparticles, the intrinsic Raman enhancement factors are on the order of 10^{14} to 10^{15}, much larger than the ensemble-averaged values derived from conventional measurements. This enormous enhancement leads to vibrational Raman signals that are more intense and more stable than single-molecule fluorescence. The excitation-polarization data is obtained from two particles under identical conditions in Fig. 5.9. These two particles were selected to show that intense Raman signals can be observed with either s- or p-polarized light (parallel or perpendicular to the plane of incidence), depending on how a particular particle is oriented relative to the polarization axis. According to theory of SERS, these two nanoparticles should have orthogonal orientations: when one is maximally excited at the direction of s polarization, the other is minimally excited, and vice versa.

Resonant Raman scattering could study the electronic elementary excitations in semiconductor nanostructured systems. Since it allows one to distinguish different types of excitations by polarization selection rules,

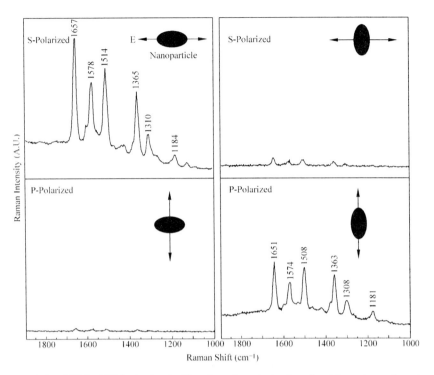

Figure 5.9 SERS of R6G obtained with a linearly polarized confocal laser beam from two Ag QDs. The R6G concentration was an average of 0.1 analysable molecule per particle. The direction of laser polarization and the expected particle orientation are shown schematically for each spectrum. Laser wavelength, 514.5 nm; laser power, 250 nW; laser focal radius, ~250 nm; integration time, 30 s. All spectra were plotted on the same intensity scale in arbitrary units of the CCD detector readout signal (adapted form Nie, et al., 1997)

collective charge-density excitations (CDEs) can be observed if the polarization directions of the incident and scattered photons are parallel to each other (polarized geometry), and collective spin-density excitations (SDEs) are measured for perpendicular polarization (depolarized geometry). In Fig. 5.10, polarized and depolarized spectra of electronic excitations in a quantum-wire (Fig. 5.10(a)) and a quantum-dot sample (Fig. 5.10(b)) are shown. Alivisatos et al. (Shiang, et al., 1993) reported resonance Raman studies of the ground and lowest electronic excited state in CdS QDs. They demonstrate that as the size is decreased the coupling of the optically generated electron and hole to the lattice decreases. These results are in contradiction to the predictions of theories based on quantum confinement with an infinite spherically symmetric potential barrier. The width of the observed longitudinal optical (LO) mode broadens with decreasing size, indicating that the resonance Raman process is intrinsically multimode in its nature. The frequency of the observed LO mode at 305 cm^{-1} has a very weak dependence

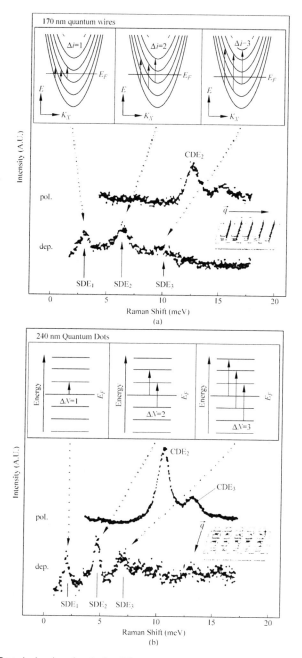

Figure 5.10 Depolarized and polarized Raman specters of electronic excitations in (a) a quantum-wire and (b) a quantum-dot sample. The inserts show schematic diagrams of the single-particle transitions, which predominantly contribute to the observed excitations. The wave vector transfer q in the experiments was $q \sim 1.3 \times 10^5$ cm^{-1} (adapted from Schuller, 1998)

on size, in contrast to results obtained from multiple quantum well systems. The temperature dependence of the frequency and linewidth of the observed LO mode is similar to the bulk and indicates that the LO mode decays into acoustic vibrations in 2.5 ps. CdS QDs of 4 nm diameter have been studied under high pressure up to 10 GPa by optical absorption and resonance Raman scattering (Haase and Alivisatos, 1992). The solid-solid phase transition from the zinc blende to the rock salt phase is observed at pressures far in excess of the bulk phase transition pressure of 3 GPa.

Transmission FTIR spectroscopy is mainly to investigate the physicochemical properties of the surface bonded or adsorbed molecules on the nanoclusters and nanocrystals. In general, the commercial FTIR instruments operate only to in medium IR region; therefore some molecular vibration in lower frequency cannot be observed. The FTIR spectrum of TiO_2 QDs (2nm) was obtained in this region (Zou, et al., 1993). TiO_2 QDs coated with a monolayer of stearate, in which there are strong peaks at 1264, 1100, 1022, and 806 cm^{-1}, respectively, out of single and combined modes of the normal polar modes. TiO_2 (rutile type) has three longitudinal optical polar modes parallel to the c-axis of 373, 458, and 806 cm^{-1}, whose IR absorption intensities were proportional to the respective electron-phonon coupling coefficients, as the continuum polarization model predicts. The coupling coefficients for LO_1 (806 cm^{-1}), LO_2 (458 cm^{-1}), and LO_3 (373 cm^{-1}) were 0.967, 0.024, and 0.007, which were exhibited clearly in the IR spectrum of TiO_2 QDs, i.e., LO_1 phonons appear dominantly in the vibration spectrum. The peak at 1264 cm^{-1} originates from the combination of LO_1 and LO_2 phonons, and its strong intensity was related to the participation of the LO_1 phonon. The chemisorbed stearate molecules are formed mololayers on the surface TiO_2 QDs, which introduces an "impurity potential" of about 1000 cm^{-1} (two peaks). This structure was due to its dielectric nature, arising in the absence of phonon dispersion. The electron and phonon localization in TiO_2 QDs coated with stearate induce significant nonlinear electron-phonon interactions, producing a long-lived polaronic state including bipolaron with small dispersion, because of the quantum confinement and dielectric confinement.

5.2.3 High Resolution Spectroscopy of Individual Quantum Dots

Current spectroscopic research observed some nanostructured systems of semiconductor QDs in which significant confinement energies exist, size fluctuations to inhomogeneous broadening of the spectral lines. This blurring of the spectra severely reduces the amount of information obtainable from spectroscopy. The finding has initiated an effort to isolate optically and study spectroscopically individual QDs. These experiments demonstrate the elegance and potential of single-QD spectroscopy (Gammon, 1998).

Important research of an artificial two-dimensional system concerned a

GaAs/AlGaAs quantum well investigated by means of low-temperature near-field scanning optical microscopy (NSOM). That is a heterostructured sample and consisted of four GaAs/Al$_{0.32}$Ga$_{0.68}$As single quantum wells (SQW) of thicknesses 31 Å, 23 Å, 45 Å, and 90 Å, each separated by 250 Å barriers and centered at depths of 266 Å, 542 Å, 826 Å, and 1144 Å, respectively (Hess, et al., 1994). An NSOM system is shown in Fig. 5.11. In brief, light from a tunable dye laser at excitation wavelength λ_{exc} was coupled into a tapered, aluminum-coated optical fiber having a subwavelength diameter aperture at its apex. After positioning this probe to <20 nm from the sample, photons emitted from the aperture and absorbed in the sample create electron-hole pairs (excitons) which migrate in the plane of the quantum well prior to recombination. The resultant luminescence was collected on the same side of the sample with a lens and transported via an optical fiber bundle to a 0.5-m spectrometer with a cooled charge-coupled device (CCD) camera for detection. Luminescent centers with sharp (<0.07 meV), spectrally distinct emission lines were imaged in this system. Temperature, magnetic field, and line width measurements establish that these centers arise from excitons laterally at interface fluctuations. Spatial variations in the PL spectrum are measured by raster scanning the sample laterally under the near-field probe in the plane of the SQW with a scan head operating in vacuum that was surrounded by an 8 T superconducting magnet and immersed in liquid helium. Here showed only near-field, far-field, and spatially averaged near-field PL spectra from an SQW sample (thickness = 23 Å, λ_{exc}=694 nm, T=2 K) in Fig. 5.12. For sufficiently narrow wells, virtually all emission originates from

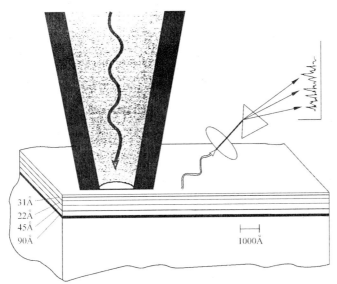

Figure 5.11 Cutaway schematic of the illumination and collection geometry used for near-field PL microscopy/spectroscopy. The probe and the sample portions are drawn to scale (adapted from Hess, et al., 1994)

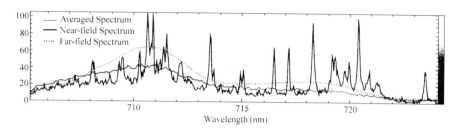

Figure 5.12 Near-field, far-field, and spatially averaged near-field PL spectra from a 23 Å single quantum well (SQW) produced with 696 nm excitation at 2 K (adapted from Hess, et al., 1994)

such centers. NSOM provides a means to access energies and homogeneous line widths for the individual eigenstates of these centers, but in far-field it is only broad PL spectral lines.

This breakthrough in the fabrication of defect-free QDs provides the opportunity for experimental verification of the effects of 3-dimensional (3D) quantum confinement in semiconductor nanostructures. Peng, et al., (1998) have reported kinetics of II-VI and III-V colloidal semiconductor nanocrystal growth: "focusing" of size distributions. For a smaller number of uniformly sized inorganic nanocrystal systems, PL narrower band shape should be nearly homogeneous forms of line broadening.

Ensembles of defect-free InAlAs QDs embedded in AlGaAs have been grown by molecular beam epitaxy. Cathodoluminescence (CL) was used to directly image the spatial distribution of the QDs by mapping their luminescence and to spectrally resolve very sharp peaks from small groups of dots, thus providing experimental verification for the discrete density of states in a QDs system (Leon, et al., 1995). CL is used to obtain spectral decomposition of the broader PL QDs ensemble into lines narrower than 1 meV. This spectroscopic resolution is a direct consequence of the reduced dimensionality of these structures and permits, in some cases, direct imaging of single QDs.

Figure 5.13 shows the CL spectra of this sample taken under different conditions. The luminescence of a thick sample measured using defocused electron beam (Fig. 5.13, spectrum a) and the same broad Gaussian shape obtained the PL measurements. In both CL and PL the AlGaAs peak at 623 nm is seen. This broad peak (full width at half maximum, 46 meV) corresponds to the excitation of large number of dots, and it is attributed to slight variation in the 0D confining potentials caused mostly by size dissimilarities. Spike-like feature appear over the broader background in spectrum a of Fig. 5.13 because of the excitation of a smaller number of dots in CL than in PL, even with a defocused electron beam.

Figure 5.13, spectra b and c, displays very sharp peaks, a fraction of a meV in width, originating from either individual dots or a few dots with equal or almost equal dimensions. These spikes are reproducible and are therefore

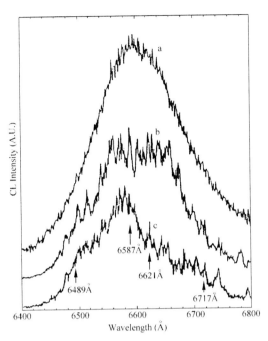

Figure 5.13 Low-temperature (10 K) CL spectra for QD ensembles of different sizes in the InAlAs/AlGaAs sample. These were obtained at low magnification with (spectrum a) a defocused electron beam on a thick sample and (spectra b and c) a focused electron beam on a thin sample. Spectra b and c were obtained from different areas of the sample. A 120 keV, 82 mA electron beam was used (adapted from Leon, et al., 1995)

not simply noise. These spectra show different sets of peaks; while the sharpness of the features is maintained, the peak have different intensities because they were taken from different areas in the sample, thus probing different dot ensembles. The peak width in the CL experiment is 0.6 meV, slightly larger than the results obtained with lower temperature (2 K) PL experiments of etched meshes in the same sample, which gave a peak width as small as 0.4 meV.

The PL and CL peaks are Gaussian (inhomogeneous line shapes) for an infinite (or very large) dot ensemble. The luminescence signal exhibits larger fluctuations from this Gaussian envelope. When the number of dots probed gets smaller, Fig. 5.14 shows a non-Gaussian shape, resulting from the statistics of a small number of dots. The small signal collected from this type of measurement decreases the spectroscopic resolution of the experiment because larger slit widths are required.

An example of single-QD PL spectra appears in Fig. 5.15. The spectra shown were obtained at temperature of 6 K by successively reducing the size of the laser spot on a GaAs quantum-well sample through the use of small apertures in a metal mask. The bottom trace is a PL spectrum obtained with a

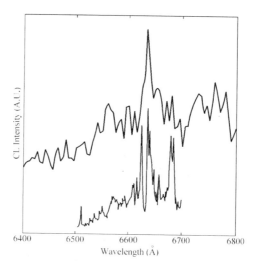

Figure 5.14 Low- and high-resolution CL spectra from the same area obtained with a very thin sample and a small condenser aperture. The spectrum is no longer Gaussian because of the small number of QDs probed. This spectrum corresponds to the excitation of only a few tens of dots (adapted from Leon, et al., 1995)

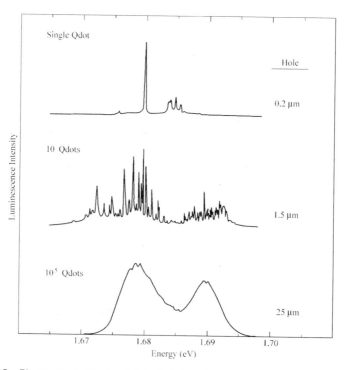

Figure 5.15 PL spectra excited and detected through apertures in an aluminum mask with diameters ranging from 25 μm to 0.2 μm. In this way, the emission from individual QDs is resolved (adapted from Gammon, 1998)

macroscopic laser spot diameter of 25 μm. The spectrum shows two broad peaks corresponding to the recombination of excitons in parts of the quantum well that are either 10 or 11 monolayers wide (2.8 nm or 3.1 nm). The spectrum is strongly inhomogeneously broadened as shown most directly by a reduction in the aperture size. The relatively broad lines break up into a decreasing number of extraordinarily narrow PL spikes as the aperture is reduced to submicroscopic dimensions. These PL spikes arise from excitons localized in individual QD potentials. These phenomena include exciton excited states, fine-structure splitting, hyperfine shifts, and time-dependent spectral jumping. The QD system originates in narrow GaAs/Al_xGa_{1-x}As quantum wells (Fig. 5.16). Here it will consider the bound electron-hole pair known as an exciton. An exciton in a QD interacts with the nuclei through an electric or a magnetic field. In narrow quantum wells with thicknesses of the order of nanometers, changes in the quantum-well width of a single monolayer caused by interface fluctuations lead to large changes in confinement energy (of the order of 10 meV). As a result in some regions of the quantum wells, the excitons are localized in all three dimensions instead of just one.

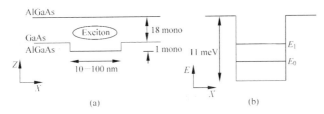

Figure 5.16 (a) Schematic diagram of an exciton's lateral localization in a QD potential formed by monolayer-high interface steps in a narrow GaAs quantum well. (b) The resulting energy diagram of the QD (adapted from Gammon, 1998)

One of the most valuable probes of the QDs is PLE spectroscopy. In this technique, the intensity of one of the PL spikes is recorded as the laser energy is scanned (Fig. 5.17). In the simplest interpretation of this experiment, the resulting spectrum maps out the excited states of the QD. For each QD PL line, an excitation spectrum serves as a "fingerprint" of that particular a rough estimate of the lateral size of the QD can be calculated from the energy separations (E_1-E_0). For the example in Fig. 5.17, one could estimate a lateral dimension of approximately 40 nm (Gammon, 1998).

In PL the energy spectra of excitons are measured optically with high sensitivity and selectivity, even down to the single-QD level. With Raman scattering, the spectra of optical phonon can be measured optically. However Raman scattering is usually weak, and phonons spectroscopy at the individual QD level would be considered hopeless. But the novel use of resonant Raman scattering can actually provide the sensitivity and selectivity to measure the optical-phonon spectrum associated with individual QDs. Figure 5.18 shows the results of an experiment that probes both the phonon and the exciton

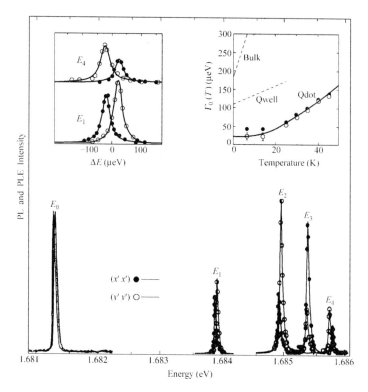

Figure 5.17 PL and PLE spectra taken in two polarization configurations. Both the PL line (E_0) and the associated PLE resonances are linearly polarized with a doublet structure, as magnified in the left inset for the first and fourth excited-state transitions. The right inset shows the temperature dependence of the PL lines (E_0) as compared to measurements on wide quantum-well and bulklike samples (adapted from Gammon, 1998)

spectra of individual QDs. Gammon et al. first demonstrated the Raman spectroscopy of a single QD. The PL spectrum from the GaAs/AlAs sample with an aperture size of 25 μm (Fig. 5.18(a)) provides an ensemble measurement of QDs, whereas the PL spectrum from a 0.8 μm aperture (Fig. 5.18(c)) shows a single QD PL line at E_0 (ground state energy). For the Raman scattering experiment, it is necessary to reduce the PL in order to measure the resonant Raman signal. For this purpose, they used thin AlAs barriers, which increased the tunneling rate out of the quantum well and reduced the PL intensity. An excitation spectrum of the QD PL line at energy E_0 exhibits sharp resonances within about 10 meV of the QD PL line corresponding to direct absorption into QD states followed by relaxation into the luminescing by emission of acoustic phonons (Fig. 5.18(b)). At higher energies (Fig. 5.18(d)), starting about one transverse optical (TO) phonon energy above the luminescing state, there is additional sharp structure, although about an order of magnitude weaker than the lower energy excitation resonances.

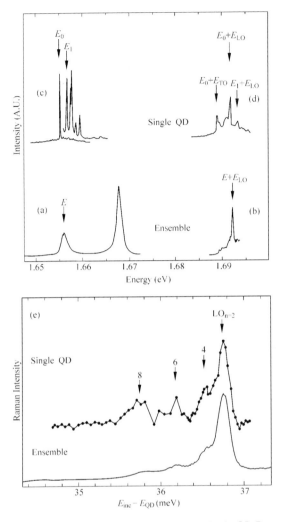

Figure 5.18 GaAs/AlAs emission spectra demonstrating single QD Raman spectroscopy. (a) An ensemble PL measurement through a 25 μm aperture, showing emission at 1.656 and 1.668 eV from the 19- and 18-monolayer parts of the quantum well, respectively. (b) Ensemble PL excitation spectrum (25 μm). (c, d) PL and PLE spectra, respectively, obtained through a 0.8 μm aperture. The PL spectrum shows a single QD-PL line at E_0. The excitation spectrum shows sharp electronic resonances within the first 10 meV and Raman resonances starting about 34 meV above the luminescing states. (e) The top spectrum is a high-resolution LO phonon spectrum from a QD, obtained from the intensity of the QD emission as a function of the energy difference between the QD (E_{QD}) and the laser (E_{EXC}) as the laser is scanned. The bottom (ensemble) spectrum is the resonant Raman spectrum obtained in the conventional way from the 25 μm aperture (adapted from Gammon, et al., 1997)

The structure between the energies labeled $E_0 + E_{TO}$ and $E_0 + E_{LO}$ represents the optical phonon spectrum modulated by the QD exciton phonon interaction. This strong enhancement corresponds to resonance Raman scattering from a single QD. The resonance linewidths are extremely sharp, allowing the selective probing of individual QDs through individual phonons. There is also a peak at higher energies ($E_1 + E_{LO}$) that corresponds to a two step process in which the 2D exciton makes a transition into the excited state of the QD at E_1 accompanied by the emission of an LO phonon, and then makes a transition to the QD luminescing state by emitting an acoustic phonon. The Raman spectrum from individual QDs can be acquired by plotting the intensity of that QD PL line as a function of the difference between the exciting and detected light energy ($E_{exc} - E_{QD}$) (Fig. 5.18(e)) for the LO_n phonon region. The structure in the ensemble spectrum (bottom trace in Fig. 5.18(e)) arises from confined and interface LO phonons due to the vertical confinement.

5.2.4 Ultrafast Spectroscopy in Quantum Confined Structures

QDs exhibit unique electronic, magnetic, and optical properties because of their small size and larger surface-to-volume (S/V) ratio. They offer an intriguing possibility to study electron behavior in nanoscopic quantum structures. They also provide an opportunity to investigate problems related to surfaces or interfaces because of their interfacial nature. Advances in optical spectroscopic techniques have revealed the distinctive features of these nanostructures. Recently much interesting work by El-Sayed and his group has been done on ultrafast time-resolved spectroscopy.

Using time-resolved optical hole (oh) burning techniques with femtosecond (fs) laser, the time dependence of the spectral diffusion of the oh is examined by Little et al. (1998) for both the CdS QDs and the CdS/HgS/CdS quantum dot quantum well (QDQW) nanoparticles. It is found that the nonradiative relaxation of the optical hole is at least 3 orders of magnitude slower in the QDQW than in the QD system. Analysis of the second derivative of the broad transient bleach spectrum of the QDQW system in the 1.6—2.5 eV energy region at 50 fs delay time is found to have a minimum at 2.1 eV, corresponding to a minimum in the radiative probability. Around this energy, the rise and decay times of the transient bleach in the spectrum are found to change greatly. These results suggest that spectral diffusion in the QDQW is a result of relaxation from high- to low-energy exciton states involving an intervening dark state at an energy of ~2.0 eV. The energies of the maxima and minimum of the second-derivative curve are found to be in good agreement with recent theoretical calculations of the energies of the radiative and dark charge-separated state, respectively. In the latter, the hole is in the CdS clad and the electron is in the HgS well. The slow nonradiative relaxation processes involving this state are expected to be slow owing to the large change in the charge carrier effective masses as they cross from the CdS clad to the HgS well. Burda et al. (1999) reported on the fs charge carrier dynamics of CdS nanoparticles (NPs)(quinone composites). They observed that in the absence of surface quinones, the excited charge carriers get trapped at the surface within 30 ps. Subsequent electron-hole recombination takes place on a much longer time scale ($>10^{-7}$s). However, if quinones are adsorbed on the CdS NP surface, photoinduced electron transfer across the NP interface to the adsorbed quinone takes place in 200—400 fs. Subsequent back electron transfer to the NP valence band occurs within 3 ps, which is 5 orders of magnitude farster than the inherent electron-hole recombination rate in the CdS itself. The quinone bypasses the CdS core and the surface trap states and shuttles the electron from the conduction band to the hole in the valence band.

Electron dynamics in gold NP with an average diameter between 9 and 99 nm have been studied by fs transient absorption spectroscopy (Link, et al., 1999; Link and El-Sayed, 1999a). Following the plasmon bleach recovery after low power excitation indicates that a non-Fermi electron distribution thermalizes by electron-electron relaxation on a time scale of 500 fs to a Fermi distribution. It can therefore be considered the different interactions, such as electron-electron, electron-phonon, electron-defect, and surface scattering. By applying a simple two-level model the dephasing time of the coherent plasmon oscillation is calculated to be less than 5 fs. This is consistent with the fact that the dominant electronic dephasing mechanism involves electron-electron interaction rather than electron-phonon coupling.

In recent years, research on the QDs assembly (especially for diameters less than 10 nm) has attracted widespread attention. These materials exhibited a strong nonlinear optical response in the vicinity of the exciton resonance, which is enhanced by the quantum confinement and dielectric confinement effects. In reality, many effects have interconnected roles in the enhancement of the nonlinear optical response of the QDs assembly. These effects are related to the electronic structure of the interface especially. We have reported the optical nonlinearity for 3.5 nm α-Fe_2O_3 QDs (Li, et al., 1993). The third-order optical susceptibilities χ^3 of hydrosol and organosol are 1.2×10^{-11} and 2.1×10^{-9} esu, respectively. This measurement is performed by the degenerate four-wave mixing (DFWM) technique. To study the population lifetime of excited states in hydrosol and ogranosol, the time-delayed four-wave mixing with incoherent light (TDFWM-IL) technique has been used. In the α-Fe_2O_3 QDs hydrosol system, the population lifetime of the electronic state $\tau_1 = 9$ ps, if α-Fe_2O_3 is encapsulated as an organosol system, an even faster nonlinear optical response has been measured. The population lifetime of the electronic state was less than 70 fs.

Electron-hole pairs can be generated promptly by light, followed by their diffusion to the surface. In a nanoscale QDs system, this kind of diffusion process is very quick and charges on surface happen to separate. This accumulation of charge separation is the origin of optical nonlinearities. For α-

Fe_2O_3 QDs hydrosol, the electronic wave function in the cluster extends to the water distance, so the dielectric confinement effect is not considerable. This will reduce the optical nonlinearities. In an α-Fe_2O_3 QDs organosol assembly, the situation is different. The dielectric constant of ST (stearate) or DBS (dodecylbenzene sulfonate) is much smaller than that of α-Fe_2O_3 QDs, and its electronic wave function is blocked by the organic-coated layer. Therefore the quantum and dielectric confinement effects are significant. The surface dipole layer formed by surface coating improves the diffusion speed of electron-hole pairs and enhances the surface charge separation, which leads to the accumulation of charge separation in the surface. This process occurs in 70 fs. It can be verified that the luminescence is quenched in the organosol. The high density of carriers and the correlation of electrons in the surface lead to the large and fast optical nonlinearities of α-Fe_2O_3 QDs organosol assemblies.

In the early days of semiconductor materials, researchers considered only linear effects important, and they investigated theoretically only the limit of very dilute exciton population. Excitons were considered to be "approximate" bosons. A dense exciton population enhances excitonic effects in quantum confined structured system. These effects originate from the unique properties of interacting many-body systems. With the development of ultrashort pulse lasers it became possible to investigate such coherent nonlinear optical effects. They occur before relaxation processes are able to destroy the phases of the populations and polarizations. An example of this effect, known as the excitonic Stark effect. This effect has attracted a lot of attention, because it can be viewed as a Bose condensation of population in all the excitonic states stimulated by the "symmetry breaking" pump field. In QDs, because of the exciton-exciton Coulomb coupling, coherent polarization waves can interact with one another, and the corresponding nonlinearity also contributes to wave mixing (Chelma, 1993).

From the viewpoint of chemistry, recent studies have confirmed that the surface of QDs plays as important a role as size effect in the equilibrium and dynamic properties of both semiconductor and metal QDs (Zhang, 1997). In particular, direct measurements of charge carrier dynamics in QDs using time-resolved techniques could give a fundamental understanding of how charge carriers interact with phonons, other carriers, and the surface. The large percentage of surface atoms, which often have dangling bonds, as well as surface defects and adsorbates introduce a high density of "surface states". These states can fall within the bandgap for a semiconductor and function to trap charge carriers (electron and hole). They are often termed trap states and significantly influence the charge carrier behavior.

A popular approach is transient absorption spectroscopy. In this approach, a short laser pluse ($\sim$100 fs) is used to create excited-state electrons and a second laser pulse is used to probe the excited population. The dynamics are measured from the change in absorbance of the probe pulse due to excited-state absorption, as a function of delay time between the two pulses. A related approach based on transient bleaching measurement, in which the probe pulse monitors the ground-state recovery, has also been used to study charge carrier dynamics and nonlinear optical properties of QDs systems. The large bleach at the first exciton energy may arise from a decrease in oscillator strength of the excitonic transitions due to trapped electrons and holes or band filling due to excess conduction band electrons. The time-resolved differential absorption spectra could observe for a series of time delays Δt between the pump and the probe before and just after excitation of a QDs system. Time-resolved photoluminescence measurements have also been applied to characterize excited charge carriers in QDs. Charge carriers may radiatively recombine, emitting near band-edge and sub-bandgap light. The near band-edge emission is mainly attributed to band-edge excitons. The sub-bandgap emission is proposed to arise from recombination of surface trapped carriers. Thus, the rise and decay of the time-resolved luminescence yield information on the formation and depletion of band-edge and trap states. For CdS colloids, electron trapping is found to be dominant in the early dynamics, which occurs in <100 fs (Zhang, 1994). A longer trapping time (500 fs to 8 ps) is deduced for CdSe QDs from photon echo experiments (Mittleman, 1994). The photon echo is an optical analogue of something observed in magnetic-resonance spectroscopy, namely spin echo. A pulse of radiofrequency radiation is applied to the system under investigation. This pulse has a duration and amplitude. A time τ later, a second pusle with $\theta_2 = \pi$ is applied. After an additional period τ, the magnetization of the system, which has decayed to thermal equilibrium following the application of each pulse, suddenly rises to a large value and then decays again. This appears as an "echo" of the first pulse on an NMR spectroscopy. That is spin echo. The photon echo proceeds in a very similar manner. Shank and Alivisatos et al. (Schoenlein, 1993) reported the first direct measurements of fs electronic dephasing (~ 85 fs) in CdSe QDs using three-pulse photon echoes and novel mode-suppression technique.

Studies of the surface effect on electron dynamics provided further insight into the electronic relaxation mechanism. One way to change the surface characteristics is to chemically modify the surface, appropriate surface electronic passivation, which often results in removal of certain trap states and enhanced luminescence. For CdS, Zhang (1997) have observed that the early electron dynamics are relatively insensitive to surface modification even though luminescence is substantially enhanced. As shown in Fig. 5.19, the dynamics remained essentially unchanged while luminescence increased 200-fold upon modifying the surface with excess Cd^{2+} ion.

A major difference between a metal and a semiconductor is in their electronic band structure. In metals with Fermi levels lying in the band center, the relevant energy level spacing is very small, even in relatively small particles, whose properties resemble those of the bulk. The quantum confinement effect is expected to occur at much smaller sizes for metal than for semiconductor nanoparticles. This difference has fundamental consequences

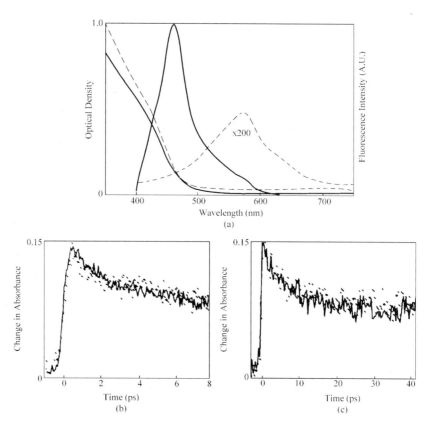

Figure 5.19 Effect of surface modification in CdS colloids. (a), Electronic absorption (left) and fluorescence (right) spectra for modified (solid lines) and unmodified (dashed lines) samples. (b, c) Dynamics measured (solid lines) and unmodified (dotted lines) samples on short (b) and longer (c) times scales (adapted from Zhang, 1997)

for electronic relaxation processes. Zhang have measured the electronic relaxation to be 1 ps for Au_{55} and 300 ps for Au_{13}. The relaxation in Au_{55} ($2R = 2$ nm) is much faster than in large particles, 14—40 nm, which show a 7 ps lifetime, as shown in Fig. 5.20.

Gammon, et al. (1997), Bonadeo, et al. (1998) apply the methodology developed in atomic and molecular systems to extend the studies to the ultimate quantum limit of a single zero-dimensional QD. Picosecond optical excitation has been used to coherently control the excitation in a single QD on a time scale that is short compared with the time scale for loss of quantum coherence. The results demonstrate coherent optical control of an exciton wave function and take the first step toward wave function engineering in these systems. The excitonic wave function is manipulated by controlling the optical phase of two-pulse sequence through timing and polarization. Wave function engineering techniques, developed in atomic and molecular systems, are used to monitor and control a nonstationary quantum mechanical state composed of

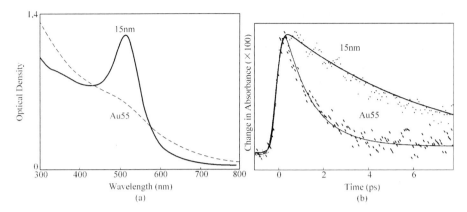

Figure 5.20 (a) Electronic absorption spectra for the two different sizes of Au particles: solid line for 15 nm and dashed line for 2 nm. (b) Comparison of electron relaxation dynamics in different sized Au particles [(a) 15 nm and (b) 2 nm]. Dotted lines are experimental data, and solid lines are fits using a single exponential with time constants of 7 ps (a) and 1 ps (b), respectively (adapted from Zhang, 1997)

a superposition of eigenstates. In particular, they coherently control the state of excitation in a QD, measure the autocorrelation of the wave function using wave packet interferometry, and manipulate the relative phase of the eigenstates in a quantum superposition of states to generate a simple target wave function.

5.3 The Control of Nanostructures by Spectroscopic Diagnosis

5.3.1 Processing on the Nanostructures

Recently, many scientists and engineers have looked for methods to control the sizes of QDs and make possible the formation of ordered lateral two-dimensional superlattices or vertical superlattices for heterojunction thin films. One approach is "top-down" method, molecular beam epitaxy nanolithographic technology, which has been developed with the development of microelectronics and processing techniques for traditional inorganic semiconductors. This technique of nanoscale manipulation can reach only the upper limits of sizes defined by nanostructure physics, and has successfully manipulated artificial atoms and molecules (Kastner, 1992; Livermore, et al., 1996). "Bottom-up" method is based on molecular and supramolecular assembly techniques that have been proposed by chemists in recent years (Whitesides, et al., 1991). With this method, it is possible to prepare monodispersed defect-free nanocrystal QDs 1—10 nm in size and to control easily QDs coupling to form nanocrystal molecules, even quantum dot superlattices in two or three dimensions.

Why are QDs with sizes less than 10 nm important? It is possible to observe correlated single-electron tunneling phenomena of QDs assemblies at room temperature. If one electron is added to an isolated QD, the electrostatic energy ($e^2/2C$) will increase. The single-electron effect can be observed when the electrostatic energy exceeds the thermal excitation energy (K_{BT}). In QDs around 10 nm in circumference, the capacitance can be reduced to the order of 10^{-18} F. This means that the Coulomb staircase on single-electron tunneling can be displayed at room temperature (thermal excition energy ~ 0.0258 eV). Andres et al. (1996a) investigated successfully Coulomb staircase phenomena of assemblies of Au QDs with diameter of 1.8 nm at room temperature. The QDs, each encapsulated by a monolayer of amphiphilic molecules with short alkyl chains, were cast from a colloidal solution dropped onto a flat substrate to form a close-packed QD monolayer with intermolecular weak interaction forces. It was demonstrated by Andres et al. (1996b) that quantum dot superlattices using Au QDs could be constructed. Bawendi et al. (Murry, et al., 1995) reported that CdSe QDs with photoactive properties had been assembled into three-dimensional superlattices. The size and spacing of the QDs within the superlattices are controlled with near atomic precision. This control is a result of synthetic advances that provide CdSe QDs that are monodisperse within the limit of atomic roughness. So, weak interaction forces are exhibited among the coated QDs. In solution, the slower growth rates produce larger colloidal crystal three-dimensional (3-D) superlattices with more regular geometries. Comparison of fluorescence spectra for QDs close packed in the solid with dots in a dilute matrix reveals that the emission line shape of the dots in the solid is modified and red shifted. These results show that weak interaction forces of quantum dots exist in nanostructured supramolecular assemblies.

Mirkin et al. (1996) and Alivisatos et al. (1996) successfully constructed supramolecular self-assemblies of QDs and oligonucleotides by precise molecular recognition of nucleotides and Watson-Crick base-pairing interactions, respectively. These methods can precisely control the coupling distance of two quantum dots of a species. Similarly, artificial molecules of homostructures may carry out the coupling of quantum dots of heterostructures to form donor-acceptor assemblies, even 2-D or 3-D QD superlattices. Peng et al. (1997a; 1997b) investigated in detail, wet-chemical methods for the synthesis of higher order structures of metal and semiconductor QDs; moreover we describe herein the successful preparation and isolation of a linked homodimer of CdSe QDs and highly luminescent CdSe/CdS core/shell QDs.

QDs can be formed within alternating multilayers of amphiphilic molecules using the Langmuir-Blodgett (LB) method. This is a kind of Langmuir-Blodgett multilayer in which nanoparticles are inserted between organic layers of periodic spacing. In general, this system can be classified according to two kinds of methods. One is called order synthesis and involves cluster formation of Cu_2S, CdS, CdSe, and PbS, which are generated in situ between the polar headgroups from the LB films (Peng, et al., 1992a). We have synthesized Q-size PbS monolayers consisting of a two-dimensional domain or linear form of PbS in the polar planes of stearic acid (SA) LB films by chemical reaction. We synthesized also the PbS within PMAO LB films in different X/Y ratios. Polymaleic acid with octadecanol monoester (PMAO) is an amphiphilic oligomer that can be used for ordered synthesis and ordered assembly of the inorganic nanoparticles. As a polymer, PMAO should be better than common fatty acids in terms of thermal and mechanical stability in LB films. The molecular structure is shown below.

$$-[(-CH)_x-(-CH-)_y]_N- \quad N \sim 20, X > Y$$

with COOH on the first CH, and C=O—O—$C_{18}H_{37}$ on the second CH.

$X/Y = 2.0 : 1.0$ PMAO1; $X/Y = 3.7 : 1.0$ PMAO2; $X/Y = 6.0 : 1.0$ PMAO3.

From the molecular structure, it can be seen that the ratio of carboxylic groups to hydrocarbon chains is always larger than one. Our previous work reported a ratio of 2.0 (PMAO1), a value twice as large as that in common fatty acids, and the PbS or CdS within PMAO LB films showed a larger blue shift of the optical absorption edge than those within stearic acid LB films (Peng, et al., 1994). By controlling the ratio of carboxylic groups and hydrocarbon chains, we can efficiently control the size of inorganic nanoparticles.

A series of uv-vis spectra for PbPMAO with different ratios of carboxylic groups and hydrocarbon chains that reacted with H_2S and lasted 150 h is shown in Fig. 5.21. We can see that the absorption threshold of the spectrum was observed at about 435, 470, 515 nm, corresponding to PMAO1, PMAO2, and PMAO3, respectively. According to the absorption threshold and band gap energy versus particle diameter curve of Henglein (1995), the PbS size in PMAO LB films was less than 2.5 nm. Also, a direct and simple method for determining the band gap value is according to the exciton peak position. A much larger blue shift of the optical absorption edge with long tails near the absorption edge is shown in Fig. 5.21. The long tail may be due to defect states, particle size distribution, and/or indirect transitions. The structural absorption spectrum of PbS we observed is close to the 0-0 transition assigned to the first allowed excited state. The exciton peaks appear at slightly shorter wavelengths of 280, 295, and 370 nm corresponding to PMAO1, PMAO2, and PMAO3, respectively.

Compared with PMAO1—3, the absorption energy is shifted to lower

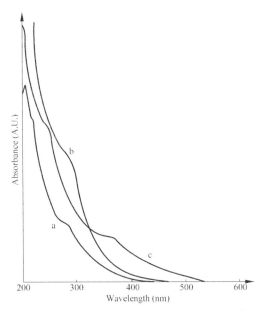

Figure 5.21 UV-vis absorption spectra of PbS monolayers formed by the chemical reaction of H_2S with (a) PMAO1 LB films with PbS monolayers. (b) PMAO2 LB films with PbS monolayers and (c) PMAO3 LB films with PbS monolayers (adapted from Peng, et al., 1992a; 1994)

values for particles prepared at higher ratios of carboxylic groups and hydrocarbon chains for PMAO molecules. High ratios of carboxylic groups and hydrocarbon chains apparently favor the formation of larger particles. It is true that the C=O stretching vibration band is located at 1720—1730 cm^{-1} when the COOH groups form a sideways dimmer structure as reported by Peng et al. (1994). The adjacent COOH groups within the individual PMAO molecules are separated by only about 0.27 nm. However, the distance between PMAO molecules is approximately 0.5 nm, the same as for fatty acid. Therefore, it is easy to form the aggregate from the oligomeric molecular unit, indicating that the size of the aggregate of the PbS within PMAO LB films is smaller than that formed in fatty acid LB films. This result is consistent with the fact that the blue shift in the PMAO matrix is greater than in the case of fatty acid salts. Moreover, the number of Pb^{2+} ions bound to one PMAO molecule apparently increases with increasing ratio of carboxylic groups and hydrocarbon chains. Therefore, the size of the PbS QDs increases with increasing ratio of carboxylic groups and hydrocarbon chains.

The preliminary results reported here indicate that PbS forms smaller QDs in PMAO LB films than in stearic acid LB films. This implies that the PMAO LB matrix provides not only dense metal ions for the preparation of inorganic nanoparticulate monolayers but also the possibility of controlling the size of the nanoparticles within the monolayers by controlling the ratio of carboxylic groups and hydrocarbon chains.

In another method, nanoparticles have been inserted between the headgroups of LB films. Bawendi et al. (Dabbousi, et al., 1994) used the Langmuir-Blodgett technique to deposit monolayers of CdSe nanocrystallites on a surface. Fendler et al. (Kotov, et al., 1994) reported the preparation of a TiO_2 monoparticulate layer based on the "membrane-mimetic approach to advanced materials". We have also reported the preparation of α-Fe_2O_3-stearate alternating Langmuir-Blodgett multilayers using a monodisperse nanoparticle α-Fe_2O_3 hydrosol directly as the subphase. Furthermore, the amphiphilic molecular orientation of 7 and 2 nm nanoparticulate α-Fe_2O_3-stearate alternating Langmuir-Blodgett films was studied by means of FTIR absorption spectra.

5.3.2 Spectroscopic Diagnosis

Some extremely high-quality inorganic nanocrystals can be prepared by 'bottom-up' wet-chemical methods. Further, the presence of the organic molecules on the surface of a nanocrystal enables extensive chemical manipulation after the synthesis. One important issue for QDs concerns the nature of surface passivation. It is essential in a confined semiconductor structure to epitaxially passivate dangling bonds at the surface (Alivisatos, 1998). For the control of nanostructures, in the process of preparing, the spectroscopic diagnosis is absolutely necessary. To explain clearly the efficacy of spectroscopic diagnosis, it is best to give some practical examples.

5.3.2.1 Metal QDs

Alvarez et al. (1997) reported the optical absorption spectra of a series of nanoscale crystalline Au clusters that are passivated by a compact monolayer of n-alkylthiol(ate)s and have been measured across the electronic energy range (1.1—4.0 eV) in dilute solution at ordinary temperature. Each of the $\sim$20 samples range in effective core diameter from 1.4 to 3.2 nm ($\sim$70 to $\sim$800 Au atoms) and has been purified by fractional crystallization and has undergone a separate composition by mass spectrometry. With decreasing crystallite size the spectra uniformly show a systematic evolution (Fig. 5.22), specifically (i) a broadening of the so-called surface-plasmon band until it is essentially unidentifiable for crystallites of less than 2.0 nm effective diameter, (ii) the emergence of a distinct onset for strong absorption near the energy ($\sim$1.7 eV) of the interband gap (5d—6sp), and (iii) the appearance in the smallest crystallites of a weak steplike structure above this onset, which is interpreted as arising from a series of transitions from the continuum d-band to the discrete level structure of the conduction band just above the Fermi level.

Pileni et al. (Taleb, et al., 1998) compared the optical properties of

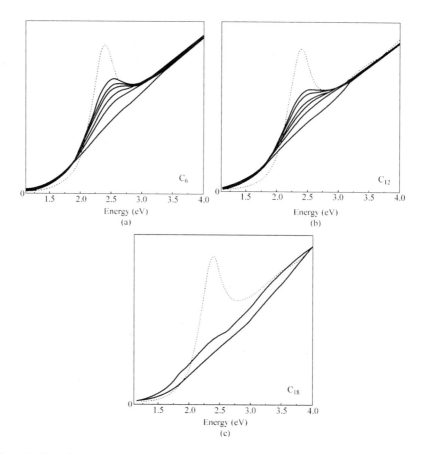

Figure 5.22 Optical absorption spectra (absorbance vs. photon energy) for dilute solutions of several purified fractions of QD gold molecules passivated by (a) hexyl-, (b) dodecyl-, and (c) octadecylthiol(ates). The spectra are scaled to unity at 4 eV and are compared to an aqueous solution of commercial colloidal Au particles (dotted lines) of 9 nm mean size. The peak amplitude (near 2.5 eV) descends with the metallic core diameters: (a) 3.2, 2.5, 2.4, 2.2, 2.0, and 1.7 nm (SC_6 passivant); (b) 2.5, 2.4, 2.2, 2.1, 2.0, and 1.7 nm (SC_{12} passivant); (c) 1.7 and 1.4 nm (SC_{18} passivant) (adapted from Alvarez, et al., 1997)

nanosized Ag particles dispersed in hexane solution and self-assembled in a 2D or 3D network. When the particles form monolayers organized in a hexagonal network, the plasmon peak of Ag nanoparticles is shifted toward low energy, with an increase in bandwidth compared to that observed with free coated particles dispersed in hexane solution. Such a shift is attributed to an increase in the dielectric constant of the matrix environment of the nanoparticles.

The linear and nonlinear ($\chi^{(2)}$) optical responses of Langmuir monolayers of organically functionalized Ag QDs were measured by Heath et al. (Collier, et al., 1997) as a continuous function of inter-particle separation under near-ambient conditions. As the distance between metal surfaces was decreased from 12 to ~5 Å, both quantum and classical effects were observed in the optical signals. When the separation was less than 5 Å, the optical second-harmonic generation (SHG) response exhibited a sharp discontinuity, and the linear reflectance and absorbance began to resemble those of a thin metallic film, indicating that an insulator-to-metal transition occurred. This transition was reversible.

The 40 Å silver cores capped with propanethiol are denoted 40 Å/C_3. It is characterized by a size distribution width of ~10%. Several spectra collected from a Langmuir film of 40 Å/C_3 particles are shown in Fig. 5.23. Upon initial compression, the film becomes more reflective. In this region, $\chi^{(2)}$ is increasing exponentially. This decrease in reflectivity is accompanied by an equally large increase in absorptivity. The final reflectance spectrum is similar to that reported for thin, metallic silver films and indicates that the Langmuir

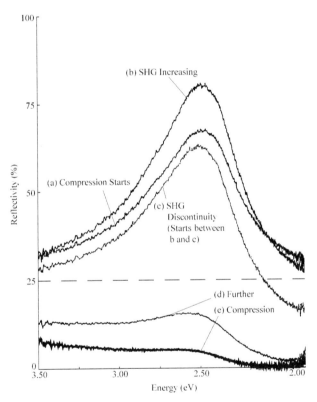

Figure 5.23 UV-vis reflectance spectra from a Langmuir monolayer of 40 Å particles passivated with n-propanethiol, collected in situ as the film compressed. The dashed line corresponds to the reflectance of a clean water surface. Correlations with the nonlinear optical response are indicated. Below ~5 Å separation between particles, the reflectance spectra begin to resemble that of a thin, metallic film. The decrease in reflectance and disappearance of the strong surface plasmon resonance correspond to a discontinuity in the SHG signal. All data were collected from an uncollapsed film, and the transition was reversible (adapted from Collier, et al., 1997)

film has become metallic. Heath et al. (Shiang, et al., 1998) show that metal QDs can be treated as "artificial atoms" and crystallized into "artificial solids" that can be tuned by controlling interparticle coupling through the application of pressure.

5.3.2.2　III—V semiconductor QDs

The synthesis of colloidal III—V QDs is more difficult than for II—VI QDs because (1) the synthesis must be conducted in rigorously air-free and water-free atmospheres, (2) it generally requires higher reaction temperatures, (3) it requires much longer reaction times, and (4) it involves more complicated metalorganic chemistry. The best results to date have been obtained for InP QDs (Nozik, et al., 1998). Nozik et al. (1998) reported the spectroscopic behavior of InP QDs. The emission and absorption features shift to higher energy with decreasing QD size. Resonant PL spectra (size-selective excitation into the tail of the absorption onset) show increasing FLN with increasing excitation wavelength; PL and PLE were used to derive the PL red shift as a function of QD size (Fig. 5.24 and Fig. 5.25). An

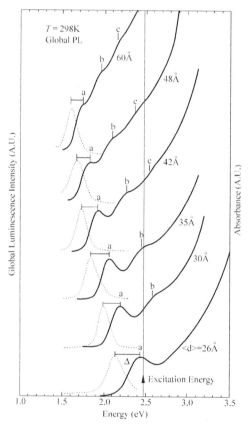

Figure 5.24 Absorption (solid line) and global PL (dotted line) spectra at 298 K for colloidal ensembles of nip QDs with different mean diameters. All QD colloidal samples were photoexcited at 2.48 eV (adapted from Nozik, et al., 1998)

analysis of the single-dot resonant red shift (difference between PL peak and the first absorption peak) as a function of the single QD diameter indicates that results are consistent with a model in which the emission occurs from an intrinsic, spin-forbidden state, split from its singlet counterpart, due to screened electron-hole exchange (Micic, et al., 1997).

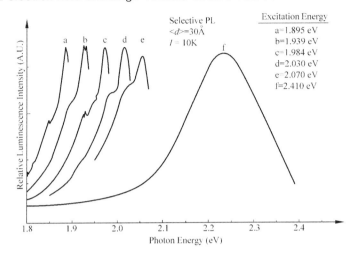

Figure 5.25 PL spectra at 10 K for an ensemble of InP QDs with a mean diameter of 32 Å for different excitation energies. The first absorption peak for this QD ensemble is at 2.17 eV, so that PL curves a—e result from excitation (1.895—2.070 eV) in the red tail of the onset region of the absorption spectrum and are FLN spectra; curve f is a global PL spectrum since its excitation was at 2.41 eV and is well to the blue of the first absorption peak (adapted from Nozik, et al., 1998)

Alivisatos and Heath et al. (Guzelian, et al., 1996) synthesized the dodecylamine capped InP QDs from 20 Å to 50 Å in diameter. Raman spectroscopy reveals TO and LO modes near the characteristic bulk InP positions as well a surface mode resulting from finite size. Figure 5.26 shows resonance Raman spectra for several QDs sizes. The TO and LO modes are found at approximately 311 and 347 cm^{-1}, respectively. The surface mode (S) is near 330—335 cm^{-1}, although for the smallest particles the LO and S features are indistinguishable.

Landin et al. (1998) studied PL from individual InAs QDs buried in GaAs. At low excitation power density, the spectra from these QDs consist of a single line. At high excitation power density, additional emission lines appeared at both higher and lower energies, separated from the main line by about 1 meV. At even higher excitation power density, this set of lines was replaced by a broad emission peaking below the original line.

These results rule out electron- or hole-excited states as the source of the observed extra emission peaks. Because of time reversal invariance, the confined states in the dot are doubly degenerate. There, the dot can be filled with two electrons and two holes without occupying any of the single-particle

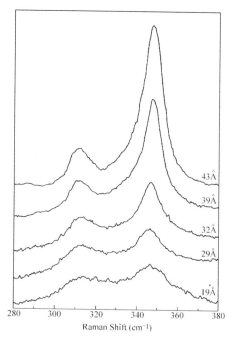

Figure 5.26 Resonance Raman spectra for several nanocrystal sizes. The TO and LO modes are found at approximately 311 and 347 cm^{-1}, respectively. The surface mode is near 330—335 cm^{-1} (adapted from Guzelian, et al., 1996)

excited states. These states can undergo recombination to produce a photon, there are four possibilities involving the single-particle ground states: e + h → photon, e + 2h → h + photon, 2e + h → e + photon, and 2e + 2h → e + h + photon, where e and h each stand for a single electron or hole, respectively. For noninteracting particles, all four of the resulting photons would have the same energy. However, the Coulomb interaction changes the energies of the multiparticle states such that they are no longer the sum of single particle energies. The relative shifts can be positive or negative, depending on details of the well shape. Hence, four distinct lines should be seen, and the number of observed lines should depend on the excitation power.

The resulting shifts with respect to single recombining electron-hole pair (e + h) are −1.6 meV for an additional electron (2e + h), +4.3 meV for an additional hole (e + 2h), and +2.6 meV for an additional exciton (2e + 2h). Figure 5.27 shows the PL spectra at different excitation power densities for four different QDs. The continuous emission from the QDs appears simultaneously with the occurrence of band filling in the wetting layer (Fig. 5.27(e)). The continuous emission from the QDs is attributed to an interaction between the band-filled wetting layer and the state-filled dot. The high-resolution spectra of QDs is indicated in Fig. 5.28.

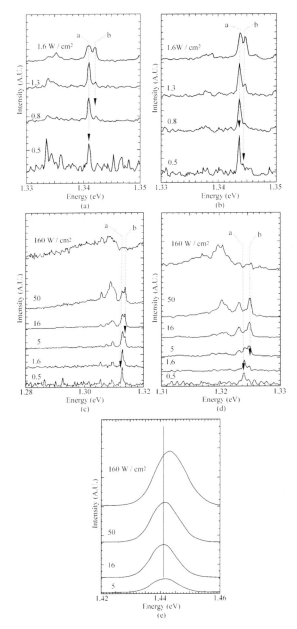

Figure 5.27 Spectra at different excitation power densities for four different QDs and for the wetting layer. (a, b) Evolution of the initial single line "a", as well as the evolution of the extra line "b" at higher energy, for excitation power densities between 0.5 and 1.6 W/cm^2. (c, d) Evolution of the spectra for excitation power densities between 0.5 and 160 W/cm^2. Additional emission peaks appear at lower energies; at the highest excitation power density, a continuum-like emission is seen. (e) Evolution of the emission from the wetting layer as a function of excitation power density (adapted from Landin, et al., 1998)

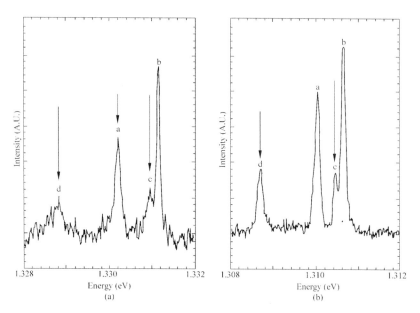

Figure 5.28 High-resolution spectra of QDs showing fine structure, generated with an intermediate excitation power density (16 W/cm²). Four peaks are shown for each island and are labeled "a" to "d", where a is the original single peak, b is the same as in Fig. 5.27, c is the shoulder peak just below b, and d is the low-energy peak. For these two islands, the lines a and d have the same splitting of ~1.4 meV with respect to one another, and the lines b and c have the same splitting of 0.25 meV. The a-d and b-c pair have different relative spacings between the islands, where the a-b splitting of the higher-energy island is 1.00 meV and of the lower-energy island is 0.65 meV (adapted from Landin, et al., 1998)

5.3.2.3 II—VI core/shell Structured QDs

Peng et al. (1997c) reported the synthesis of epitaxially grown, wurtzite CdSe/CdS core/shell nanocrystals. The core absorption spectrum is characteristic of nearly monodisperse nanocrystals with an absorption onset significantly from the bulk value of wurtzite CdSe (1.74e). After refluxing the TOPO (tri-n-octylphosphine) capped cores overnight in pyridine, no change was detected in their absorption spectra. These core nanocrystals were capped with amine and used as a standard sample for comparison to core/shell samples in all experiments. During shell growth, the absorption spectrum roughly maintained its overall shape, with a slight broadening of feature, while shifting to lower energy. For the smaller cores a shift of approximately 0.05 eV accompanied each CdS stock injection, while smaller shifts of 0.02 eV per injection were observed for the larger cores. Smaller cores also showed a total absorption shift of near 0.17 eV, while larger cores showed a total shift of about 0.06 eV before the reaction mixture became turbid. These changes in the absorption spectra are caused by the closing together of

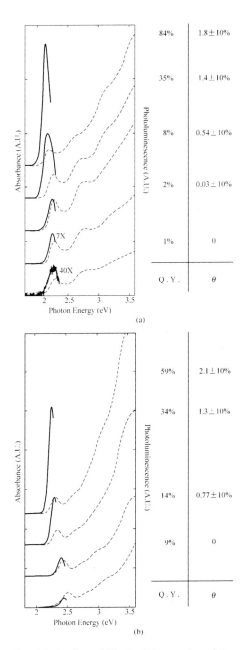

Figure 5.29 Absorption (dashed) and PL (solid) spectra of two series of core/shell QDs. Spectra were taken after successive injections of CdS stock solution. The increase in quantum yield and of coverage of CdS with each injection is also shown. Q.Y.: quantum yield of PL. θ: number of monolayers of shell growth. All spectra were taken at a concentration corresponding to an optical density (OD) of roughly 0.2 at the peak of the lowest energy feature in the absorption spectrum. Fig. 5.3(a): 30 Å CdSe core diameter series; Fig. 5.3(b): 23 Å CdSe core diameter series (adapted from Alivisatos, 1998; Peng, et al., 1997c)

electronic levels that accompanies a size increase and by a slight worsening of the size distribution. The PL and absorption spectra for two series of nanocrystals having two different CdSe core sizes and varying shell thicknesses are given in Fig. 5.29. These core/shell systems could increase PL quantum yield up to at least 50%. The experimental results indicate that in the excited state the hole is confined to the core and the electron is delocalized throughout the entire structure. The photostability can be explained by the confinement of the hole, while the delocalization of the electron results in a degree of electronic accessibility that makes these QDs attractive for use in optoelectronic display devices.

5.3.2.4 Heterogeneous Nanostructures

The Si nanocrystals (diameter 2—5 nm) showed a broad PL spectrum, peaking at 880 nm, attributed to the recombination of quantum confined excitons. But the PL peak wavelength blue shifts and PL intensity were depended on oxidation depths of the Si nanocrystals (Brongersma, et al., 1998). FTIR measurements could be performed to diagnose of Si QDs surface oxides. After oxidization, the stretching mode diminishes, accompanied by a broad oxygen absorption due to the Si-O-Si antisymmetric and symmetric stretching. The band consists of a single asymmetric peak extending from 1000 to 1250 cm^{-1} with the peak at 1100 cm^{-1}. That is due to TO antisymmetric stretching Si-O-Si modes. The symmetric stretching appears centered at $\sim$840 cm^{-1} (Thompson, et al., 1998). The Raman spectrum of the surface-oxidized Si QDs displayed a sharp peak at about 510 cm^{-1} (Li, et al., 1997). Porous Si is a high surface area material produced by electrochemically etching single-crystal silicon is HF-containing electrolytes. It consists of a large number of interconnected Si nanocrystallites with a surface that is almost entirely covered with hydrogen atoms (Harper, et al., 1997). The surface of the resulting porous Si material is H-terminated and is characterized by infrared absorptions assigned to Si-H_x stretches around 2100 cm^{-1} and a Si-H_2 scissor mode at 915 cm^{-1} (Fig. 5.30).

In "bottom-up" assembly nanostructured materials, an important interim goal is the self-assembly and self-organization of QDs arrays or superlattices in solution. In these complex, which are refered to as heterosupermolecules, the intrinsic properties of the QDs and the molecule persist and there exists cooperative hetrosupramolecular function. For instance, the associated function for the TiO_2 QDs-organic supramolecular assembly is light-induced electron transfer (Cusack, et al., 1997). TiO_2 can act as a sensitizer for light-reduced redox processes because of its electronic structure, which is characterized by a filled valence band and an empty conduction band (Hoffmann, et al., 1995; Linsebigler, et al., 1995). The quantum-sized semiconductor TiO_2 particles may result in increasing photoefficiencies for systems in which the rate-limiting step is charge transfer (Konenkanp, et al., 1993). The widespread application of titanium dioxide (TiO_2) as a

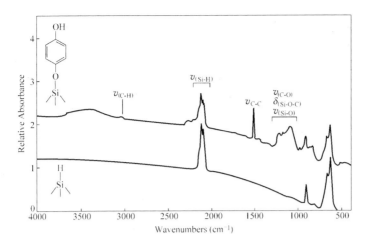

Figure 5.30 FTIR spectra of a porous Si sample before and after derivatization. Bottom: As-formed (HF etched) porous Si showing bands assigned to $\nu_{(Si-H)}$ of the surface hydrides. Top: Sample after treatment with benzoquinone, showing the appearance of bands characteristic of a surface-bound hydroquinone moiety; $\nu_{(C-C)}$ ring stretching mode at 1518 cm^{-1}, broad $\nu_{(O-H)}$ band at 3414 cm^{-1}, aromatic $\nu_{(C-H)}$ at 3052 cm^{-1}, and bands at 1229, 1176, and 1091 cm^{-1} assigned to $\nu_{(C-O)}$ stretching, $\delta_{(Si-O-C)}$ bending and $\nu_{(Si-O)}$ stretching vibrations. The top spectrum is offset from the bottom one by 1.0 absorbance unit for clarity. The spectra were obtained in a spectroelectrochemical cell constructed of Teflon with the appropriate holes and mounting to allow acquisition of transmission IR spectra without altering the sample position (adapted from Harper, et al., 1997)

photoactive species led us to prepare a Langmuir-Blodgett-type multilayer of TiO$_2$ (anatase type) inserted between the organic monolayers in order to construct lateral and vertical two- or three-dimensional organic-inorganic alternating multilayers.

A nanoparticulate TiO$_2$-stearate (TiO$_2$-St) monolayer was obtained directly using TiO$_2$ hydrosol as the subphase. The surface pressure versus surface area isotherm showed that the monolayer could be compressed to a mean molecular area of 0.25 nm^2. The monolayer was transferred onto hydrophobic and hydrophilic n-type, p-type Si substrates at a dipping speed of 18 cm/min under a surface pressure of 25 mN/m. The transfer ratio was 1.0± 0.1. The transmission electron microscopic (TEM) images of TiO$_2$-stearate monolayers showed relatively densely packed nanoparticles. A higher surface coverage was obtained, but a great number of pinholes were observed on the film surface due to the irregular shape of the particles. The AFM images of the monolayer showed that a high-quality lateral structure was formed (Li, et al., 1997).

Lian et al. (Ghosh, et al., 1998) reported femtosecond IR spectra of injected electrons in colloidal TiO$_2$ nanoparticles. The direct detection of electrons in the QDs with subpicosecond time resolution provides a new approach to study ultrafast interfacial electron transfer between semiconductor QDs and molecular adsorbates. The dynamics of electron injection from sensitized to QDs and the subsequent back-transfer and relaxation dynamics of the injected electrons correspond to the rise and decay of the transient IR signal of injected electrons. Using the technique, the injection time for coumarin 343 sensitized TiO_2 QDs in D_2O is determined to be 125±25 fs. The subsequent decay dynamics of the injected electrons in QDs are found to be different from conduction band electrons in a bulk TiO_2 crystal. Kamat et al. (1997) synthesized composite nanoclusters of Au/CdS with core/shell geometry. These QDs are photoactive and exhibit transient bleaching in the 400—600 nm region when subjected to 355 nm laser pulse excitation. Capping of gold colloids with ultrasmall CdS nanoclusters (particle diameter ∼4 nm) significantly alters the picosecond dynamics of the gold core. The bleaching of the surface plasmon absorption of the gold core is achieved by exciting the CdS shell. The major fraction of the interparticle electron transfer between CdS and Au nanoclusters is completed within the laser pulse duration of 18 ps.

5.3.3 Photovoltage Spectroscopy of Surface and Interface

Surface-photovoltage spectroscopy (SPS) is a powerful tool for the investigation of the influence of band gap and sub-band illumination on the electronic properties of semiconductor surfaces and interfaces with other materials. The spectra are obtained by monitoring the work function of a semiconductor as a function of the wavelength of the incident light. Kelvin probe technique, as an established direct, noncontact method in semiconductor surface electronics to determine the work function of a conducting (or semiconducting) solid, is widely used in surface photovoltage measurements (SPV) (Moons, et al., 1997a; 1997b). This technique has also been proven useful for the identification: (1) The surface potential of organized molecular films on solid substrates, and to evaluate the effect of adsorbates on the work function of semiconductor; (2) The position of the Fermi level of extended organized molecular films and evaluate the electric fields present in inorganic/organic heterostructures. Additionally, SPV can be carried out using the Kelvin probe technique and yield information about band bending near the surface of a semiconductor and/or the built-in voltage at the buried interface inside a heterojunction, and SPS can provide information about the energetic position of localized levels inside the band gap.

The nanoscale TiO_2-St monolayers deposited on the n-type, p-type Si substrate are shown in Fig. 5.31. For comparison, surface-photovoltage spectra of blank n-type and p-type Si substrates in contact with n-ITO glass are also shown in Fig. 5.31. According to the diagrams, we could analyze in detail the relation between interfacial charge of heterostructures and photoinduced charge transfer.

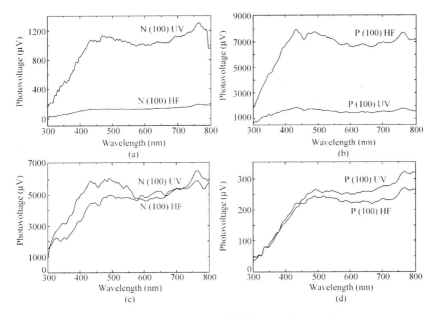

Figure 5.31 Surface photovoltage spectrum of TiO_2-stearate monolayers on an n-type, p-type Si substrate: (a) clean n-type silicon substrate, (b) clean p-type silicon substrate, (c) n-type silicon substrate coated with a TiO_2-St LB monolayer, (d) p-type silicon substrate coated with a TiO_2-St LB monolayer. UV, hydrophilic Si surface; HF, hydrophobic Si surface (adapted from Li, et al., 1997)

In the n-Si/n-ITO heterostructural system, the response values of SPS should be decreased because of the n/n interface. If the n-Si is treated with an HF solution, forming a hydrophobic surface, a higher electrostatic potential barrier is formed at the space charge layer of the n-Si(100)HF/ITO interface. So the response value of SPS is the smallest one. In an n-Si(100)UV/ITO cell, due to a thinner layer of SiO_2 grown on the n-Si surface, a higher electrostatic potential can form at the space charge layer, the response value of SPS will be larger than that of the former. In contrast, the hydrophilic and hydrophobic p-Si(100)/ITO cells have large response values of SPS because of the p/n interface. Furthermore, it is of interest to note that the response value of the SPS for the p-Si(100)UV/ITO cell has proved to be smaller than that of the p-Si(100)HF/ITO cell. Because photoinduced electron transfer can result from enhanced electrostatic potential at an interface, there is a larger response value for the SPS of the p-Si(100)HF/ITO cell.

After the deposition of a TiO_2-St monolayer on the Si substrate, the response values of SPS are clearly changed. Two kinds of proposed energy band diagrams, neglecting any SiO_x interfacial layer, for n-Si(100)/TiO_2-St/ITO and p-Si(100)/TiO_2-St/ITO heterostructures are shown in Fig. 5.32. Because of the high conductivity of ITO, it is assumed that the Fermi level

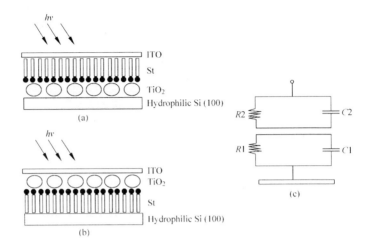

Figure 5.32 (a) TiO_2-St monolayer is deposited on a hydrophilic Si(100) surface, i.e., Si(100)UV; (b) The TiO_2-St monolayer is deposited on a hydrophobic Si(100) surface, i.e., Si(100)HF; (c) equivalent circuit (adapted from Li, et al., 1997)

coincides with the conduction band edge in ITO.

The n-Si(100)/TiO_2-St/ITO heterostructural cell appears to have two interfaces, n-Si(100)/TiO_2-St and TiO_2-St/ITO. This double-junction model can be represented by the equivalent circuit in terms of a pair of tunnel capacitors (R1, C1; R2, C2) in series (Livemore, et al., 1996). A schematic diagram of the Si(100)/TiO_2-St/ITO heterostructures is shown in Fig. 5.33. We have measured the SPS of the n-type and p-type silicon substrate with a TiO_2-St LB monolayer (Fig. 5.32) deposited. In an n-Si(100)/TiO_2-St/ITO cell, the electrostatic potential has the same orientation whether the junction is from an interface of n-Si(100)/TiO_2-St or TiO_2-St/ITO, and the electrostatic potential should be increased in the series $C1$ and $C2$, which causes enhancement of SPS. The SPS of four heterostructural cells are shown in Fig. 5.32.

When the cell is illuminated by light with a wavelength greater than 600 nm, the indirect band gap of the n-Si substrate display only an absorption transition and caused large response values of SPS. The response value of SPS is further increased with illumination by light of shorter wavelengths. This is because the light is absorbed by both n-Si(100) and TiO_2. In this case, the diffuse direction of the photogenerated electron-hole pairs is advantageous for enhancement of the interfacial electric field. At the same time, it can be seen that the values of SPS for TiO_2-St monolayers deposited on the hydrophobic n-Si(100)HF surface are smaller than for those on the hydrophilic n-Si(100)UV surface. The photovoltage response of the n-Si(100) substrate is increased about 25 and 5 times after a monolayer of TiO_2-St LB films was transferred onto hydrophobic and hydrophilic n-Si(100) surfaces, respectively.

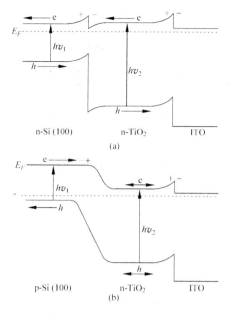

Figure 5.33 Equilibrium energy band diagrams for Si/TiO$_2$-St/ITO cells. (a) n-Si/TiO$_2$-St/ITO cell; (b) p-Si/TiO$_2$-St/ITO (adapted from Li, et al., 1997)

On the contrary, in the p-Si(100)/TiO$_2$-St/ITO cell, the electrostatic potential is in the opposite direction, so the SPS values should be decreased in a series of junction capacitors. If TiO$_2$ can absorb photons and cause photogenerated electron-hole pairs, these electrons can be confined in the lower potential well, and the response values of SPS can be further decreased. The photovoltage response is decreased by about 40 and 5 times due to the deposition of the TiO$_2$-St monolayer on the hydrophobic and hydrophilic p-Si(100) surfaces, respectively.

References

Alivisatos, A. P., MRS Bulletin. **23**, 19 (1998)
Alivisatos, A. P., Jour. Phys. Chem., **100**, 13226(1996b)
Alivisatos, A. P., Science. **271**, 933 (1996a)
Alivisatos, A. P. K. P. Johnson, X. G. Peng, T. E. Wilson, C. J. Loweth, M.

Acknowledgment The author thanks the National Climbing Program and the National Natural Science Foundation of China (NNSFC) for financial support.

P. Bruchez Jr, and P. G. Schultz. Nature. **382**, 609 (1996)
Alvarez, M. M., J. T. Khoury, T. J. Schaaff, M. N. Shafigullin, I. Vezmar, and R. L. Whetten. J. Phys. Chem., B **101**, 3706 (1997)
Andres, R. P., T. Bein, M. Dorogi, S. Feng, J. I. Henderson, C. P. Kubiak, M. Mahoney, R. G. Osifchin, and R. Rdifenberger. Science. **272**, 1323 (1966a)
Andres, R. P., J. D. Bielefeld, J. I. Henderson, D. B. Janes, V. R. Kolaqunta, C. P. Kubiak, W. J. Mahoney, and R. G. Osifchin. Science. **273**, 1690 (1966b)
Bhargava, R. N., D. Gallagher, X. Hong, And A. Nurmikko, Phys. Rev. Lett. **72**, 416 (1994)
Bonadeo, N. H., D. Erland, Gammon, D. Park, D. S. Katzer, and D. G. Steel Science. **282**, 1473 (1998)
Brongersma, M. L., A. Polman, K. S. Min, E. Boer, T. Tambo, and H. A. Atwater. Appl. Phys. Lett., **72**, 2577 (1998)
Burda, C., T. C. Green, S. Link. and M. A. El-Sayed. J. Phys. Chem., B **103**, 1783 (1999)
Chelma, D. S., Phys. Today, June, 46 (1993)
Collier, C. P., R. J. Saykally, J. J. Shiang, S. E. Henrichs, and J. R. Heath, Science. **277**, 1978 (1997)
Cusack, L., R. Rizza, A. Gorelov, and D. Fitzmaurice. Angew Chem. Int. Ed. Engl. **36**, 848 (1997)
Dabbousi, B. O., C. B. Murray, M. F. Rubner, and M. G. Bawendi, Chem. Mater., **6**, 216 (1994)
Empedocles, S. A., and M. G. Bawendi. Science. **278**, 2114 (1997)
Haase, M. and A. P. Alivisatos. J. Phys. Chem., **96**, 6756 (1992)
Hess, H. F., E. Betzig, T. D. Harris, L. N. Pfeiffer, and K. W. West, Science. **264**, 1740 (1994)
Gammon, D., MRS Bulletin. **23**, 44 (1998)
Gammon, D., S. W. Brown, E. R. Snow, T. A. Kennedy, D. S. Katzer, and D. Park. Science. **277**, 85 (1997)
Ghosh, H. N., J. B. Asbury, and T. Q. Lain, J. Phys. Chem. B **102**, 6482 (1998)
Guzelian, A. A., J. E. B. Katari, A. V. Kadavanich, U. Banin, K. Hamad, E. Juban, A. P. Alivisatos, R. H. Wolters, C. C. Arnold, and J. R. Heath. J. Phys. Chem., **100**, 7212 (1996)
Harper, T. H. and M. J. Sailor. J. Am. Chem. Soc., **19**, 6943 (1997)
Henglein, A., Ber. Bunsenges. Phys. Chem., **99**, 903 (1995)
Hoffmann, M. R., S. T. Martin, W. Choi, and D. W. Bahnemman. Chem. Rev., **95**, 69 (1995)
Kamat, P. V. and B. Shanghavi. J. Phys. Chem., B **101**, 7675 (1997)
Kastner, M. A., Rev. Mod. Phys., **64**, 849 (1992)
Konenkanp, R., R. Henninger, and P. Hoyer. J. Phys. Chem., **97**, 7328 (1993)

Kotov, N. A., F.C. Meldrum, and J.H. Fendler. J. Phys. Chem.. **98**, 8827 (1994)

Landin, L., M.S. Miller, M.E. Pistol, C.E. Pryor, and L. Samuelson. Science. **280**, 262 (1998)

Lehn, J. M., and Angew. Chem. Int. Ed. Engl.. **27**, 89 (1988)

Lehn, J. M. and Angew. Chem. Int. Ed. Engl. **29**, 1304 (1990); Lehn, J. M., Supramolecular Chemistry, Concepts and Perspectives, 1995 (V Weinheim).

Leon, R., C. R. Petroff, D. Leonard, and S. Farard. Science. **267**, 1966 (1995)

Li, L. S., J. Zhang, L. J. Wang, Y. M. Chen, Z. Hui, L. F. Chi, H. Fuchs, and T. J. Li. J. Vac. Sci. Technol.. **15**, 1618 (1997)

Li, S. T., S. J. Silvers, and M. S. El-Shall. J. Phys. Chem.. B **101**, 1794 (1997)

Li, T. J., L. Z. Xiao, X. G. Peng, Y. Zhang, B. S. Zou, D. J. Wang, H. S. Fei, X. N. Bao and Z. Q. Zhu. Photophysical studies on nanoscale clusters and cluster-assembled materials. In: Photochemical and Photoelectrochemical Conversion and Storage of Solar Energy. Proceedings of the Ninth International Conference on Photochemical Conversion and Storage of Solar Energy, IPS-9, 23—28 August 1992, ed. by Z. W. Tian and Y. Cao (International Academic Publishers Beijing, China 1993) pp. 318—329

Link, S., C. Burda, Z. L. Wang, And M. A. El-Sayed. J. Chem. Phys.. **111**, 1255 (1999)

Link, S. and M.A. El-Sayed. J. Phys. Chem.. B **103**, 4212 (1999)

Linsebigler, A. L., G. Lu, and Jr. J. T. Yates. Chem.. Rev. **95**, 735 (1995)

Little, R. B., C. Burda, S. Link, S. Logunov, and M. A. El-Sayed. J. Phys. Chem.. A **102**, 6581 (1998)

Livermore, C., C.H. Crouch, R.M. Westervelt, K.L. Campman, and A.C. Gossard. Science. **274**, 1332—1335 (1996)

Micic, O. L., H. M. Cheong, H. Fu, A. Zunger, J. R. Sprague, A. Mascarenhas, and A. J. Nozik. J. Phys. Chem.. B **101**, 4904 (1997)

Mirkin, C. A., R. L. Letsinger, R. C. Mucic, and J. J. Storhoff. Nature. **382**, 607 (1996)

Mittleman, C. B., R. W. Schoenlein, J. J. Shiang, V. L. Colvin, and A. P. Alivisatos. Phys. Rev. B: Condens. Matter. **49**, 14435 (1994)

Moon, E., A. Goossen, and T. Savenije, J. Phys. Chem.. B **101**, 8492 (1997a)

Moon, E., M. Eschle, and M. Gratzel. Appl. Phys. Lett.. **71**, 3305. Yoffe, A. D., 1993, Advance (1997b)

Murray, C. B., C. R. Kagan, and M. G. Bawendi. Science. **270**, 1335 (1995)

Nie, S. and S.R. Emory, Science. **275**, 1102 (1997)

Norris, D. J., M. G. Bavendi, and L. E. Brus.: "Optical properties of semiconductor nanocrystals (quantum dots). In: Molecular electronics. A "chemis-

try for the 21st century" monograph ed. by Jortner, J. and Rather, M. (IUPAC and Blackwell Science Ltd 1997) pp.281—323 Phys. Lett. **59**, 1826.
Nozik, A. J. and O.I. Micic. MRS Bulletin. 23,24 (1998)
Peng, X. G. T. E. Wilson, A. P. Alivisatos, and P. Z. Schultz. Angew. Chem.. Int. Ed. Engl. **36**, 145 (1997a)
Peng, X., J. Wickham, and A. P. Alivisatos. J. Am. Chem. Soc.. **120**, 5343 (1998)
Peng, X. G., C. S. Michael, A. V. Kadavanich, And A. P. Alivisatos. J. Am. Chem. Soc.. **119**,7019 (1997b)
Peng, X. G., S.Q. Guan, X.D. Chai, Y.S. Jiang, and T.J. Li. Jour. Phys. Chem.. **96**, 3170 (1992a)
Peng, X. G., R. Lu, Y.Y. Zhao, L.H. Qu, H.Y. Chen, and T.J. Li. Jour. Phys. Chem.. **98**, 7052 (1994)
Peng, X. G., M. C. Schlamp, A. V. Kadavanich, and A. P. Alivisatos. J. Am. Chem. Soc.. **119**, 7019 (1997c)
Peng, X. G., Y. Zhang, J. Yang, B. S. Zou, L.Z. Xiao, and T. J. Li. J. Phys. Chem.. **96**, 3412 (1992b)
Schoenlein, R.W., C.B. Mittleman, J.J. Shiang, A.P. Alivisatos, and C. V. Shank. Phys. Rev. Lett.. **70**, 1014 (1993)
Schuller, C.. Physica E. **3**, 121 (1998)
Shiang J. J., J. R. Hearh, C. P. Collier, and R. J. Saykally. J. Phys. Chem.. B **102**, 3425 (1998)
Shiang, J. J., S.H. Risbud, and A.P. Alivisatos. J. Chem. Phys.. **98**,8432 (1993)
Taleb, A., C. Petit, and M.P. Pileni. J. Phys. Chem.. B **102**, 2214 (1998)
Thompson, W. H., Z. Yamani, L. AbuHassan, O. Gurdal, and M. Nayfeh. Appl. Phys. Lett. **73**,841 (1998)
Whitesides, G. M., J. P. Mathias, and C. T. Seto. Science. **254**, 1312 (1991)
Yang, J., X.G. Peng, Y. Zhan, H. Wang, and T.J. Li. J. Phys. Chem.. **97**, 4484 (1993)
Yoffe, A. D.. Advance in physics. **42**, 173 (1993)
Zhang, J. Z. Acc. Chem. Res.. **30**, 423 (1997)
Zhang, J. Z., R.H. O'Neil, and T.W. Roberti. J. Phys. Chem.. **98**, 3859 (1994)
Zou, B. S., L. Z. Xiao, T. J. Li, J. L. Zhao, Z. Y. Lai, and S. W. Gu. Appl. Phys. Lett.. **59**, 1826 (1991)
Zou, B. S., Y. Zhang, L.Z. Xiao, And T.J. Li. J. Appl. Phys.. **73**, 4689 (1993)

6 Dynamic Properties of Nanoparticles

Jin Z. Zhang

6.1 Introduction

Nanoparticles (NPs) refer to materials with size dimension on the length scale of a few to a few hundred nanometers. Both equilibrium and dynamic properties of nanomaterials can be very different from those of their corresponding bulk materials or isolated atoms and molecules (Henglein, 1989; Gratzel, 1989; Bawendi, et al., 1990a; 1990b; Wang, 1991; Schmid, 1994; Tolbert and Alivisatos, 1995; Miller, et al., 1995; Fendler and Meldrum, 1995; Alivisatos, 1996a; 1996b; Liu, et al., 1977; Kamat, 1997; Zhang, 1997; Collier, et al., 1998; Heath and Shiang, 1998). Their properties are often strongly dependent on the particle size, shape, and surface properties. It is fundamentally interesting to understand how properties of nanoparticles vary with these parameters. For example, spatial confinement is expected to lead to changes in the density of states (DOS) for both electrons and phonons and the rate of electron-hole recombination. The possibility to control the materials properties by varying these parameters is significant to many technological applications ranging from microelectronics to non-linear optics, opto-electronics, catalysis and photoelectrochemistry. For instance, the color and redox potential of semiconductor nanoparticles can be tuned by changing particle size. This can be very useful for applications in optics, non-linear optics and photocatalysis (Henglein, 1989; Kamat, 1993). The possibility to form surperlattices from NPs presents a whole new class of nanomaterials for photonics and other applications.

One complication involved with the study of nanoparticles is that the various parameters are not all independent. It is thus often difficult to study the dependence of the materials properties on just one parameter. For example, as the particle size decreases, the surface-to-volume (S/V) ratio increases and the particle shape can also change. However, with careful control of synthesis conditions, it is possible in many cases to obtain high quality samples that allow investigation of the dependence of the nanoparticle properties on some of the parameters.

Nanoparticles are certainly not new and have been around for a long time. Recently, nanomaterials have received considerable renewed attention because of their promise in many current and potential applications in areas such as microelectronics, non-linear optics, detectors, sensors, and catalysis. Properties of nanoparticles have been characterized using a variety of techniques including spectroscopy, microscopy and X-ray techniques. Most studies have focused on their equilibrium properties, such as optical absorption and emission, particle shape, surface structure, interparticle interaction, self-assembly and formation of surperlattices (Fendler and Meldrum, 1995; Heath, 1992; Weller, 1996; Andres, et al., 1996; Alivisatos, 1997; Collier, et al., 1998; Sarathy, et al., 1999).

Direct study of the dynamic properties of charge carriers in nanoparticles is relatively new. Dynamic studies not only help to gain important fundamental insight into the charge carrier properties but also provide complementary information to equilibrium studies, since dynamic properties are intimately coupled with equilibrium properties. For example, electronic relaxation reflects interaction of the electron with phonons, surface, and other charge carriers. Most dynamic studies have been performed using time-resolved laser spectroscopic techniques. In this chapter, I will provide an overview of some of the recent studies of dynamics properties of semiconductor and metal nanoparticles. The first part of the chapter will cover experimental techniques for synthesis and characterization. Since synthesis of nanoparticles is dealt with extensively in other chapters of this book, I will only briefly discuss synthesis of a few nanoparticle systems that will be used as examples for dynamic studies later in this chapter. For the same reason, discussion of characterization and related techniques will also be brief. The discussion of experimental techniques will focus on time-resolved laser spectroscopy, which is the primary technique for dynamic studies. The second part of the chapter will concentrate on dynamic studies of semiconductor nanoparticles. A number of important examples are discussed in some detail, including CdS, AgI, PbI_2, PbS, TiO_2, Fe_2O_3, ZnO, and Si. The third part of the chapter will focus on metal nanoparticles, with Au as a primary example. For both semiconductor and metal nanoparticles, I will concentrate on examination of possible dependence of charge carrier dynamics on parameters such as particle size, surface, and shape. Comparison and contrast will be made between different nanoparticle systems. A summary will be given at the end of the chapter.

6.2 Experimental Techniques

6.2.1 Synthesis of Semiconductor Nanoparticles

There are many different ways to make semiconductor and metal nanoparticles. Detailed discussions of synthesis of nanoparticles are presented in other chapters of this book. I will mention only a few examples of nanoparticle synthesis that are relevant to this chapter. Colloidal nanoparticles are usually prepared by wet chemistry methods in solution. Semiconductor NPs can be prepared by arrested precipitation from homogeneous solution by controlled release of ions or forced hydrolysis in the presence of surfactants. Surfactants or stabilizers are often used to stabilize NPs by stopping or controlling the crystal growth following nucleation, since, otherwise, NPs are thermodynamically unstable and tend to coalesce and grow to large size or bulk crystals (Spanhel, et al., 1987; Gallardo, et al., 1989; O'Neil, et al., 1990a; 1990b; Colvin, et al., 1992). In some cases the stabilization is carried out by ions or charges on the surface of the NPs.

A number of metal chalcogenide semiconductor NPs can be easily prepared using wet colloidal chemistry methods. For example, CdS can be synthesized by controlled mixing of Cd^{2+} with sulfide ions in the presence of surfactants (Duonghong, et al., 1982; Colvin, et al., 1992). To passivate the surface of NPs to remove surface defects and control surface properties, organic and biological molecules can be used, e.g. TOPO used for CdS and CdSe (Murray, et al., 1993; Bowen Katari, et al., 1994) and cysteine and glutathione for CdS and ZnS (Bae and Mehra, 1998; Nguyen, et al., 1999). Passivation of surface defects often leads to enhanced photoluminescence (Spanhel, et al., 1987; Roberti, et al., 1998). ZnS and CdSe NPs can be synthesized using similar techniques as used for CdS. Mixed or doped colloidal particles such as $Zn_xCd_{1-x}S$ or Mn-doped ZnS can be prepared by mixing the desired metal ions as the starting material in synthesis (Khosravi, et al., 1995; Roberti, et al., 1998). PbS NPs can be prepared with reaction of $Pb(NO_3)_2$ with H_2S in the presence of surfactant polymers such as poly(vinyl) alcohol (Nenadovic, et al., 1993; Zhou, et al., 1993). NPs of MoS_2, a layered semiconductor, can be prepared using inverse micelle methods (Wilcoxon and Samara, 1995; Parsapour, et al., 1996].

Metal oxides NPs can be prepared using similar techniques. For instance, TiO_2 can be prepared by hydrolyzing Ti(IV) salt, e.g. $TiCl_4$ (Serpone, et al., 1995) or Ti(IV) tetraisopropoxide (Choi, et al., 1994; Smith, et al., 1997b). γ-Fe_2O_3 NPs can be synthesized by selective reaction and oxidation of Fe^{2+} and Fe^{3+} ions (Kang, et al., 1996; Cherepy, et al., 1998). α-Fe_2O_3 NPs can be prepared by hydrolysis of Fe^{3+} (Faust, et al., 1989; Cherepy, et al., 1998).

Several metal iodide semiconductors can be prepared conveniently. PbI_2 NPs can be prepared based on reaction between $Pb(NO_3)_2$ with KI in organic solvents such as alcohols and acetonitrile (Sandroff, et al., 1986; Sengupta, et al., 1999). In water, a surfactant such as poly(vinyl alcohol) (PVA) is needed for making PbI_2 NPs (Artemyev, et al., 1997; Sengupta, et al., 1999). AgI can be prepared by reacting NaI with $AgNO_3$ in the presence of a stabilizing polymer such as poly(N-vinyl)pyrrolidone (PVP) (Henglein, et al., 1989; Brelle and Zhang, 1998).

Preparation of Si NPs is more involved (Fojtik, et al., 1994; Heinrich, et al., 1992; Bley, et al., 1996; Zhang, et al., 1998; Takagi, et al., 1990; Li, et al., 1997; Littau, et al., 1993; Wang, et al., 1996; Cao and Hunt, 1994) and there is no simple wet chemistry methods are available for making Si NPs at this point.

6.2.2 Synthesis of Metal Nanoparticles

Metal nanoparticles are usually prepared by reducing appropriate metal ions with reducing agents in solution with or without surfactant molecules to stabilize the particles (Henglein, 1993; Zhang, 1997; Smith, et al., 1997b). For example, Au NPs can be prepared by reducing $HAuCl_4$ with sodium citrate (Frens, 1973). Ag and Pt NPs can be synthesized by reducing platinum or silver ions with borohydride (Suh, et al., 1983; Wilenzick, et al., 1967). Metal NPs such as Au capped with organic molecules, e.g., n-alkylthiol(ate)s, have also been prepared (Alvarez, et al., 1997; Chen and Murray, 1999).

It should be pointed out that most colloidal NPs prepared have a finite size distribution. The average size and size distribution are strongly dependent on the experiment conditions of preparation, including reactant concentration, temperature, mixing rate, impurities, and surfactants. The size distribution can be made quite narrow in some systems with carefully controlled experimental conditions or use of separation methods following synthesis (Murray, et al., 1993; Bowen Katari, et al., 1994; Schaaff, et al., 1997). To date, synthesis of truly single-sized NPs have been achieved only in a few cases, such as CdS (Lee, et al., 1988; Herron, et al., 1993; Vossmeyer, et al., 1991) and Au and Pt (Benfield, et al., 1989; Fauth, et al., 1991; Schmid, 1991). It remains an interesting and challenging problem to develop robust synthetic methods for making single-sized NPs.

6.2.3 Characterization of Nanoparticles

The characterization of nanoparticles involve various microscopy, spectroscopic and x-ray techniques. Microscopy techniques such as TEM, STM and AFM are the most direct methods for determining particle size and shape (Reetz, et al., 1995; Fendler and Meldrum, 1995; Liu, et al., 1977). With high resolution microscopy such as HRTEM, particle shape, crystal lattice structures and grain boundaries of nanoparticles can be clearly determined, as shown for Si (A) and Au (B) nanoparticles in Fig. 6.1. With high resolution microscopy, it is found that most nanoparticles are not spherical but can be formed in different interesting shapes such as cubes, prisms, or needles, often with well-defined facets (Fendler and Meldrum, 1995; Petroski, et al., 1998; Link, et al., 1999; Patel et al., 2000). The shapes of particles are sensitive to the preparation methods and conditions, e. g. reaction temperature and surfactants used to stabilize the particles. Surface properties of NPs have been studied using various type of spectroscopy, such as UV/visible (Weller, et al., 1986; Zhang, et al., 1994a), fluorescence (Chestnoy, et al., 1986; Colvin, et al., 1992; Herron, et al., 1993; Roberti, et al., 1998), IR (Sengupta, et al., 1999), and Raman (Bowen Katari, et al., 1994). These spectroscopic techniques are very useful in characterizing structural, optical and electronic properties of nanoparticles. In particular, fluorescence spectroscopy is sensitive to the electronic properties of surface trap states that are usually difficult to examine by other methods. Raman and IR are sensitive to lattice structure and species adsorbed on the particle surface. In addition, X-ray diffraction (Lee, et al., 1988; Schmid, 1991; Tolbert and Alivisatos, 1994; Vossmeyer, et al., 1991) and X-ray absorption and photoemission techniques (Van Buuren, et al., 1998) have also been employed to characterize nanoparticles. In particular XAFS (X-ray absorption fine structure) measurements have been found useful in determining local and surface structures of NPs, as demonstrated successfully for TiO_2 (Chen, et al., 1997) and CdTe (Rockenberger, et al., 1998).

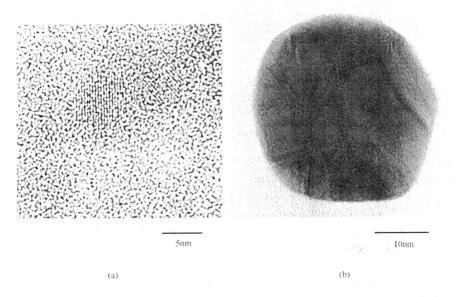

Figure 6.1 High resolution transmission electron microscopy (HRTEM) images of Si (a) and Au (b) NPs

6.2.4 Dynamics Measurements with Time-Resolved Techniques

For measurements of charge carrier dynamics in nanoparticles, time-resolved laser techniques, including transient absorption, transient bleach, and time-resolved fluorescence, have been employed. The basic idea is to first excite the nanoparticles with a short femtosecond or picosecond light pulse generated from an ultrafast laser system. The excited charge carriers are then monitored with a second laser pulse that is delayed with respect to the excitation (pump) pulse. The change in absorption of the second (probe) pulse reflects change in excited state charge carrier population. Alternatively, the ground state population can be monitored as a transient bleach signal, which indicates depletion and recovery of the ground state population. Another method to monitor excited state population is to time-resolve the luminescence from the excited state. This technique is only useful for luminescent nanoparticles. A schematic diagram of a typical laser setup used for time-resolved measurements is shown in Fig. 6. 2. In the system used in our lab, femtosecond laser pulses generated from a Ti-sapphire oscillator are amplified in a Ti-sapphire regenerative amplifier (Zhang, et al., 1994a). The final output (250 μJ/pulse, 100 fs) centered at 780 nm is doubled in a KDP crystal to yield 390 nm pulses (30 μJ/pulse), which are used to excite the sample. The remaining 780 nm light following doubling is used to generate a white light continuum (500—950 nm) in a quartz or sapphire window, from which a single wavelength can be selected as a probe pulse to interrogate the excited state population. The time delay between the probe and excitation (pump) pulses can be varied with a delay line, e. g. a translation stage. The excited state dynamics can be monitored by measuring changes of absorbance of the probe pulse as a function of delay time between the pump and probe. To obtain multiwavelength spectra, the white light can be used as a probe and dispersed after the sample with multiwavlength detectors such CCD or dual diode array detectors.

  The excited state population can also be monitored by detecting fluorescence following excitation. To time-resolve the fluorescence on the ultrafast time scales requires special techniques such as fluorescence up-conversion. In this method, fluorescence is mixed with a "gating" pulse in a non-linear optics crystal to generate a up-converted signal and the gating pulse is time-delayed with respect to the excitation pulse. The up-converted signal with frequency equal to the sum of the gating pulse frequency and fluorescence frequency, is linearly proportional to the fluorescence intensity and contains dynamic information about the emitting species (Iwai, et al., 1998). This can provide fs time resolution, however, the signal is expected to be weak due to the typically low luminescence yield of nanoparticles. So far, no studies have been reported on the use of fluorescence up-conversion to study charge carrier dynamics in nanoparticles on the fs time scale. But it is anticipated that such experiments will be conducted in the future, since it is an excellent way to directly probe different trap states versus bandedge states. This is because the wavelength of emission is different for the bandedge states compared to trap states. By selecting the emission wavelength being up-converted, one

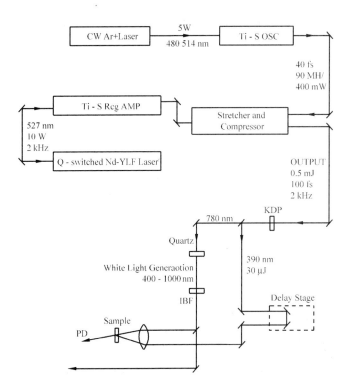

Figure 6.2 Schematic illustration of a fs Ti-sapphire laser system used for transient absorption measurements. The abbreviations used are: CW for continuous wave, Ti-S OSC for Ti-sapphire oscillator, Ti-S Reg AMP for Ti-sapphire regenerative amplifier, KDP for potassium dihydrogen phosphate, IBF for interference bandpass filter, and PD for photodiode

can unambiguously determine dynamics of the emitting states. The recent advent of OPA (optical parametric amplifier) technology has made it easy to generate tunable excitation wavelengths from laser systems such as the one shown in Fig. 6.2. Tunable excitation wavelength is desirable for studying samples absorbing at different wavelengths.

6.3 Dynamic Properties of Semiconductor Nanoparticles

6.3.1 Theoretical Considerations

Semiconductors are generally divided into two classes: direct and indirect bandgap according to their electronic band structure. Optical absorption and emission as well as electronic properties of these two types of semiconductors can be very different. For direct bandgap semiconductors, electronic transitions from the valence band to the conduction band is electrical dipole allowed and the electronic absorption as well as emission is usually strong. The absorption cross section depends on the bandgap according to:

$$\alpha(hv) = A(hv - E_g)^\beta \qquad (6.1)$$

where E_g is the bandgap of the semiconductor, A is a constant depending on the index of refraction and reduced masses of the electron and hole of the semiconductor. β is either 1/2 for allowed direct transitions, all momentum (k) —— conserving transition, or 3/2 for forbidden direct transitions, all k except $k = 0$ (Pankove, 1991). For indirect bandgap semiconductors, the valence band to conduction band electronic transition is electrical dipole forbidden and the transition is phonon-assisted, i.e., both energy and momentum of the electron-hole pair are changed in the transition. The absorption cross section depends on the bandgap according to:

$$\alpha(hv) = A'(hv - E_g + E_p)^2 / \exp(E_p/kT) - 1 \qquad (6.2)$$

for $hv > E_g - E_p$ or

$$\alpha(hv) = A'(hv - E_g - E_p)^2 / 1 - \exp(-E_p/kT) \\ + A'(hv - E_g + E_p)^2 / \exp(E_p/kT) - 1 \qquad (6.3)$$

for $hv > E_g + E_p$, where A' is a constant depending on the semiconductor and E_p is the phonon energy (Pankove, 1991). Both the absorption and emission are weaker compared to direct bandgap semiconductors. Figure 6.3 shows a comparison of the electronic absorption and emission spectra of CdS, a direct bandgap semiconductor, and Si, an indirect bandgap material. The peak near 420 nm for CdS NPs is known as the exciton peak. Such excitonic features are absent in indirect semiconductor materials.

For NPs of semiconductors, quantum confinement effects play an important role in their electronic and optical properties. The quantum confinement effect can be qualitatively explained using the effective mass approximation (Brus, 1984; 1986; 1990). For a spherical particle with radius R, the effective bandgap, $E_{g.\,eff}(R)$, is given by:

$$E_{g.eff}(R) = E_g(\infty) + \frac{h^2 \pi^2}{2R^2}\left(\frac{1}{m_e} + \frac{1}{m_h}\right) - \frac{1.8e^2}{\varepsilon R} \qquad (6.4)$$

where $E_g(\infty)$ is the bulk bandgap, m_e and m_h are the effective masses of the electron and hole, and ε is the bulk optical dielectric constant or relative permittivity. The second term on the right hand side shows that the effective bandgap is inversely proportional to R^2 and increases as size decreases. On the other hand, the third term shows that the bandgap energy decreases with decreasing R due to the increased Coulombic interaction. The second term

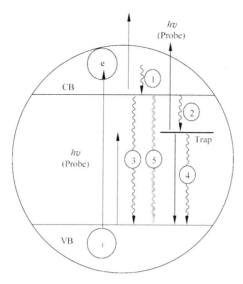

Figure 6.3 Schematic illustration of charge carrier relaxation in semiconductor NPs and the pump-probe scheme for monitoring the carrier dynamics. The long solid line with upward arrow indicates excitation and the short solid lines with arrows indicate probe of the excited or ground state. The curved lines with downward arrows indicate different relaxation processes: (1) electronic relaxation within the conduction band, (2) trapping, (3) bandedge electron-hole recombination (radiative and non-radiative), (4) trapped electron-hole recombination (radiative and non-radiative), and (5) exciton-exciton annihilation

becomes dominant with small R, thus the effective bandgap is expected to increase with decreasing R. The quantum size confinement effect becomes significant especially when the particle size becomes comparable to or smaller than the Bohr exciton radius, a_B, which is given by:

$$a_B = \varepsilon_0 \varepsilon_r h^2 / \pi \mu e^2 \qquad (6.5)$$

where ε_0 and ε_r are the permittivity of vacuum and relative permittivity of the semiconductor, μ is the reduced mass of the electron and hole, $m_e m_h / (m_e + m_h)$, and e the electron charge. The Bohr radius of CdS is around 2.4 nm (Gratzel, 1989) and particles with radius smaller or comparable to 2.4 nm show strong quantum confinement effects as indicated by a significant blue shift of their optical absorption relative to that of bulk (Duonghong, et al., 1982; Colvin, et al., 1992; Zhang, et al., 1994a).

The study of dynamic properties of charge carriers in semiconductor nanoparticles has received considerable attention in the last few years. Issues of interest include charge carrier recombination, trapping, carrier-carrier interaction, and their dependence on particle size, shape, surface characteristics. Systems that have been studied include CdS, TiO_2, CdSe, Fe_2O_3, AgI, CuS, PbI_2, Ag_2S, PbS, Si, MoS_2, and coupled colloids such as CdS/TiO_2 (Evans, et al., 1994) or core/shell structures such as CdS/HgS

(Kamalov, et al., 1996). In almost all these systems, the dynamics of photogenerated charge carriers seem to be dominated by trapping and recombination of trapped carriers caused by a high density of surface trap states. Figure 6.4 shows schematically the major pathways for charge carrier relaxation following above bandgap photoexcitation. The first step of relaxation should be electronic relaxation in the conduction band and hole relaxation in the valence band. This is mainly due to electron-phonon interaction and is expected to be on the time scale of 100 fs or less. Once the electron is relaxed to the bottom of the conduction and the hole to the top of the valence band, they can recombine radiatively or non-radiatively. If there are few or no bandgap states, the recombination should be primarily radiative, i.e., strong bandedge luminescence should be observed, and the lifetime should be on the order of nanoseconds or longer. When there are states in the bandgap due to surface defects, they act to trap the charge carriers on time scales faster than radiative recombination, typically a few picoseconds or shorter. The trapped charge carriers can recombine non-radiatively or radiatively, producing trap state emission that is red shifted with respect to bandedge emission. The trap states have lifetimes on the time scale from tens of picoseconds to nanoseconds or microsecond or even longer, depending on the nature of the trap states. The trap states also vary significantly in their energy levels or trap depth, which in turn determines how fast the trapping occurs and how long the trap state lives. It should also be pointed out that the excitonic state, formed as a result of Coulombic interaction between the electron and hole, usually lies slightly below the bottom of the conduction band (few tens of meV) but above the trap states. This excitonic state can also be populated during the relaxation process. However, since the binding energy is typically small, e.g., 30 meV for CdS, the excitonic state is at thermal equilibrium with the bottom of the conduction band at room temperature. It is thus generally difficult to experimentally distinguish between the bottom of the conduction band and the excitonic state. We will simply refer them to as bandedge states.

In terms of direct experimental probes of the excited charge carrier dynamics, three major techniques based on time-resolved laser spectroscopy are used, including transient absorption, transient bleaching and time-resolved fluorescence. These techniques have been described in detail in Section 6.2.4. The following sections present several specific examples to illustrate how time-resolved experiments can be used to investigate the relaxation pathways of charge carriers in semiconductor nanoparticles. We will attempt to examine how the charge carrier dynamics depend on various parameters such as surface, size and shapes of nanoparticles.

6.3.2 CdS, CdSe and Related Systems

CdS is among the most extensively studied semiconductor nanoparticle

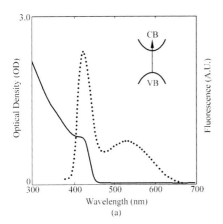 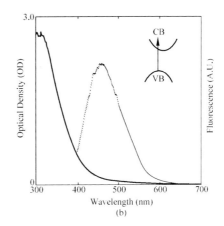

Figure 6.4 Comparison of electronic absorption spectra of a direct bandgap, CdS (top), and an indirect bandgap, Si (bottom), semiconductor. The solid lines are absorption and the dotted lines are fluorescence with excitation at 390 nm. The difference in band structure in momentum space representation is shown schematically in the inset

systems. Earlier dynamics studies on the picosecond time scale identified a strong transient bleach feature near the excitonic absorption region of the spectrum (Dimitrijevic and Kamat, 1987; Haase, et al., 1988; Hilinski, et al., 1988; Kamat, et al., 1989a; 1989b; Wang, et al., 1990; Eychmuller and Weller, 1994). It was noticed that the peak of the bleach feature shifts with time and one explanation proposed was increased screening by charge carriers for the particles (Henglein, et al., 1986). Later, femtosecond measurements were carried out and found a power dependence of the bleach recovery time (Ernsting, et al., 1990). The bleach recovery follows a double exponential rise with the fast component increasing with power faster than the slower component. Recent work from our lab using transient absorption found a similar power dependence of the electronic relaxation dynamics featuring a double exponential decay behavior with a fast (2—3 ps) and slow (50 ps) decay components (Zhang, et al., 1994a; 1994b; Roberti, et al., 1998). As shown in Fig. 6.5, it can be clearly seen that amplitude of the fast decay component increases with excitation intensity faster than that of the slow component. It grows nonlinearly, slightly subquadratic, with excitation intensity. This non-linear fast decay was first attributed to nongeminate electron-hole recombination at high excitation intensities (Zhang, et al., 1994a). We have subsequently carried out a more detailed investigation using fs transient absorption in conjunction with ns time-resolved fluorescence and found that the bandedge fluorescence was also power dependent (Roberti, et al., 1998). These results led us to propose that the fast decay is due to exciton-exciton annihilation upon trap state saturation, as suggested previously (Zheng, et al., 1988), and the slow decay is due to trapped charge carrier recombination. Therefore, the transient absorption signal observed seems to have contributions from both bandedge electrons (excitons) and trapped electrons. At early times, the bandedge electrons have significant contribution, especially when trap states are saturated at high excitation intensities, while as time progresses the contribution from trapped electrons becomes more dominant. On long time scales (hundreds of ps to ns), the signal is essentially all from trapped charge carriers. We believe that this is true for many other colloidal semiconductor nanoparticles (Zhang, 1997; Brelle and Zhang, 1998; Cherepy, et al., 1998; Sengupta, et al., 1999; Patel et al., 2000).

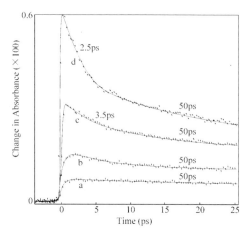

Figure 6.5 Excitation intensity dependent photoinduced electron relaxation dynamics of CdS NPs probed at 780 nm, following excitation at 390 nm

One important issue involved in describing charge carrier relaxation in semiconductor NPs is the charge carrier trapping rate. For CdS NPs, an electron trapping time constant of about 100 fs has been suggested (Zhang, et al., 1994a; Skinner, et al., 1995). A longer trapping time (0.5—8 ps) was deduced for CdSe NPs based on time-resolved photon echo experiment (Mittleman, et al., 1994). An even longer trapping time of 30 ps has been reported for CdS NPs based on measurement of trap state emission (O'Neil, et al., 1990b). A similar 30 ps electron trapping time for CdS NPs has been reported based on study of the effects of adsorption of electron acceptors such as viologen derivatives on the particle surface (Logunov, et al., 1998). The hole trapping was found to be faster, a 1 ps hole trapping time has been reported for CdS based on time-resolved photoluminescence measurements (Klimov, et al., 1996). The difference in trapping times reported could be either due to a difference in the samples used or different interpretations of the data obtained. It can be concluded, however, that the trapping time is on the order of a few hundred of fs to tens of ps, depending on the nature of the NPs and quality of the sample.

Charge transfer dynamics from CdS and CdSe NPs to electron acceptors, e.g. viologen derivatives, adsorbed on the particle surface have been studied using transient absorption, transient bleach and time-resolved fluorescence (Logunov, et al., 1998; Burda, et al., 1999). Electron transfer was found to take place on the time scale of 200—300 fs and and competes efficiently with trapping and electron-hole recombination. These results are important to understanding interfacial charge transfer involved in photocatalysis and photoelectrochemistry applications.

6.3.3　Metal Oxide Nanoparticles: TiO_2, Fe_2O_3, ZnO, SnO_2

Metal oxides play an important role in catalysis and photocatalysis. Their nanoparticles can usually be prepared easily by hydrolysis. Among the different metal oxide nanoparticles studied, TiO_2 has received the most attention because of its stability, availability, and promise for applications, e.g. solar energy conversion (O'Regan and Gratzel, 1991). Studies of charge carrier dynamics have been performed for TiO_2 nanoparticles alone and, more popularly, with dye sensitization. For TiO_2 NPs alone with excitation at 310 nm, the photoinduced electrons were found to decay following second-order kinetics with a second-order recombination rate constant of 1.8×10^{-10} cm^3/s (Colombo Jr., et al., 1995). The electron trapping was suggested to occur on the time scale of 180 fs (Skinner, et al., 1995).

Dye sensitization of TiO_2 has been studied extensively, partly because of its potential for solar energy conversion (O'Regan and Gratzel, 1991) and photocatalysis (Serpone and Pelizzetti, 1989). Since TiO_2 alone does not absorb visible light, dye sensitization can extend the absorption into the visible region. In dye sensitization, the electron is injected from a dye molecule on the TiO_2 nanoparticle surface, as illustrated in Fig.6.6. There are several requirements for this to work efficiently. First, the excited state of the dye molecule needs to lie above the bottom of the conduction band of the TiO_2 nanoparticle. Second, strong binding of the dye onto the TiO_2 nanoparticle surface is desired for fast injection. Third, back electron transfer to the dye cation following injection should be minimal. Fourth, the dye molecule must have strong absorption in the visible region of spectrum for solar energy conversion. A number of dye molecules have been studied and tested for solar energy conversion applications over the years (Kalyanasundaram and Gratzel, 1998). To date, the dye molecule that shows most promise for applications is a Ru complex, Ru(4,4'-dicarboxyl-2,2'-bipyridine)$_2$(NCS)$_2$(N$_3$), which showed the highest reported light-to-electricity conversion efficiency of 10% (O'Regan and Gratzel, 1991; Nazeeruddin, et al., 1993). This work has stimulated strong interest in understanding the mechanism of charge injection and recombination in such dye-sensitized nanocrytalline systems.

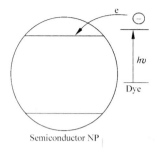

Figure 6.6 Schematic illustration of charge injection in dye sensitization of a semiconductor NP

The rates of electron injection and subsequent recombination or back electron transfer in dye sensitization are expected to be dependent on the nature of the dye molecule and the NPs, especially the surface characteristics of the NPs. The interaction between the dye and the NP surface will determine the rates and yields of forward and reverse electron transfer (Gratzel, 1989; Miller, et al., 1995). The shape (facets) and size of the particles could also be important and tend to vary from sample to sample depending on the preparation methods used. Recent studies of the injection rate have found that electron injection (forward electron transfer) is generally extremely fast, ~ 100 fs. For instance, for the coumarin 343 dye on TiO_2, the electron injection rate was found to be around 200 fs (Rehm, et al., 1996). For the N_3 dye on TiO_2, the first direct fs measurement reported a hot electron injection time of $<$25 fs (Hannappel, et al., 1997). However, there has been some debate over possible degradation of the dye sample used (Moser, et al., 1998; Hannapel, et al., 1998). We have recently studied an anthocyanin dye adsorbed on TiO_2 NPs and found an electron injection time of $<$100 fs (Cherepy, et al., 1997). The assignment is made simple in this system since the dye alone has a stimulated emission signal (similar to a transient bleach), while the dye on TiO_2 has a transient absorption signal. By carefully performing several control experiments, we were able to unambiguously assign the transient absorption signal to electrons injected into TiO_2, which has a rise time of $<$ 100 fs. This suggests that the electron injection from the excited dye molecule to the TiO_2 NP takes place in $<$100 fs. The most recent work by Ellingson, et al. on N_3 on TiO_2 reported an injection time of $<$ 50 fs based on transient infrared measurements (Ellingson, et al., 1998). While forward electron transfer has generally been found to be very fast, back electron transfer was found to occur on a range of time scales, from about 10 ps to μs, depending on the nature of the dye and the nanoparticle (Kamat, 1997; Cherepy, et al., 1997; Martini, et al., 1998). For example, both forward and reverse electron transfer have been studied in the case of anthracenecarboxylic acids adsorbed on different types of TiO_2 NPs and were found to be dependent on the dye molecular structure and the method used to synthesize the TiO_2 particles (Martini, et al., 1998).

Another important metal oxide system is iron oxide (Fe_2O_3). Fe_2O_3 can exist in the γ phase (maghemite) or α phase (hematite). In its α phase it can be used as a photocatalyst and in its γ phase it can be used as a component in magnetic recording because of its magnetic properties (Leland and Bard, 1987; Faust, et al., 1989; Ziolo, et al., 1992; Cornell and Schwertmann, 1996; Kang, et al., 1996). We have recently carried out fs dynamics studies of electronic relaxation in both γ and α phased Fe_2O_3 nanoparticles with 390 nm excitation. The relaxation dynamics were found to be very similar between the two types of nanoparticles, despite their difference in magnetic properties and particle shape: γ being mostly spherical with 1—2 nm average diameter and α being mostly spindle-shaped with average dimension of 1×5 nm (Cherepy, et al., 1998). The relaxation featured a multiexponential decay with time constants of 0.36 ps, 4.2 ps and 67 ps. The overall fast relaxation, in conjunction with very weak fluorescence, indicates extremely efficient nonradiative decay processes, possibly related to the intrinsic dense band structure or a high density of trap states. The fast relaxation of the photoinduced electrons is consistent with the typically low photocurrent efficiency of Fe_2O_3 electrodes, since the short lifetime due to fast electron-hole recombination does not favor charge transport that is important for photocurrent generation.

Similar studies have been done on other metal oxides such as SnO_2 and ZnO. For ZnO NPs with 310 nm excitation, the photoinduced electron decay was found to follow second-order kinetics (Cavaleri, et al., 1995), similar to TiO_2 (Colombo Jr., et al., 1995). It was also found that the initial electron trapping and subsequent recombination dynamics were size dependent. The trapping rate was found to increase with increasing particle size, which was explained with a trap-to-trap hopping mechanism. The electron-hole recombination is faster and occurs to a greater extent in larger particles because there are two different types of trap states. A different explanation, based on exciton-exciton annihilation upon trap state saturation, has been recently proposed by us for similar excitation intensity dependent and size dependent relaxation observed in CdS and CdSe NPs (Roberti, et al., 1998). This explanation would also seem to be consistent with the results observed for TiO_2 and ZnO.

For SnO_2 NPs, most dynamics work has focused on dye sensitization. For instance, fs transient absorption and bleach studies have been performed on cresyl violet H-aggregate dimers adsorbed on SnO_2 colloidal particles (Martini, et al., 1997). It was found that the electron injection from the higher energy state, resulting from exciton splitting, of the dimer to SnO_2 NPs occurs in <100 fs and back electron transfer occurs with a 12 ps time constant.

6.3.4 Other Semiconductor Nanoparticle Systems: Si, AgI, Ag$_2$S, PbS

Silicon NPs are interesting because of their luminescence properties. Since bulk silicon is an indirect bandgap semiconductor with a bandgap of 1.1 eV, it is very weakly luminescent. For opto-electronics applications, it is highly desirable to develop luminescent materials that are compatible with the current existing silicon technology developed and matured for the electronics industry. The weak luminescence of bulk silicon presents a major obstacle to its use for opto-electronics. The discovery that porous and nanocrystalline Si emit visible light with high quantum yield in 1990 (Canham, 1990) has raised hopes for new photonic devices based on silicon and stimulated strong research interest in porous silicon and Si nanoparticles (Littau, et al., 1993; Fojtik and Henglein, 1994; Brus, 1994; Cao and Hunt, 1994; Wang, et al., 1996; Bley, et al., 1996; Zhang, et al., 1997; Li, et al., 1997). Various methods have been used to make Si nanoparticles, including slow combustion of silane (Fojtik, et al., 1994), reduction of SiCl$_4$ by Na (Heath, 1992), separation from porous Si following HF acid electrochemical etching (Heinrich, et al., 1992; Bley, et al., 1996; Zhang, et al., 1998), microwave discharge (Takagi, et al., 1990), laser vaporization/controlled condensation (Li, et al., 1997), high pressure aerosol reaction (Littau, et al., 1993), laser-induced chemical vapor deposition (Wang, et al., 1996), and chemical vapor deposition (Cao and Hunt, 1994). Representative absorption and luminescence spectra of Si NPs prepared from sonicating porous Si following HF etching are shown in Fig. 6.3. HRTEM image of the same NPs are shown in Fig. 6.1, which clearly shows lattice fringes and particle size of about 3 nm. We have performed preliminary fs studies of charge carrier dynamics in these nanoparticles and found that the relaxation time is a few picoseconds and the lifetime is longer for particles with higher luminescence yield (Wu, et al., 1999). A recent dynamics study on ion-implanted Si nanocrystals using fs transient absorption identified two photoinduced absorption features, attributed to charge carriers in nanocrystal quantized states with higher energy and faster relaxation and Si/SiO$_2$ interface states with lower energy and slower relaxation (Klimov, et al., 1998). Red emission observed in this sample was shown to be from surface trap states and not from quantized states.

Silver halide nanoparticles play an important in photography and their synthesis is relatively easy. We have recently investigated the ultrafast charge carrier dynamics in AgI and core/shell structured AgI/Ag$_2$S and ArBr/Ag$_2$S NPs (Brelle and Zhang, 1998). We found that the electronic relaxation follows a double exponential decay with time constants of 2.5 ps and >0.5 ns, which are independent of excitation intensity at 390 nm. The fast decay was attributed to trapping and non-radiative electron-hole recombination dominated by a high density of trap states, as indicated by extremely low luminescence.

The slow decay was assigned to reaction of deep trapped electrons with silver cations to form silver (Ag) atom, which is the basis for latent image formation in photography. The slow decay agrees with early nanosecond studies (Micic, et al., 1990). When we compared two core/shell systems, AgI/Ag$_2$S and AgBr/Ag$_2$S, we observed a new 4 ps rise component with AgBr/Ag$_2$S. This was taken as an indication of electron transfer from Ag$_2$S to AgBr (Brelle and Zhang, 1998). At the time of experiments, we were unable to make pure Ags$_2$S nanoparticles. However, very recently we have succeeded in synthesizing pure Ag$_2$S nanoparticles capped with cysteine and glutathione. The dynamics observed for pure Ag$_2$S NPs are very similar to that of the core/shell structured AgBr/Ag$_2$S (Brelle and Zhang, 1999). Therefore, the 4 ps rise feature attributed to electron transfer previously is most likely incorrect. A more likely explanation is that the rise is due to contribution from a transient bleach signal, since there is noticeable ground state absorption at the probe wavelength. Other supporting evidence of this new assignment will be published shortly (Brelle and Zhang, 1999). Even though we need to change the assignment of the 4 ps rise for Ag$_2$S/AgBr, the original assignment for the AgI and AgI/Ag$_2$S NPs should still be valid.

PbS NPs are another interesting system in that their particle shapes can be readily varied by controlling synthetic conditions (Nenadovic, et al., 1993; Gao, et al., 1995; Schneider, et al., 1997; Ai, et al., 1999; Patel and Zhang, 1999). Also, since the Bohr radius of PbS is relatively large, 18 nm and bulk bandgap is small, 0.41 eV (Machol, et al., 1993), it is easy to prepare particles with size smaller than the Bohr radius that show strong quantum confinement effects and still absorb in the visible part of the spectrum. We have attempted to study the shape dependence of electronic relaxation in different shaped PbS NPs. While it is possible to observe significant changes in the ground state electronic absorption spectrum when particle shapes are changed from mostly spherical to needle and cube shaped, the electronic relaxation dynamics remain about the same for the apparently different shaped particles (Patel and Zhang, 1999). We attributed this to the dominance of the surface properties on the electronic relaxation, similar to other systems studied. While the shapes are different, the different samples may have similar surface properties. Therefore, if the dynamics are dominated by the surface, change in shape may not affect the electronic relaxation dynamics substantially. In order to study the possible intrinsic effect of size or shape, the surface needs to be better controlled so that the relaxation is not dominated by the surface, which is a challenging problem.

6.3.5 Nanoparticles of Layered Semiconductors: MoS$_2$, PbI$_2$

Layered semiconductors such as PbI$_2$ and MoS$_2$ are an interesting class of semiconductors with some unique properties (Peterson and Nozik, 1992). Nanoparticles of layered semiconductor can be prepared using techniques similar to those used for other semiconductors. Dynamic studies of layered semiconductor NPs have been limited. A picosecond transient emission study of charge carrier relaxation in MoS_2 NPs has been reported (Doolen, et al., 1998) and found that the relaxation was dominated by trap states. The relaxation from shallow traps to deep traps is fast (40 ps) at room temperature and slows down to 200 ps at 20 K.

We have very recently conducted a fs study of charge carrier relaxation dynamics in PbI_2 nanoparticles and found that the relaxation was dominated by surface properties and was independent of particle size in the size range (3—100 nm) studied (Sengupta, et al., 1999). The relaxation was found to be strongly dependent on the solvent used. The early time dynamics was found to show some signs of a quantum beat signal, as shown in Fig. 6.7. The exact origin of the oscillation is not completely clear at this point. Such features are rarely observed for semiconductor NPs. There have been some controversies over the nature of the optical absorption spectrum, whether it is from PbI_2 nanoparticles or from some kind of iodine complexes, and if the three major absorption peaks are due to different sized "magic" numbered particles (Sandroff, et al., 1986; Micic, et al., 1987; Peterson and Nozik, 1992). We have addressed these two questions in detail (Sengupta, et al., 1999). The short answer is that there is no evidence for "magic" numbered particles correlating with the three absorption peaks and the optical absorption is dominated by PbI_2 NPs in the samples used in our study.

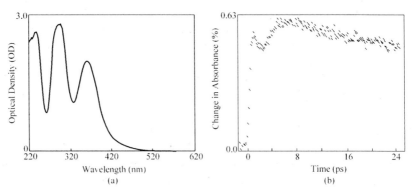

Figure 6.7 Electronic absorption spectrum (a) and photoinduced electron relaxation dynamics (b) of PbI_2 NPs in acetonitrile. The dynamics data were collected with probing at 720 nm and excitation at 390 nm. The dynamics show an oscillation at early times with a period of 6 ps

6.3.6 Effects of Particle Surface, Size and Shape

The examples discussed above seem to suggest that electronic relaxation in semiconductor nanoparticles is dominated by surface properties. This is largely because a high density of surface and defect trap states lie within the bandgap and serve to trap photogenerated charge carriers (electrons and holes). These states are usually not observed in optical absorption but in photoluminescence or dynamics measurements that are sensitive to charge carriers in trap states. The density and distribution of trap states are strongly dependent on the nature of the NPs. For example, exciton-exciton annihilation due to trap state saturation has been observed only for some NPs such as CdS, CdSe, TiO_2, and ZnO (Colombo Jr., et al., 1995; Roberti, et al., 1998) but not for other NPs such as AgI (Brelle and Zhang, 1998), Fe_2O_3 (Cherepy, et al., 1998), PbS (Patel and Zhang, 1999), and PbI_2 (Sengupta, et al., 1999). The absence of exciton-exciton annihilation in the latter systems is attributed to a higher density of trap states and thereby a higher threshold of trap state saturation. This explanation is consistent with the very weak fluorescence from these NPs, which indicates a high density of trap states. Due to difficulty in preparing single-size, single-shaped particles with uniform and well-defined surface properties, there have been no systematic studies on the effects of these parameters on the electronic relaxation dynamics. All the results reported so far seem to indicate that the surface effect is predominant. Therefore, to study the intrinsic size or shape effect, the surface needs to be better controlled, which is possible but usually nontrivial to do.

Given these difficulties, however, many interesting optical and electronic properties of semiconductor nanoparticles can still be unraveled from comparative studies with relatively high quality samples. Furthermore, for many applications, e.g., the nanocrystalline solar cells mentioned earlier or photocatalysis, it is not always necessary to use particles all with the same size, shape and surface properties. In addition, for surface chemistry application such as catalysis, the surface states are highly reactive and often necessary for conducting chemical reactions on the surface. Therefore, surface states can be useful or harmful, depending on the nature of the applications. For fundamental understanding of the effects of these parameters, however, it is highly desirable to synthesize particles with well-controlled size, shape, and surface characteristics or to develop characterization techniques that allows investigation of individual particles (Empedocles and Bawendi, 1999; Klar, et al., 1998; Perner, et al., 1998).

6.4 Dynamic Properties of Metal Nanoparticles

6.4.1 Background and Theoretical Considerations

Metal nanoparticles have interesting similarities and differences compared to semiconductor nanoparticles (Henglein, 1993; Schmid, 1994; Alivisatos, 1996a; Zhang, 1997). Similar to semiconductor NPs, metal NPs are expected to exhibit quantum effects when the size is sufficiently small. The difference is, however, that the size at which quantum effect can be observed is expected to be much smaller than that of semiconductor nanoparticles. This is due to the fundamental difference in their electronic band structures. As the particle size increases from isolated atoms (Fig. 6.8), the center of the band develops first and the edge last (de Heer, 1993; Alivisatos, 1996a). For semiconductors the Fermi level lies between the valence band and conduction band and the edges of the bands determine the low energy optical and electronic properties. Thus bandedge optical excitation strongly depends on particle size. However, for metals the conduction band is half filled with the Fermi level lying in the band center, the relevant energy level spacing is small near the Fermi level even in very small particles. Thus, quantum confinement is expected to be only significant for much smaller sized metal particles compared to semiconductor nanoparticles. This makes it more difficult to observe discrete energy levels, as suggested by some theoretical work (Halperin, 1986), in

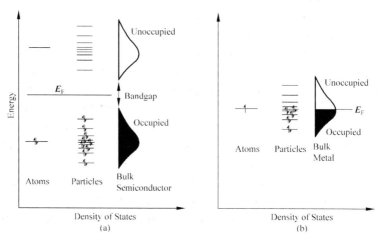

Figure 6.8 Comparison of electronic band structure and density of states (DOS) as a function of particle size between semiconductor and metal NPs

metal NPs than in semiconductor NPs. Of course, inhomogeneous size distributions and surface impurities may also make experimental examination of discrete energy levels difficult.

The difference in electronic band structure between metals and semiconductors is also expected to be reflected in their electron relaxation processes. Electrons in metal are free to move even at very low temperature and thermal excitation or photoexcitation in the UV to IR region only causes redistribution of the electron population near the Fermi level and does not result in bandgap excitation. The relaxation mechanism of hot electrons in metals is thus quite different from those in semiconductors with bandgap excitation.

Various theoretical models have been developed over the years to explain the optical properties of metal colloidal nanoparticles (Mie, 1908; Ganiere, et al., 1975; Creighton, 1982; Halperin, 1986; Barber and Chang, 1988). One key issue of interest is the size dependence of the optical absorption, e.g. the width and peak position of the surface plasmon band. In 1908, Mie proposed the first classical theoretical model based on Maxwell's equations to explain optical absorption of metal particles (Mie, 1908). The surface plasmon band of small metal particles in an insulating matrix can be calculated using the Mie theory. Reasonable agreement between the Mie theory calculation and experimentally measured spectrum had been found for Au particles >10 nm. However, for smaller particles, e.g., <10 nm, the discrepancy between theory and experiment becomes larger (Ganiere, et al., 1975). Kawabata and Kubo later developed a quantum mechanical model to calculate the absorption as a function of particle size (Kawabata and Kubo, 1966). Instead of treating the particle surface as a source of scattering, they regard the surface as a boundary condition that causes the normally quasi-continuous conduction electron energy levels to become discrete. Neither one of these two models was sufficiently good to predict the shift of the plasmon peak to longer wavelength with decreasing size for particles <10 nm (Ganiere, et al., 1975). Later, Kreibig introduced the "free path effect", the influence of conduction electron collision with particle surface, to describe differences in optical constants between metal particles and bulk metal (Kreibig and Gragstein, 1969; Genzel, et al., 1975). It was suggested that the electronic relaxation time should be dependent on particle size:

$$\tau^{-1} = \tau_{bulk}^{-1} + v_F/R \qquad (6.6)$$

where τ^{-1} is the electronic relaxation time of metal particles which is an average relaxation time caused by phonons, impurities, and lattice defects, τ_{bulk}^{-1} is the relaxation of the bulk metal, v_F is the Fermi velocity and R is the radius of the particle. This surface collision effect appears to be supported by some of the more recent dynamics studies which found size-dependent electronic relaxation in ways that are consistent with (6.6). The following presents some specific examples of dynamics studies on metal nanoparticles using time-resolved laser techniques.

6.4.2 Gold (Au) Nanoparticles

The most intensively studied metal nanoparticle system is gold (Au), which is a classic example of metal nanoparticles. The optical properties of Au NPs have been studied extensively both experimentally and theoretically (Vijayakrishnan, et al., 1992; Leff, et al., 1996; Genzel, et al., 1975; Halperin, 1986; Henglein, 1993). The surface plasmon band near 560 nm is caused by collective electron oscillation over the particle surface. Dynamics properties of Au NPs were first investigated in a picosecond study and the electronic relaxation time was reported to be >2 ps (Heilweil and Hochstrasser, 1985). Later, femtosecond studies have been performed by a number of groups. The first study by our group found an electronic relaxation of 7 ps plus a longer time offset for Au NPs (14—40 nm diameter) in aqueous solution using fs transient absorption measurements, as shown in Fig. 6.9 (Faulhaber et al., 1996; Smith, et al., 1997a). When the excitation intensity was increased, the 7 ps time constant did not change but the amplitude of the offset increases non-linearly, which was proposed to be due to a phonon bottleneck effect (lattice heating) or photoejected electrons. The 7 ps relaxation time, attributed to electron-phonon interaction, was slower than that of bulk Au films, 1 ps (Brorson, et al., 1987) but faster than what would be predicted based on a theoretical model (Belotskii and Tomchuk, 1992; Gorban, et al., 1991). The slower relaxation of NPs compared to bulk was attributed to weakening of electron-phonon coupling with decreasing size, while the discrepancy with the theoretical prediction was suggested to be due to imperfection of the particle surface (Faulhaber, et al., 1996). Later on when smaller sized Au NPs were studied under similar conditions, it was found that the relaxation time was shorter (1 ps) for very small Au clusters (2 nm), which contained 55 Au atoms (Smith, et al., 1997b). This result, in conjunction with other theoretical and experimental work, led us to propose that the electronic relaxation in metal particles are determined by both electron-phonon interaction and surface scattering (Smith, et al., 1997b). As the particle size decreases, the electron-phonon interaction becomes weaker, which results in a longer relaxation time, while the surface collision frequency increases, resulting in shorter relaxation time. These two factors have competing effects in determining the relaxation time.

Independent studies by others have found somewhat different relaxation times for similarly prepared Au NPs. Ahmadi and co-workers found a 2.5 ps relaxation time for both large (15 nm) colloidal particles in aqueous solution and small Au particles (1.9—3.2 nm) passivated with a monolayer of alkylthio (ate) groups using femtosecond transient bleach measurements (Ahmadi, et al., 1996; Logunov, et al., 1998). The relaxation time was found to increase to as long as 5 ps with increasing excitation intensity. More recently, Hodak,

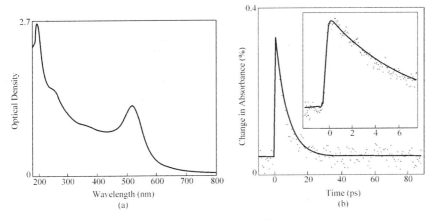

Figure 6.9 Electronic absorption spectrum (a) and photoinduced electron relaxation dynamics (b) of Au NPs in aqueous solution. The dynamics data were measured with 790 nm probe pulses following 390 nm excitation

et al. reported a relaxation time of ~0.6—0.8 ps for 11 nm Au NPs in aqueous solution using transient bleach measurements (Hodak, et al., 1998a; 1998b). They also found the relaxation time to increase with excitation intensity and suggested that at low intensity the relaxation time was similar to that of bulk gold. Theoretical calculation based on the two-temperature model (Anisimov, et al., 1974) was carried out and provided semiquantitative agreement with the experimental results and the agreement became better for lower excitation intensity results (Hodak, et al., 1998b). Most recently the same authors have found evidence of quantum beats due to coherently prepared acoustic vibrational modes in Au NPs and the beat frequency was inversely proportional to the particle size within the size range of 14—17 nm in diameter (Hodak, et al., 1998c). An independent study of Au NPs in glass matrix found a relaxation time of 4 ps (Perner, et al., 1996). Another interesting study was recently conducted on Au nanoshells with an Au_2S dielectric core and found an electronic relaxation time of 1.65 ps (Averitt, et al., 1998), smaller than that observed in bulk Au films (0.8 ps) (Groeneveld, et al., 1995).

The difference in the electronic relaxation time reported for Au NPs could be due to a number of reasons. First, the samples and the dispersion environment can be different from one lab to another. There is evidence that the relaxation time is dependent on the solvent used for the same sized particles (Zhang, et al., 1996). The surface properties and other defects including grain boundaries, which could depend sensitively on the preparation methods used, can also influence the relaxation. Second, the excitation intensity is different in the different measurements, especially given the evidence of power-dependent relaxation times mentioned above. Third, it is also possible that the fs measurement methods are different. Some labs used transient absorption and others have used transient bleach as a way to monitor the relaxation process. If the electron simply relaxes from the excited state, which is responsible for the transient absorption observed, to the ground state, which accounts for the bleach recovery, the two measurements should give the same relaxation time. However, if there are complications involved such as "intermediate states" or phonon bottleneck, the two measurements may not give the same result. These different possibilities should be examined carefully in future investigations in order to fully clarify the issues of whether the electronic relaxation is size-dependent and if the relaxation is the same or different from that of the bulk.

6.4.3 Other Metal Nanoparticles: Ag, Cu, Sn, Ga and Pt

Several other metal NPs have been studied in terms of their electronic relaxation dynamics using ultrafast laser techniques. The first femtosecond study of metal nanoparicles was carried out in our lab on silver (Ag) nanoparticles (Roberti, et al., 1995). It was found that the electronic relaxation follows a double exponential decay with time constants of 2.5 ps and 40 ps, independent of excitation intensity and particle size (4—10 nm in diameter). The time constants are attributed to electron-phonon interaction and phonon cooling through phonon-solvent interactions, respectively. An independent study of CuNPs has been reported and the electronic relaxation was found to be 1.3 ps and 0.7 ps with the probe wavelength on and off resonance, respectively, with the surface plasmon peak at 2.22 eV (Bigot, et al., 1995). Subsequently, a study of Sn nanoparticles was reported and suggested a size-dependent relaxation time that decreases with decreasing size in the 4—12 nm range, which was attributed to enhanced surface collision (Stella, et al., 1996). Recently, Ga NPs have been studied and found to have an electronic relaxation time varying from 0.6 ps to 1.6 ps with particle size increasing from 5 to 9 nm (Nisoli, et al., 1997). We have also conducted studies of Pt nanoparticles and found a faster relaxation for smaller particles, 1.7 ps for 3 nm particles and 3.5 for 35 nm particles (Smith, et al., 1997a), similar to hat observed by us for Au NPs. There are much less comparative studies performed on these metal nanoparticles compared to gold NPs.

6.4.4 Effects of Surface, Size and Shape

The examples given above indicate that electronic relaxation in metal NPs may be dependent on particle size. However, just like in semiconductor NPs, it is often difficult to study the size effect because complications caused by other competing factors such as surface and shape variations with size. It appears that the surface effect for metal nanoparticles is not as significant or dominant as for many semiconductor NPs in determining the electronic relaxation dynamics. However, it is clear from the above discussion that the surface or environment is important. Similar to semiconductor NPs, there have been no systematic studies of surface or size effect on the electronic relaxation of metal NPs. This is partly because it is not a trivial problem to try to control the size and surface properties at will. There is also very little theoretical guidance in this issue, despite some models or conjectures proposed in the literature. This is partly due to the many-body nature of the system, in conjunction with often ill-defined interface and embedding environments.

One further complicating factor is the shape of particles. Most NPs are not spherical as one may first expect. They often have different shapes, such as cubes, prisms, needles, and rods, with different facets. Just like size and surface properties, the shape is sensitive to the preparation methods and synthesis conditions used. Recently, some exciting progress has been made in some cases in controlling the shape of metal NPs, e.g. Pt (Petroski, et al., 1998) and Au (Link, et al., 1999). However, it is still difficult to obtain a single shape for a particular sample: i.e., the sample is typically a mixture of shapes and reproducibility is often a problem. Because of these problems mentioned, the samples studied are usually an ensemble average of particles with variations in size, shape and surface properties. The data obtained are thus averaged over these different particles. Even given that, some properties of metal particles can still be studied when they are not particularly sensitive to these variations. For example, the optical absorption spectrum of metal NPs, e.g., Au, prepared in different labs at different times is highly reproducible. Some properties may be more sensitive to these variations, e.g., the electronic relaxation time. In that case, it is important to try to control these parameters the best one can to make comparison of different samples and studies meaningful. To date, no dynamics studies have been reported on shape-selected metal nanoparticles. The shape could be important since surface scattering affects electronic relaxation and the different facets of different shaped particles could very well affect surface scattering.

6.5 Summary and Prospects

Nanoparticles show great promise for a number of emerging technological applications because of their novel properties. The control over size, shape, and surface properties presents a challenge in the synthesis of nanoparticles. It is these variables that make NPs interesting for studying many fundamental issues related to quantum size effects and interfaces. For applications, one of the major limitations with NPs is that charge transport between particles is very limited compared to bulk single crytalline materials. Thus, it is important to understand interaction between particles, e. g. , in surperlattice structure, that is critically related to charge transport. The study of interparticle interactions has received growing attention recently. Some alternative systems include nanowires, nanotubes or nanorods that have enhanced charge transport properties compared to nanoparticles. This is another area of growing interest.

In terms of characterization techniques, surface sensitive methods are needed to better characterize the surface properties of NPs. It might be possible to develop single particle characterization techniques based on spectroscopy or microscopy or both (Empedocles and Bawendi, 1999; Klar, et al. , 1998; Perner, et al. , 1998). This would eliminate the need to prepare single-sized or truly monodisperse particles, which is often difficult to do. Only when the surface properties are better controlled, can the intrinsic size and shape effects be studied effectively, since otherwise the surface properties tend to dominate the optical and electronic properties of the NPs.

Applications of NPs are not yet fully explored. Some areas of current or potential applications include light-emitting diodes, solar cells, lasers, sensors, fluorescence imaging, and detectors. Nanocomposite materials of NPs with organic materials or biological molecules, such as conjugated polymers, are another interesting class of materials worth exploring for both fundamental studies and new applications. Such composite materials may take advantage of both the inorganic NPs and the organic or biological materials when properly combined. The possibilities of applications as well as interesting issues concerning nanomaterials are tremendous and require further research and exploration.

References

Ahmadi, T. S. , S. L. Logunov, and M. A. El-Sayed. J. Phys. Chem. . **100**,

---

Acknowledgments This work was supported in part by grants from the Petroleum Research Fund administered by the American Chemical Society. Collaborative University of California/ Los Alamos Research Fund, University of California Energy Institute, University of California Faculty Research Fund, and National Science Foundation Summer Undergraduate Research Fellowship program. I am grateful to my students, postdoctors, and co-workers who have contributed to the work described in this chapter, including Dr. Nerine Cherepy, Dr. Trevor Roberti, Brian Smith, Archita Sengupta, Fanxin Wu, Chris Grant, Mike Brelle, Melissa Kreger, Dr. K. C. Mandal, Dr. H. Deng, B. Jiang, Dr. M. Gratzel, and Dr. Greg Smestad. The Si samples used for Fig. 6.1A and Fig. 6.3 (bottom) were kindly provided to us from Prof. Jeff Coffer's lab at Texas Christian University. The HRTEM was done at the Central Microscopy Facility at Lawrence Berkeley National Lab with help from Dr. C. Song.

8053(1996)
Ai, X., L. Guo, Y. Zou, Q. Li and H. Zhu. Mater. Lett., **38**, 131(1999)
Alivisatos, A.P., J. Phys. Chem., **100**, 13226(1996a).
Alivisatos, A.P., Science, **271**, 933(1996b)
Alivisatos, A.P., Endeavour, **21**, 56(1997)
Alvarez, M.M., J.T. Khoury, T.G. Schaaff, M.N. Shafigullin, I. Vezmar and R.L. Whetten. J. Phys. Chem., B **101**, 3706(1997)
Andres, R.P., J.D. Bielefeld, J.I. Henderson, D.B. Janes, V.R. Kolagunta, C.P. Kubiak, W.J. Mahoney and R.G. Osifchin. Science, **273**, 1690 (1996)
Anisimov, S.I., B.L. Kapeliovich, T.L. Perelman. Zh. Eksp. Teor. Fiz. **66**, 776 (1974, Sov. Phys., JETP 39, 375)
Artemyev, M.V., Yu.P. Rakovich and G.P. Yablonski. J. Cryst. Growth, **171**, 447(1997)
Averitt, R.D., S.L. Westcott and N.J. Halas. Phys. Rev., B **58**, 10203 (1998)
Bae, W. and R.K. Mehra. J. Inorg. Chem., **70**, 125(1998)
Barber, P.W. and R.K. Chang. Optical effects associated with small particles (World Scientific, Singapore 1988)
Bawendi, M.G., M.L. Steigerwald and L.E. Brus. Ann. Rev. Phys. Chem., **41**, 477(1990a)
Bawendi, M.G., W.L. Wilson, L.J. Rothberg, P.J. Carroll, T.M. Jedju, M.L. Steigerwald and L.E. Brus. Phys. Rev. Lett., **65**, 1623(1990b)
Belotskii, E.D. and P.M. Tomchuk. Int. J. Electronics, **73**, 955(1992)
Benfield, R.E., J.A. Creighton, D.G. Eadon and G. Schmid. Z. Phys., D **12**, 533(1989)
Bigot, J.Y., J.C. Merle, O. Cregut, A. Drunois. Phys. Rev. Lett., **75**, 4702 (1995)
Bley, R.A., S.M. Kauzlarich, J.E. Davis and H.W.H. Lee. Chem. Mater., **8**, 1881(1996)
Bowen Katari, J.E., V.L. Colvin and A.P. Alivisatos. J. Phys. Chem., **98**, 4109(1994)
Brelle, M. and J.Z. Zhang. J. Chem. Phys., **108**, 3119(1998)
Brelle, M.C. and J.Z. Zhang. Chem. Phys. Lett., (1999) (submitted)
Brorson, S.D., J.G. Fujimoto and E.P. Ippen. Phys. Rev. Lett., **59**, 1962 (1987)
Brus, L.E., J. Chem. Phys., **80**, 4403(1984)
Brus, L.E., J. Phys. Chem., **90**, 2555(1986)
Brus, L.E., Phys. Rev. Lett., **65**, 1623(1990)
Brus, L., J. Phys. Chem. **98**, 3575(1994)
Burda, C., T.C. Green, S. Link and M.A. El-Sayed; J. Phys. Chem., B **103**, 1783(1999)
Canham, L.T., Appl. Phys. Lett., **57**, 1046(1990)
Cao, W. and A.J. Hunt. Appl. Phys. Lett., **64**, 2376(1994)

Cavaleri, J. J., D. E. Skinner, D. P. Colombo, Jr. and R. M. Bowman. J. Chem. Phys. 103, 5378(1995)

Chen, L. X., T. Rajh, Z. Wang and M. C. Thurnauer. J. Phys. Chem.. B 101, 10688(1997)

Chen, S. and R. W. Murray. Langmuir. 15, 682(1999)

Cherepy, N. J., G. P. Smestad, M. Gratzel and J. Z. Zhang. J. Phys. Chem.. B 101, 9342(1997)

Cherepy, N. J., D. B. Liston, J. A. Lovejoy, H. Deng and J. Z. Zhang. J. Phys. Chem.. B 102, 770(1998)

Chestnoy, N., T. D. Harris, R., Hull and L. E. Brus. J. Phys. Chem.. 90, 3393(1986)

Choi, W., A. Termin and M. R. Hoffman. J. Phys. Chem.. 98, 13669(1994)

Collier, C. P., T. Vossmeyer and J. R. Heath. Ann. Rev. Phys. Chem.. 49, 371(1998)

Colombo, Jr. D. P., K. A. Roussel, J. Saeh, D. E. Skinner, J. J. Cavaleri and R. M. Bowman. Chem. Phys. Lett.. (1995)

Colvin, V. L., A. N. Goldstein and A. P. Alivisatos. J. Am. Chem. Soc. 114, 5221(1992)

Cornell, R. M. and U. Schwertmann. The Iron Oxides. (VCH, New York 1996).

Creighton, J. A.. in Surface enhanced Raman scattering. ed. Chang, R. K. and T. E. Furtak (Plenum Press, New York 1982)

de Heer, W. A.. Rev. Mod. Phys.. 65, 611(1993)

Doolen, R., R. Laitinen, F. Parsapour and D. F. Kelley. J. Phys. Chem.. B 102, 3906(1998)

Dimitrijevic, N. M. and P. V. Kamat.. J. Phys. Chem.. 91, 2096(1987)

Duonghong, D., J. J. Ramsden and M. Gratzel. J. Am. Chem. Soc.. 104, 2977(1982)

Ellingson, R. J., J. B. Asbury, S. Ferrere, H. N. Ghosh, J. R. Sprague, T. Lian and A. J. Nozik. J. Phys. Chem. B 102, 6455(1998)

Empedocles, S. A. and M. G. Bawendi. J. Phys. Chem.. B 103, 1826(1999)

Ernsting, N. P., M. Kaschke, H. Weller and L. Katsikas. J. Opt. Soc. Am.. B 7, 1630(1990)

Evans, J. E., K. W. Springer and J. Z. Zhang. J. Chem. Phys.. 101, 6222 (1994)

Eychmuller, A. and H. Weller. J. Phys. Chem.. 98, 7673(1994)

Faulhaber, A. E., B. A. Smith, J. K. Andersen and J. Z. Zhang. Mol. Cryst. Liq. Cryst.. 283, 25(1996)

Faust, B. C., M. R. Hoffmann and D. W. Bahnemann. J. Phys. Chem.. 93, 6371(1989)

Fauth, K., U. Kreigib and G. Schmid. Z. Phys.. D 20, 297(1991)

Fendler, J. H. and F. C. Meldrum. Adv. Mater.. 7, 607(1995)

Fojtik, A. and A. Henglein, Chem. Phys. Lett.. 221, 363(1994)

Frens, G.. Nature. Physical Science. 241, 20(1973)

Gallardo, S., M. Gutierrez, A. Henglein and E. Janata, Ber. Bunsenges. Phys. Chem.. **93**, 1080(1989)
Ganiere, J.-D., R. Rechsteiner and M.-A. Smithard Sold State Communications. **16**, 113(1975)
Gao, M., Y. Yang, B. Yang and J. Shen. Langmuir. **91**, 4121(1995)
Genzel, L., T.P. Martin and U. Kreibig. Z. Phys.. B **21**, 339(1975)
Gorban, S.A., N.P. Nepijko and P.M. Tomchuk. Int. J. Electronics. **70**, 485(1991)
Gratzel, M.. Heterogeneous photochemical electron transfer. (CRC Press, Boca Raton 1989)
Groeneveld, R.H.M., R. Sprik, A. Lagendijk. Phys. Rev.. B **51**, 11433 (1995)
Haase, M., H. Weller and A. Henglein. J. Phys. Chem.. **92**, 4706(1988)
Halperin, W.P.. Rev. Mod. Phys.. **58**, 533(1986)
Hannappel, T., B. Burfeindt, W. Storck and F. Willig. J. Phys. Chem.. B **101**, 6799(1997)
Hannappel, T. C. Zimmermann, B. Meissner, B. Burfeindt, W. Storck and F. Willig. J. Phys. Chem.. **102**, 3651(1998)
Heath, J.R.. Science. **258**, 1131(1992)
Heath, J.R. and J.J. Shiang. Chem. Soc. Rev.. **27**, 65(1998)
Heilweil, E.J. and R.M. Hochstrasser. J. Chem. Phys.. **82**, 4762(1985)
Heinrich, J., C. Curtis, G. Credo, K. Kavanagh and M. Sailor. Science. **255**, 66(1992)
Henglein, A., E. Kumar, E. Janata and H. Weller. Chem. Phys. Lett.. **106**, 9869(1986)
Henglein, A., M. Gutierrez, H. Weller, A. Fojtik and J. Jirkovsky. Ber. Bunsenges. Phys. Chem.. **93**, 593(1989)
Henglein, A.. Chem. Rev. **89**, 1861(1989)
Henglein, A.. Israel J. Chem.. **33**, 77(1993)
Herron, N., J.C. Calabrese, W.E. Farneth and Y. Wang. Science. **259**, 1426(1993)
Hilinski, E.F., P.A. Lucas and Y. Wang. J. Chem. Phys.. **89**, 3435(1988)
Hodak, J., I. Martini and G.V. Hartland. Chem. Phys. Lett.. **284**, 135 (1998a)
Hodak, J., I. Martini and G.V. Hartland. J. Phys. Chem.. B **102**, 6985 (1998b)
Hodak, J., I. Martini and G.V. Hartland. J. Chem. Phys.. **108**, 9210 (1998c)
Iwai, S., S. Murata and M. Tachiya. J. Chem. Phys.. **109**, 5963(1998)
Kalyanasundaram, K. and M. Gratzel. Coord. Chem. Rev.. **177**, 347(1998)
Kamalov, V.F., R. Little, S.L. Logunov and M.A. El-sayed. J. Phys. Chem.. **100**, 6381(1996)
Kamat, P.V., N.M. Dimitrijevic and A.J. Nozik. J. Phys. Chem.. **93**, 2873 (1989a)

Kamat, P. V., T. W. Ebbesen, N. M. Dimitrijevic and A. J. Nozik. J. Phys. Chem., 93, 4259(1989b)
Kamat, P. V.. Chem. Rev.. 93, 267(1993)
Kamat, P. V.. Prog. Inorg. Chem.. 44, 273(1997)
Kang, Y. S., S. Risbud, J. F. Rabolt and P. Stroeve. Chem. Mater 8, 2209 (1996)
Kawabata, A. and R. Kubo. Jpn. J. Phys. Soc.. 21, 1765(1966)
Khosravi, A. A., M. Kundu, L. Jatwa, S. K. Dehpande, U. A. Bhagwat, M. Sastry and S. K. Kulkarni. Appl. Phys. Lett.. 67, 2702(1995)
Klar, T., M. Perner, S. Grosse, G. von Plessen, W. Spirkl and J. Feldmann. Phys. Rev. Lett.. 80, 4249(1998)
Klimov, V., P. H. Bolivar and H. Kurz. Phys. Rev.. B 53, 1463(1996)
Klimov, V. I., C. J. Schwartz, D. W. McBranch and C. W. White. Appl. Phys. Lett.. 73, 2603(1998)
Kreibig, U. and C. v. Gragstein. Z. Phys.. 224, 307(1969)
Lee, G. S. H., D. C. Craig, I. Ma, M. L. Scudder, T. D. Bailey and I. G. Dance. J. Am. Chem. Soc.. 110, 4863(1988)
Leff, D. V., L. Bvandt, J. R. Heath. Langmair. 12, 4723(1996)
Leland, J. K. and A. J. Bard. J. Phys. Chem.. 91, 5076(1987)
Li, S., S. J. Silvers and M. S. El-Shall. J. Phys. Chem.. B 101, 1794 (1997)
Link, S., M. B. Mohamed and M. A. El-sayed. J. Phys. Chem.. B 103, 3073 (1999)
Littau, K. A., P. J. Szajowski, A. J. Muller, A. R. Kortan and L. E. Brus. J. Phys. Chem.. 97, 1224(1993)
Liu, J., A. Y. Kim, L. Q. Wang, B. J. Palmer, Y. L. Chen, P. Bruinsma, B. C. Bunker, G. J. Exarhos, G. L. Graff, P. C. Rieke, G. E. Fryxell, J. W. Vivden, B. J. Tarasevich, L. A. Chick. Adv. Colloid Interface Science. 69, 131 (1996)
Logunov, S., T. Green, S. Marguet and M. A. El-Sayed. J. Phys. Chem.. A 102, 5652(1998)
Machol, J. L., F. W. Wise, R. C. Patel and D. B. Tanner. Phys. Rev.. B 48, 2819(1993)
Martini, I., G. V. Hartland and P. V. Kamat. J. Phys. Chem.. B 101, 4826 (1997)
Martini, I., J. H. Hodak and G. V. Hartland. J. Phys. Chem.. 102, 9508 (1998)
Micic, O. I., L. Zongguan, G. Mills, J. C. Sullivan and D. Meisel. J. Phys. Chem.. 91, 6221(1987)
Micic, O. I., M. Meglic, D. Lawless, D. K. Sharma and N. Serpone. Langmuir. 6, 487(1990)
Mie, G.. Ann. Phys. 25, 377(1908)
Miller, R. J. D., G. L. McLendon, A. J. Nozik, W. Schmickler and F. Willig. Surface Electron Transfer Processes. VCH: New York (1995)

Mittleman, D. M., R. W. Schoenlein, J. J. Shiang, V. L. Colvin and A. P. Alivisatos. Phys. Rev. B: Condens. Matter **49**, 14435(1994)
Moser, J. E., D. Noukakis, U. Bach, Y. Tachibana, D. R. Klug, J. R. Durrant, R. Humphry-Baker and M. Gratzel. J. Phys. Chem., **102**, 3649 (1998)
Murray, C. B., D. J. Norris and M. G. Bawendi. J. Am. Chem. Soc., **115**, 8706(1993)
Murray, C. B., C. R. Kagan and M. G. Bawendi. Science, **270**, 1335(1995)
Nazeeruddin, M. K., A. Kay, I. Rodicio, R. Humphry-Baker, E. Muller, P. Liska, N. Vlachopoulos and M. Gratzel. J. Am. Chem. Soc., **115**, 6382 (1993)
Nenadovic, M. T., M. I. Comor, V. Vasic and O. I. Micic. J. Phys. Chem., **94**, 6390(1993)
Nguyen, L., R. Kho, W. Bae and R. K. Mehra; Chemosphere, **38**, 155 (1999)
Nisoli, M., S. Stagira, S. De Silvestri, A. Stella, P. Tognini, P. Cheyssac, R. Kofman. Phys. Rev. Lett., **78**, 3575(1997)
O'Neil, M., J. Marohn and G. McLendon. Chem. Phys. Lett., **168**, 208 (1990a)
O'Neil, M., J. Marohn and G. McLendon. J. Phys. Chem., **94**, 4356 (1990b)
O'Regan and M. Gratzel. Nature, **353**, 737(1991)
Pankove, J. I., Optical processes in semiconductors. Dover Publications, Inc. New York (1991)
Parsapour, F., D. F. Kelley, S. Craft and J. P. Wilcoxon. J. Chem. Phys., **104**, 4978(1996)
Patel, A., F. Wu, J. Z. Zhang, C. L. Torres-Martinez, R. K. Mehra, Y. Yang, S. H. Risbud. J. Phys. Chem., B, **104**, 11598(2000)
Perner, M., P. Bost, T. Pauck, G. von Plessen, J. Feldmann, U. Becker, M. Mennig, J. Porstendorfer, M. Schmitt and H. Schmidt. Springer Ser. Chem. Phys., **62**, 437(1996)
Perner, M., T. Klar, S. Grosse, U. Lemmer, G. von Plessen, W. Spirkl and J. Feldmann. J. Luminescence, **76**, 181(1998)
Peterson, M. W. and A. J. Nozik. in Photoelectrochemistry and photovoltaics of layered semiconductors, (ed.) A. Aruchamy (Kluwer Academic, Dordrecht), 297(1992)
Petroski, J. M., Z. L. Wang, T. C. Green and M. A. El-Sayed. J. Phys. Chem., B **102**, 3316(1998)
Reetz, M. T., W. Helbig, S. A. Quaiser, U. Stimming, N. Breuer and R. Vogel. Science, **267**, 267(1995)
Rehm, J. M., G. L. McLendon, Y. Nagasawa, K. Yoshihara, J. Moser and M. Gratzel. J. Phys. Chem., **100**, 9577(1996)
Roberti, T. W., B. A. Smith, J. Z. Zhang. J. Chem. Phys., **102**, 3860 (1995)

Roberti, T. W., N. J. Cherepy, J. Z. Zhang. J. Chem. Phys.. **108**, 2143 (1998)

Rockenberger, J., L. Troger, A. L. Rogach, M. Tischer, M. Grundmann, A. Eychmuller and H. Weller. J. Chem. Phys.. **108**, 7807(1998)

Sandroff, C. J., D. M. Hwang and W. M. Chung. Phys. Re.. B **33**, 5953 (1986)

Sarathy, K. V., P. J. Thomas, G. U. Kulkarni and C. N. R. Rao. J. Phys. Chem.. B **103**, 399(1999)

Schaaff, T. G., M. N. Shafigullin, J. T. Khoury, I. Vezmar, R. L. Whetten, W. G. Cullen, P. N. First, C. Gutierrez-Wing, J. Ascensio and M. J. Jose-Yacaman. J. Phys. Chem.. B **101**, 7885(1997)

Schmid, G., Mater. Chem. Phys.. **29**, 133(1991)

Schmid, G., (ed.). Clusters and Colloids. VCH, New York(1994)

Schneider, T., M. Haase, A. Kornowski, S. Naused and H. Weller. Ber. Bunsenges. Phys. Chem.. **101**, 1654(1997)

Sengupta, A., B. Jiang, K. C. Mandal, J. Z. Zhang. J. Phys. Chem.. B **103**, 3128(1999)

Serpone, N. and E. Pelizzetti (eds.): Photocatalysis, Fundametals and Applications. Wiley, New York (1989)

Serpone, N., D. Lawless and R. Khairutdinov. J. Phys. Chem.. **99**, 16646 (1995)

Skinner, D. E., D. P. Colombo Jr., J. J. Cavaleri and R. M. Bowman. J. Phys. Chem.. **99**, 7853(1995)

Smith, B. A., J. Z. Zhang, U. Giebel and G. Schmid. Chem. Phys. Lett.. **270**, 139(1997a)

Smith, B. A., D. M. Waters, A. E. Faulhaber, M. A. Kreger and J. Z. Zhang. J. Sol-Gel Sci. Tech.. **9**, 125(1997b)

Spanhel, L., M. Haase, H. Weller and A. Henglein. J. Am. Chem. Soc.. **109**, 5649(1987)

Stella, A, M. Nisoli, S. De Silvestri, O. Svelto, G. Lanzani, P. Cheyssac and R. Kofman. Springer Ser. Chem. Phys.. **62**, 439(1996)

Suh, J. S., D. P. DiLella and M. Moskovots. J. Phys. Chem.. **87**, 1540 (1983)

Takagi, H., Y. Yamazaki, A. Ishizaki and T. Nakagiri. Appl. Phys. Lett.. **56**, 2379(1990)

Tolbert, S. H. and A. P. Alivisatos. Science. **265**, 373(1994)

Tolbert, S. H. and A. P. Alivisatos. Ann. Rev. Phys. Chem.. **46**, 595(1995)

Van Buuren, T., L. N. Dinh, L. L. Chase, W. J. Siekhaus and L. J. Terminello. Phys. Rev. Lett.. **80**, 3803(1998)

Vijayakrishnan, V., A. Chainani, D. D. Savma, C. N. R. Rao. J. Phys. Chem.. 96,8679(1992)

Vossmeyer, T., G. Reck, L. Katsikas, E. T. K. Haupt, B. Schulz, H. Weller. Science. **267**, 1476(1995)

Wang, Y., A. Suna, J. McHugh, E. F. Hilinski, P. A. Lucas and R. D. John-

son. J. Phys. Chem., **92**, 6927(1990)
Wang, Y., Acc. Chem. Res., **24**, 133(1991)
Wang, W. X., S. H. Liu, Y. Zhang, B. B. Mei and K. X. Chen. Phys., B **225**, 137(1996)
Weller, H., H.M. Schmidt, U. Koch, A. Fojtik, S. Baral, A. Henglein, W. Kunath, K. Weiss and E. Dieman. Chem. Phys. Lett., **124**, 557(1986)
Weller, H., Angew. Chem. Int. Ed. Engl., **35**, 1079(1996)
Wilcoxon, J.P. and G.A. Samara. Phys. Rev. B **51**, 7299(1995)
Wilenzick, R. M., D. C. Russell, R. H. Morriss and S. W. Marshall. J. Chem. Phys. **47**, 533(1967)
Wu, F., M. S. El-Shall, J. L. Coffer and J. Z. Zhang. to be published. (1999)
Zhang, L. and J.L. Coffer. Chem. Mater. **9**, 2249(1997)
Zhang, L., J.L. Coffer and T.W. Zerda. J. Sol-Gel Sci. Tech., **11**, 267 (1998)
Zhang, J.Z., R.H. O'Neil, T.W. Roberti, J.L. McGowen and J.E. Evans. J. Phys. Chem., **98**, 3859(1994a)
Zhang, J.Z., R.H. O'Neil, T.W. Roberti, J.L. McGowen and J.E. Evans. Chem. Phys. Lett., **218**, 479(1994b)
Zhang, J.Z., B.A. Smith, A.E. Faulhaber, J.K. Andersen and T J. Rosales Ultrafast Processes in Spectroscopy IX, 561(1996)
Zhang, J.Z., Acc. Chem. Res., **30**, 423(1997)
Zheng, J.P., L. Shi, F.S. Choa, P.L. Liu and H.S. Kwok. Appl. Phys. Lett., **53**, 643(1988)
Zhou, H.S., I. Honma and H. Komiyama. J. Phys. Chem., **97**, 895(1993)
Ziolo, R.F., E.P. Giannelis, B.A. Weinstein, M.P. O'Horo, B.N. Ganguly, V. Mehrotra, M.W. Russell and D.R. Hoffman. Science, **257**, 219 (1992)

7 Magnetic Characterization

Adam J. Rondinone and Z. John Zhang

7.1 Introduction

Magnetic materials have been fascinating human beings for over 4000 years. Since a shepherd noticed that the iron nails in his shoe and the iron tip of his staff stuck to certain rocks in Magnesia of ancient Greece, magnetic materials have found their way into almost every part of our civilization. In our modern society we use magnetic materials daily, such as in computer disks, credit and ID cards, speakers, refrigerator door seals, and a host of other conveniences. Lodestone, the type of rock that attracted attention in ancient Greece consists mainly of magnetite (Fe_3O_4) and is probably the first known permanent magnet. Over the years, a vast number of magnetic materials have been developed and huge industries have been based upon magnetic materials. The discovery of giant magnetoresistance (GMR) in magnetic thin films with superlattice structures in 1988 has attracted broad attention (Baibich, et al., 1988; Parkin, 1995). Subsequently, GMR was observed in several magnetic granular systems consisting of the particles of Fe, Co, Ni, or their various alloys in Cu, Ag, or Au matrices (Berkowitz, et al., 1992; Xiao, et al., 1992; Chien, 1995). The discover of gigantic magnetoresistance in so called colossal magnetoresistance materials of manganate perovskites in 1993 has further rejuvenated the studies on magnetism and magnetic materials (von Helmolt, et al., 1993; Rao, et al., 1996). Despite the newly surged interests in magnetic materials, the researches on magnetic nanocrystalline materials still far lag behind the semiconductor nanocrystal researches.

Magnetic nanoparticles are of great interest in fundamental science and in technological applications. In magnetic materials, unpaired electrons on individual atoms and ions create magnetic moments. Magnetic properties of materials are fundamentally determined by the magnetic couplings at the atomic level (Leslie-Pelecky and Reike, 1996). The exchange or superexchange couplings between two adjacent magnetic moments determine the magnetic order in the materials. If such interaction is weak, the Heisenberg exchange energy is too small to overcome thermal activation. Consequently, the material is paramagnetic and there is no magnetic order in the material (Fig. 7.1(a)).

When two adjacent magnetic moments are spontaneously aligned parallel

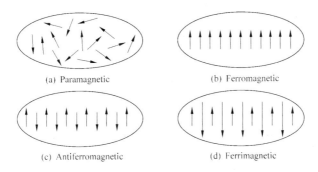

Figure 7.1 Four general types of magnetic order

below a critical temperature-the Curie temperature T_c, there is a ferromagnetic order and the material is ferromagnetic (Fig. 7.1(b)). When two neighboring magnetic moments are antiparallel aligned, there is an antiferromagnetic order in the material. The material is antiferromagnetic if these two adjacent magnetic moments have the same magnitudes and the net magnetic moment is zero (Fig. 7.1(c)). If two adjacent magnetic moments have different magnitudes, the material is called ferrimagnetic and there are net magnetic moments in the material (Fig. 7.1(d)). In addition to these couplings between electron spins, there are couplings between electron spin, S and the angular momentum, L of the orbital, in which the electron locates. Usually, magnetic materials that we use for various applications are ferromagnetic or ferrimagnetic materials. Ferrimagnetic materials can have very strong magnetic behavior similar to ferromagnetic materials. Magnetic metal oxides, in general, are ferrimagnetic materials.

Two key properties of magnetic materials are the Curie temperature and the magnetization hysteresis. The exchange coupling and therefore the Heisenberg exchange energy is directly related to the T_c of ferro- and ferrimagnetic materials. Below T_c, the net magnetic moments point to the same directions, which are along specific crystallographic zone axes. Such preferred directions are called magnetization "easy" axes, which are the results of the couplings between electron spin and the angular momentum of the electron orbital. Due to the existence of the easy axes, the material system is anisotropic to magnetization from an applied external magnetic field. When direction of an applied magnetic field is not along one of the easy axes, the applied magnetic field has to overcome the energy barrier(s) to pull all the magnetic moments away from the easy axes and point them to the field direction (Tejada, et al., 1996). This energy barrier is known as the magnetocrystalline anisotropy, E_A and is the atomic origin of magnetization hysteresis behavior of magnetic materials. It is of fundamental importance to the magnetic properties of the material, and determines the suitability of magnetic materials for specific applications.

In principle, magnetic properties of materials can be understood and controlled through magnetic couplings, and such couplings are closely related to the chemical composition and structure of magnetic materials. However, we have not acquired a clear understanding of the relationships between the magnetic properties of materials, their chemical compositions and crystal structures at the atomic level. To date, magnetic research has predominantly been conducted on bulk magnetic materials. It is difficult to directly correlate the magnetic properties of bulk materials with the magnetic couplings at the atomic level due to the formation of magnetic domain structures in bulk materials. When spontaneous magnetization occurs below T_c in magnetic materials, magnetic moments align with each other locally. However, magnetic domains appear as a consequence of reducing the magnetostatic energy, which comes from the energetically unfavorable magnetic flux leakage out of the material. Within each domain, the atomic moments align to the same direction and are spontaneously magnetized. Different domains have different magnetization directions and are separated by domain walls. The domain structure and the width of domain wall are closely related to the magnetocrystalline anisotropy of the material. When an external magnetic field is applied, the domain walls move through the material and the magnetization directions of all the domains are gradually converted to the field direction. Hysteresis behavior of magnetization in bulk materials is a process of domain wall displacement and subsequent realignment of the domain structure. The displacement of the domain wall is a very complex issue in bulk materials. In addition to the magnetocrystalline anisotropy as an intrinsic factor, the displacement of the domain wall is greatly affected by many extrinsic factors including grain boundaries, crystal defects, lattice strain and chemical impurities. Since these extrinsic factors are inevitably present in bulk magnetic materials, it is very difficult to correlate the magnetic properties with magnetic couplings and to understand them from their atomic origins in bulk magnetic materials. However, magnetic nanoparticle systems are very promising candidates for systematically studying the relationships between the magnetic properties and magnetic couplings at the atomic level.

Magnetic nanoparticles have a single magnetic domain structure. A multi-domain structure is energetically unfavorable due to the small particle size. Without the existence of the domain wall, atomic level magnetic couplings can be directly correlated with the magnetic properties of the nanoparticles, especially with the magnetocrystalline anisotropy. Surely, understanding and controlling magnetic properties of nanoparticles will elucidate the mechanisms of magnetic properties of materials and facilitate the design and control of magnetic materials.

Magnetic measurement of nanoparticles has some unique features due to the drastically reduced dimensions of materials, and consequently the occurrence of superparamagnetism (Aharoni, 1992). Superparamagnetic properties of nanoparticles are directly associated with the magnetic anisotropy in the nanoparticles. The magnetic anisotropy energy is minimized when the magnetization of the nanoparticle aligns with its easy axis. Using the Stoner-Wohlfarth theory, the magnetocrystalline anisotropy E_A of a single-domain particle can be approximated as:

$$E_A = KV\sin^2\theta \qquad (7.1)$$

where K is the effective anisotropy energy density, V is the volume of the nanoparticle, and θ is the angle between the magnetization direction and the easy axis of the nanoparticle (Stoner and Wohlfarth, 1991). In magnetic nanoparticles with a spherical shape, the magnetocrystalline anisotropy can be approximated as the total magnetic anisotropy. This anisotropy serves as the energy barrier to prevent the change of magnetization direction. When the size of ferromagnetic or ferrimagnetic nanoparticles is reduced below a threshold value, E_A becomes comparable with thermal activation energy, $k_B T$ with k_B as the Boltzmann constant. With a small anisotropy energy barrier, the magnetization direction of the nanoparticles can be easily moved away from the easy axis by thermal activation energy and/or an external magnetic field. If thermal activation is sufficiently high E_A is overcome in all directions and the magnetic moment may randomly point in any direction. Consequently, the collective behavior of the magnetic nanoparticles is the same as that of paramagnetic atoms. Although the magnetic order still exists in the nanoparticles, each particle behaves like a paramagnetic atom but with a giant magnetic moment. Such behavior is known as superparamagnetism. In a superparamagnetic state, the magnetization direction of the nanoparticle rapidly fluctuates instead of fixing along certain direction. The temperature at which the magnetic anisotropy energy barrier of a nanoparticle is perpetually overcome by thermal activation energy is known as the blocking temperature.

In this chapter, four physical methods for studying the magnetic properties of nanoparticles are discussed including superconducting quantum interference device (SQUID) magnetometry, Mössbauer spectroscopy, neutron diffraction, and Lorentz microscopy with a transmission electron microscope (TEM). Certainly, the discussion here on these methods are only intended to serve as an introduction to the magnetic measurement of nanoparticles. Interested readers are strongly encouraged to refer to books that are devoted to each method.

7.2 SQUID Magnetometry

A magnetic field is described by magnetic flux lines. The strength of the field, H corresponds to the density of the flux line in vacuum. H is a vector indicating the strength and the direction of the magnetic field. When a magnetic material is put into an external magnetic field, the material is magnetized by aligning its magnetic moments along the direction of the external magnetic field. The density of flux lines in the material is represented by magnetic induction, B that has a relationship with the magnetic field, H and the magnetization of the material, M as:

$$B = H + 4\pi M \qquad (7.2)$$

where B expresses the strength of the local magnetic field in the material, B/H is the permeability, μ of the material and M/H is the magnetic susceptibility, χ (Hirst, 1997). The units for B are Gauss (G) and Tesla (T) with 1 T = 10 000 G. H has the unit of Oersted (Oe). An Oersted and a Gauss have the same dimensions. In practice, Gauss is usually used for both B and H. The equivalent unit for M is the volume magnetization represented by the total magnetic moments of the sample divided by the volume of the sample. The reported unit for magnetic moment is called the electromagnetic unit, emu in shorthand during the measurement by using a SQUID magnetometer. Therefore, M has units of emu/cm^3, and 1 emu/cm^3 corresponds to $(1/4\pi)$ G. For the measurement on magnetic nanoparticles, it is difficult to obtain the volume of a nanoparticulate sample. The commonly encountered unit for M is emu or emu/gram. The unit is also generally converted to emu/ml by assuming the density of the bulk magnetic materials.

The SQUID magnetometer is arguably the most popular type of magnetometers. It has the distinction of being the most sensitive magnetometer. The working mechanisms of the SQUID are based on two fundamental principles of superconductivity (Gallop, 1990). One is the quantization of the magnetic flux in a superconducting ring. The other one is the Josephson effect.

In 1962, Brian Josephson at the University of Cambridge hypothesized that superconducting electron pairs, Cooper pairs, can tunnel from a superconductor through a thin layer of an insulator into another superconductor. This Josephson effect can be understood from the propagation of electron waves. Superconducting current is phase coherent, meaning that all conducting electrons within a superconducting material can be described by a single wave function. If the insulating layer between two superconductors is thin enough, the electron wave in one superconductor can penetrate the insulating layer and overlap with the electron wave in the other superconductor. Consequently, the Cooper pairs tunnel through the insulating junction.

For a relatively easy understanding of the mechanism of the Josephson junction and SQUID, we here discuss the case of a SQUID consisting of two identical Josephson junctions, probably the easiest SQUID configuration to be understood. More rigorous discussions can be found in solid-state physics and superconductivity books (Weinstock, 1996).

The total supercurrent, I_T is divided into I_A and I_B, which tunnel through junction A and B respectively (Fig. 7.2).

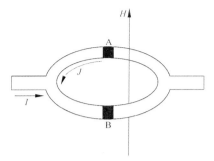

Figure 7.2 DC SQUID. Field H produces a persistent current J, which adds to current I/2 at junction B and subtracts from I/2 at junction A

$$I_T = I_A + I_B \tag{7.3}$$

When a magnetic field is applied, a magnetic flux Φ goes through the superconducting loop and induces a persistent current J. The induced current will add onto I_A while subtracting from I_B. When the current waves flow through the junctions, their phases shift by δ_A and δ_B, respectively.

$$I_A = I_o \sin \delta_A \tag{7.4}$$
$$I_B = I_o \sin \delta_B \tag{7.5}$$

where I_o is the maximum allowable current flowing through the junction. It is well known that the magnetic flux going through a superconducting loop is quantized. Consequently, there is quantum interference between I_A and I_B after passing the junctions. The total supercurrent of the SQUID, I_T changes periodically with the variation of magnetic flux. The maximum amplitude of the total supercurrent at a given flux can be identified as the critical current of the SQUID, I_c.

$$\delta_A - \delta_B = 2\pi \Phi / \Phi_o \tag{7.6}$$
$$I_c = 2I_o \cos[(\delta_A - \delta_B)/2] = 2I_o \cos(\pi \Phi / \Phi_o) \tag{7.7}$$

where Φ is the quantum of flux passing through a superconducting ring and $\Phi_o = 2.0678 \times 10^{-7}$ Gauss-cm^2. The critical current in this two-junction SQUID becomes a function of the magnetic flux passing through the loop. The SQUID becomes a magnetometer used to determine the magnetic flux by measuring the critical current in the superconducting loop.

The SQUID is coupled to the superconducting pickup coils through induction. The loop of the pickup coils encircles the magnetic sample. The magnetic field from the sample is inductively coupled to the pickup coils when the sample moves through them. The detection coils measure the local changes in magnetic flux density due to the magnetic field from the sample.

The coil is fashioned as a gradiometer and contains three opposing loops (Fig. 7.3). The first and third loops contain one clockwise turn each, the

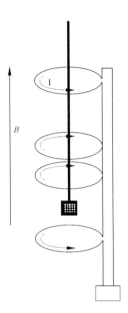

Figure 7.3 Gradiometer. External field B induces a local field in the sample, which generates signal at individual coils during movement. The signals from background and the applied field are cancelled by the 2+2 opposing coils on the top and bottom

center loop contains two counterclockwise turns. The two plus two opposing coils serve to cancel the signal from any field that is homogenous over the three loops, such as any background magnetic field that may be present (Clarke, 1994). Any field that is inhomogeneous to the three loops, such as the field from a small sample, generates a signal that is not cancelled. When coupled to a superconducting pickup coil the SQUID is a tremendously sensitive device for detecting magnetic field. Certainly, many auxiliary technologies are required for a complete magnetometer.

The magnetic properties of materials are usually determined by studying the response of materials to an applied magnetic field. A SQUID magnetometer contains a high field electromagnet capable of producing homogeneous and static fields of several Tesla. In order to generate high fields efficiently, SQUID magnetometer utilizes a superconducting coil made of either niobium-tin or niobium-titanium. Both materials require liquid helium temperatures to operate a high field. The instrument contains the electronics for both charging the superconducting magnet and measuring the exact field strength. Both the sample and the detector loop reside inside the magnetic core. The gradiometer design of the pickup coils also cancels the signal of the superconducting magnet, leaving only the sample signal.

We here use the magnetic studies on $MgFe_2O_4$ spinel ferrite nanoparticles to illustrate the typical SQUID magnetometer measurements. The $MgFe_2O_4$ nanoparticles have been prepared by coprecipitation and subsequent thermal annealing. The sample is cooled at first to 1.7 K without a magnetic field present. After a magnetic field of 100 G is applied, the magnetic susceptibility is measured as the measuring temperature slowly rises. The magnetic susceptibility of $MgFe_2O_4$ nanoparticles measured from 1.7 K to 300 K is shown in Fig. 7.4 as a function of temperature.

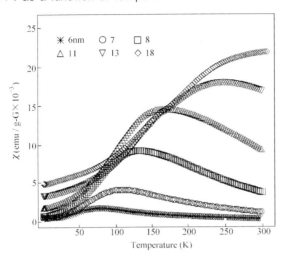

Figure 7.4 Rising blocking temperature vs. increasing particle size. Blocking temperature is encountered at maximum magnetization

Initially, the magnetic susceptibility increases. Upon reaching a certain temperature, the susceptibility starts to decrease with increasing temperature behaving as Curie-Weiss paramagnetic materials. Although these nanoparticles show a paramagnetic behavior at 300 K, they have the same antiferromagnetic order that the $MgFe_2O_4$ bulk materials possess. Therefore, these nanoparticles are typical superparamagnetic materials. The temperature at which the magnetic susceptibility starts to decrease is the blocking temperature T_B. The blocking temperature correlates with the nanoparticle size as the inset in Fig. 7.4 displays. The Curie transition temperature of this nanoparticle system is around 600 K, which has been determined by high-temperature magnetic susceptibility measurements.

The temperature dependent magnetic susceptibility of $MgFe_2O_4$ nanoparticles also depends upon the sample cooling conditions in the SQUID magnetic field. A typical magnetic behavior of these $MgFe_2O_4$ nanoparticles is represented by a sample with 11 nm size (Fig. 7.5). When the sample is cooled from room temperature to 1.7 K without an external magnetic field (zero field cooled; ZFC), its magnetic susceptibility initially increases as temperature is increased from 1.7 K, and then starts to decrease at the blocking temperature (plot a in Fig. 7.5). When the same sample is cooled under a magnetic field (field cooled; FC), its susceptibility is highest at

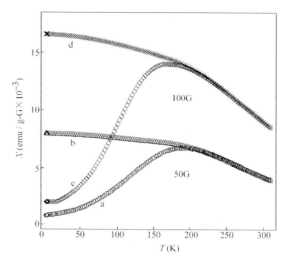

Figure 7.5 Zero-field-cooled susceptibility (a + c) and field-cooled susceptibility (b + d)

1.7 K. It decreases as temperature increases, and overlaps with the results obtained from the zero field cooling measurement when temperature rises above the blocking temperature (plot b in Fig. 7.5). The blocking temperature changes as the magnitude of applied magnetic field changes. Figure 7.5 displays the different temperature-dependent changes of magnetic susceptibility at 50 Gauss field (plot a and b) and 100 Gauss field (plot c and d).

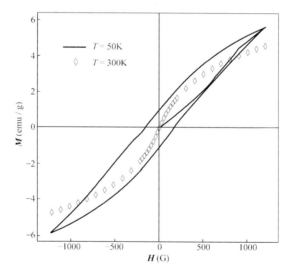

Figure 7.6 Magnetic hysteresis at 50 K and 300 K

Below the blocking temperature, the magnetization of $MgFe_2O_4$ nanoparticles displays hysteresis as the applied magnetic field changes strength and direction (plot a in Fig. 7.6). The $MgFe_2O_4$ nanoparticles with 11 nm diameter have a coercivity of 165 Gauss. When the temperature of magnetization measurement is raised above the blocking temperature, the hysteresis disappears, and the nanoparticles show a paramagnetic-like magnetization (plot b in Fig. 7.6).

The results of the magnetic measurements clearly show the superparamagnetic features of these $MgFe_2O_4$ nanoparticles. Since the blocking temperature indicates the threshold point of thermal activation, the magnetocrystalline anisotropy of the nanoparticles is overcome by thermal activation starting at T_B and the magnetization direction of each nanoparticle simply follows the applied field direction. Consequently, the nanoparticles show paramagnetic properties. Below the blocking temperature, thermal activation is no longer able to overcome the magnetocrystalline anisotropy of the nanoparticles. Hence, the magnetization direction of each nanoparticle rotates from the field direction back to its own easy axis without any movement of the nanoparticle. Since the nanoparticles and consequently their easy axes are randomly orientated, overall susceptibility is reduced with decreasing temperature as Fig. 7.5 shows. Larger particles possess a higher E_A and require a larger K_BT to become superparamagnetic. Therefore, T_B increases with increasing particle size.

When the sample is cooled under a magnetic field, the direction of the magnetic moment of each nanoparticle is frozen to the field direction as the temperature decreases below the blocking temperature. As a result the field-cooled sample shows the highest susceptibility at 1.7 K. When a stronger magnetic field is applied, the magnetization direction of the nanoparticles becomes easier to switch to the field direction, and less assistance is required from the thermal activation. Consequently, the peak (T_B) in $\chi(T)$ plot shifts to a lower value with a larger applied field (Fig. 7.4). Since the magnetic anisotropy blocks the change of magnetization direction at $T < T_B$, a certain magnetic field is required to change the magnetization direction. Therefore, the magnetic nanoparticles have shown coercivity and hysteretic behavior. When the temperature rises above the blocking temperature, the energy barrier formed by the magnetic anisotropy is overcome within the measurement time and the magnetization direction of the nanoparticles follows the field direction without any hindrance.

Clearly, the magnetometer can provide a great deal of information on the magnetic properties of nanoparticles. By combining the SQUID measurement with some mathematical models and other techniques, quantitative information such as magnetic anisotropy distribution and magnetocrystalline anisotropy constant can be obtained.

7.3 Mössbauer Spectroscopy

Resonant absorption is the fundamental working mechanism in modern spectroscopic methods. Photons from a radiation source are absorbed by samples inducing a change in state between two energy levels that matches energy of the incident photon. When the energy transitions are between the energy states of a nucleus, and the photons are in the γ-ray region, such resonant absorption is known as Mössbauer spectroscopy (Greenwood and Gibb, 1971; Bancroft, 1973). The foundation of Mössbauer spectroscopy is nuclear gamma resonance, which was discovered by Rudolph Mössbauer in 1957.

In a Mössbauer spectrometer, the source generating γ-rays is a radioactive nucleus. When a transition with an energy E_t occurs in the source nucleus from the excited state E_e to the ground state E_g, γ-rays are emitted with the energy $E_γ$. Since the energy levels are distinct in each type of nucleus, only the same type of nucleus in the sample can resonantly absorb the γ-rays from the source. Therefore each radioactive source is limited to only studying the samples containing the same nucleus as the source. For instance, Mössbauer spectroscopy studies on iron compounds require an ^{57}Fe source, and studies on tin compounds require a ^{119}Sn source. There are more than 45 elements in which nuclear gamma resonance has been observed. Practically, however, only a limited number of these elements can be used in Mössbauer spectroscopy due to extreme difficulties in experimental setup. To date, most Mössbauer spectroscopy studies have been done with ^{57}Fe, and to a lesser extent with ^{117}Sn.

During the development of nuclear gamma resonance one of the major concerns was the recoil process associated with emission and absorption of γ-ray photons. When a high-energy γ-ray photons is emitted, a free nucleus usually recoils with a recoil energy E_R. The overall energy has is conserved during the transition.

$$E_e = E_g + E_γ + E_R \qquad (7.8)$$
$$E_t = E_e - E_g = E_γ + E_R \qquad (7.9)$$

If the free nucleus has a mass M and a recoil velocity v, the recoil energy E_R can be expressed as:

$$E_R = 1/2 Mv^2 \qquad (7.10)$$

The momentum of the recoiling nucleus is Mv The γ photon has a momentum of mc, which is expressed by γ-ray energy as:

$$mc = mc^2/c = E_\gamma/c \qquad (7.11)$$

where m is the mass of photon and c is the speed of light. The momentum should be conserved for the recoil process.

$$Mv = mc \qquad (7.12)$$

Therefore, the energy of the recoil may be expressed in terms of the γ-ray energy:

$$E_R = 1/2 Mv^2 = 1/2(Mv)^2/M = 1/2(E_\gamma/c)^2/M = E_\gamma^2/2Mc^2 \qquad (7.13)$$

For the nucleus in the sample resonantly absorbing the γ-rays, the same recoil-like process also occurs. The overall energy conservation for the resonant adsorption in the sample becomes:

$$E'_e + E_R = E'_g + E_\gamma \qquad (7.14)$$
$$E'_t = E'_e - E'_g = E_\gamma - E_R \qquad (7.15)$$

When the γ-rays reach the sample, the photon may impart a large recoil energy on a sample nucleus. Consequently, the energy for the resonant transition, $E_\gamma - E_R$ may be below the threshold of the resonant adsorption E'_t. For a free atom, the nucleus can recoil without hindrance and E_R is too large to observe the nuclear gamma resonance.

The Nobel laureate Rudolph Mössbauer demonstrated that a significant fraction of γ-rays can be emitted and absorbed in solids without recoil. Therefore, E_R is negligible and a strong nuclear gamma resonant absorption occurs in the samples at a solid-state form. The recoil free nuclear emission and absorption of a high-energy photon has come to be known as the Mössbauer effect. The successful observation of the Mössbauer effect can be considered due to the solid surroundings of the nuclei. The emitting and absorbing nuclei are rigidly bound in solids, which behaves as a lattice. The lattice can vibrate and the recoil energy can excite the lattice vibration. However, the vibrational energy is quantized and lattice vibration may only occur in discrete energy levels. Hence the recoil energy has to be high, matching with these discrete levels in order to excite the lattice vibration. The recoil energy is seldom sufficient to excite quantized vibrations in the lattice. When the recoil energy may not excite quantized vibration in the lattice, the recoil mass M must then be taken as the entire crystal rather than the single atom. The large increase in M causes a proportional decrease in the recoil energy E_R, yielding a significant fraction of recoil free emissions and absorptions. This recoil free fraction of the transition is known as the Mössbauer fraction. Its value depends upon the type of nucleus and the solid. In general, it decreases as the energy of γ-ray increases.

Mössbauer spectroscopy is unique among the spectroscopic methods because it detects transitions within nuclear, rather than electronic, energy states. It therefore yields information about nuclear excited states, electronic-nuclear interactions, nuclear quadrupole moments and magnetic field effect on the nucleus. Temperature and time-resolved experiments may also yield information about magnetic relaxation times and kinetics of structural change. For instance, gravitational experiments using Mössbauer spectroscopy have given experimental support to Einstein's theory on gravitational red shift. But for the purposes of this chapter we restrict the discussion to Mössbauer spectroscopy's contribution to magnetic studies of iron-based materials.

Mössbauer spectrometers, like other types of spectroscopic techniques, require such basic components as a source, an absorber and a detector. Because of the nature of the experiment one also requires a computer and a multichannel analyzer to sort the data. Other accessories may be added for high-and low-temperature studies or X-ray backscattering experiments. An entire Mössbauer spectroscopic setup including linear motor and source (A), detector (B), amplifier (C), linear gate (D) and controller (E) is represented schematically in Fig. 7.7.

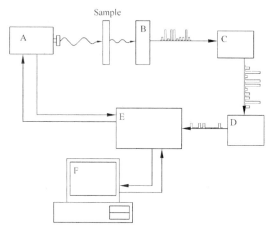

Figure 7.7 Schematic diagram of Mössbauer spectroscopy instrumentation. Linear motor with source (A) is driven by controller (E). Detector (B) and linear amplifier (C) collect data. Discriminator (D) removes unwanted data. Controller collects data and stores in computer (F)

Furthermore, a cryogenic dewar or a furnace may be used for low and high temperature experiments, respectively.

Studies involving magnetic materials containing iron species utilize ^{57}Fe Mössbauer spectroscopy. The source for ^{57}Fe studies is ^{57}Co element bound in a palladium matrix. ^{57}Co has a half-life of 272 days. It transmutes to an excited ^{57}Fe energy state through a nuclear β-decay of electron capture. The excited ^{57}Fe nucleus may decay to the ground state in one of two paths, more than 90% of the decay goes through the path that releases γ photons with an energy of 14.4 keV (Fig. 7.8).

Resonant absorption spectroscopy records the absorption peaks at

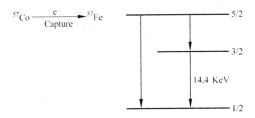

Figure 7.8 Nuclear process in the source of ^{57}Fe Mössbauer spectroscopy. ^{57}Co nucleus captures an electron to form ^{57}Fe in the 5/2 excited state. Possible decay routes include 5/2→1/2 and 5/2→3/2→1/2. 3/2→1/2 decay releases 14.4 KeV gamma ray

various energies. As in all spectroscopies, this technique requires a source to cover a distinct range of energies. The gamma photons released in nuclear decay transitions have specific energies, such as 14.4 keV from the ^{57}Fe source. The energy of gamma photons must be varied to create an energy spectrum in order to conduct spectroscopy. Energy variation is achieved by using the Döppler effect, a phenomenon originally discovered with sound waves in 1842 but applicable to light. According to the Döppler effect, the movement of a radiation source with respect to an absorber or detector will alter the energy of a photon by:

$$\Delta E = (v/c) E_\gamma \qquad (7.16)$$

where v is the velocity of the source respect to absorber. In a typical Mössbauer experiment, the source is vibrated at a velocity not exceeding 100 mm/s. Döppler shifting of the gamma photons from a ^{57}Fe source yields a change in energy of 4.8×10^{-6} eV, corresponding to a change of one part in ten billion. Although the energy difference is subtle, it is sufficient to record the change of the nuclear energy states due to the change the electron density on the nucleus.

Döppler shifting is achieved by mounting the gamma photon source on a suitable moving object of which the velocity is well known. Modern Mössbauer spectrometers utilize an electromagnetic linear motor to achieve the Döppler vibration. The linear motor is driven by a controller. It also contains a feedback loop, allowing the controller to observe the velocity profile of the motor and correlate it to resonant absorption data in a computer. The velocity profile of the motor usually is not accurate enough to be used to calibrate spectra. One must use either a standard absorber or an absolute velocity measurement device. Typically, the standard absorber is sodium nitroprusside [$Na_2Fe(CN)_5NO \cdot 2H_2O$], iron foil, or α-Fe_2O_3. The first standard gives two well-defined resonant absorption peaks and the other two standards produce six well-defined peaks. All these peaks may be used to generate an accurate velocity profile to apply to unknown samples.

Absolute velocity measurement utilizes an optical interferometer, which measures the instantaneous velocity based on destructive interference fringes. Interference fringes are generated by mounting the mobile mirror of the interferometer onto the back of the linear motor armature. Absolute velocity is easily calculated using fringes per unit time and an accurate timing circuit. By convention, all Mössbauer data are reported as absorbance versus source velocity in mm/s.

There are several types of Mössbauer detector including scintillation and gas-filled proportional counters. For the purposes of ^{57}Fe spectroscopy, one would prefer a krypton-filled gas proportional counter. This simple counter contains an anode and a cathode in the presence of krypton gas. Gamma photons ionize the krypton allowing free electrons to accelerate toward the anode. As the electrons accelerate, they strike other krypton atoms, causing more ionization and quickly forming an electron cascade, which registers at the anode as a voltage pulse. Detectors of this type maintain a positive 1—2 kV potential at the anode and must have a sensitive pre-amplifier.

Other gamma photons with higher energies are also released by the decaying ^{57}Fe nucleus (Fig. 7.8 decay diagram). Gamma absorption by the sample and by the parts of the equipment near to the source may produce stray X-ray photons. Gas proportional counters produce a pulse proportional in voltage to the energy of the incoming photon. A linear gate after the amplifier may measure the amplitude of each pulse and block signals from photons of an energy not near 14.4 keV. Without the linear gate, the signal to noise ratio would be so high that the absorption signal would be drowned out.

For a resonant absorption signal in a ^{57}Fe Mössbauer spectrum, the energy transition takes place from the ground state with a nuclear angular momentum of 1/2 to an excited state with a nuclear spin of 3/2. The energy of these nuclear states is strongly affected by the interaction between the charge distribution on the nucleus and the electrons that have a finite probability to reside on the nucleus. According to the fundamental concepts of quantum mechanics, only the electrons in s-orbital will be found on the nucleus, and therefore the energy shift of the nuclear states correlates with the density of s-electrons. Due to the strong screen effects of p- and d-electrons to the s-electrons, the variation of the electron density in valence shell also changes the energy of the nuclear states. The different extent of overlap between the nuclear charge density and the s-electron density causes the resonant absorption peak to shift as Fig. 7.9 shows. This shift is known as the isomer shift in Mössbauer spectroscopy.

The isomer shift is highly dependent upon the chemical environment, namely the oxidation state of the element and its bonding with the ligands. The magnitude of the shift is also affected by other factors such as pressure and temperature. In routine experiments, we only consider the variation of chemical environment. The isomer shift enables ^{57}Fe Mössbauer spectroscopy to be used to discriminate between ferric, ferrous cations and iron atoms. Usually, the three standard samples mentioned above are treated as zero-shift references. In general, ferrous ions produce a large shift, and ferric ions have a small shift. Iron metal gives a minute shift. Unfortunately, oxidation state of iron species may only be determined by judging the spectrum against similar samples. No atomic or ionic model exists which may simply take the magnitude of the shift and determine the oxidation state.

A classic and useful example of chemical isomer shift is the spectrum of magnetite, Fe_3O_4. Chemically, Fe_3O_4 is $FeO - Fe_2O_3$ containing both ferrous and ferric ions. The Mössbauer spectrum reflects two patterns, one for each cation. FeO has an isomer shift larger than Fe_2O_3.

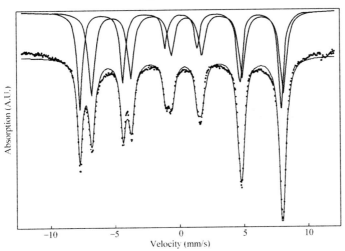

Figure 7. 9 Mössbauer spectrum of Fe_3O_4. $Fe^{2+} + Fe^{3+}$ sites overlap to generate the pattern for Fe_3O_4. Sites may be discriminated by a difference in chemical shift, as Fe^{2+} typically has a higher chemical shift

One unique and fundamental property of Mössbauer spectroscopy is the ability to detect and quantify quadrupole moments of the nucleus (Long and Grandjean, 1993). Any nucleus with a spin greater than 1/2 will have an aspherical charge distribution. If such a nucleus is in an electric field gradient (EFG), the degeneracy is removed between the 3/2 and 1/2 spin states in an excited ^{57}Fe nucleus. This translates into a difference in transition energy between the $1/2_g \rightarrow 1/2_e$ and the $1/2_g \rightarrow 3/2_e$ transition. As a result, within the Mössbauer spectrum a single absorption peak is split into two peaks, the center of which is the position of the original peak and is used to identify the isomer shift (Fig. 7.10).

The quadrupole splitting is reported as the magnitude of the split between the two peaks, in mm/s. The origin of the EFG is predominantly the asymmetric electronic orbital surrounding the nucleus. Point charges within the crystal lattice provide a minor contribution. Careful experimentation may yield information about the distribution of the electronic orbitals as a function of

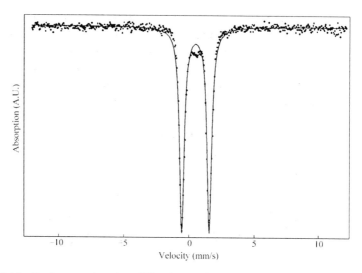

Figure 7.10 Nuclear quadrupole splitting by an electric field gradient at the nucleus

ligand field strength and ligand position. The existence of an EFG has been useful in many Mössbauer studies for helping to solve structure, ligand position and even behavior of unbound electron pairs. However, it does not have much impact on magnetic studies of materials.

The true usefulness in Mössbauer spectroscopy with respect to magnetic materials lies within the magnetic hyperfine interactions. The Zeeman effect indicates that a magnetic field at the nucleus will remove degeneracy from each of the ground and excited nuclear spin states. This magnetic hyperfine interaction between the nuclear energy states and a magnetic field creates two distinct ground ($+1/2$, $-1/2$) and four distinct excited ($+3/2$, $-3/2$, $+1/2$, $-1/2$) states with a possible eight transitions. Because transitions are only allowed when $\Delta m = 0$ or ± 1, there are six possible transitions. The spectrum for an Fe nucleus under a magnetic field will therefore have six equidistant absorption lines corresponding to the six transitions and provide the magnetic hyperfine structure. The center of the six lines is the position of the unsplit peak, and the magnitude of the sum of the six peaks in the sextet will equal to the magnitude of the unsplit peak. The presence of a sextet is an indication that the nucleus is experiencing a magnetic field (Fig. 7.11).

The magnetic field may be internal or external in nature. Internal magnetic fields come from the magnetic moments at the lattices. Magnetic moments originate from two major sources, the electronic angular momentum, or spin, and the nuclear-electronic spin interactions. The electronic angular momentum contribution is much larger than the nuclear interaction contribution. Overall, the internal field is estimated as the sum of three terms:

$$H = H_s + H_l + H_d \qquad (7.17)$$

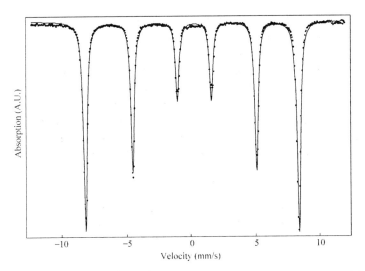

Figure 7.11 Nuclear magnetic hyperfine splitting. Presence of a magnetic field at the nucleus splits the quadrupole doublet into a hyperfine sextet

where H_s is a nuclear-s-orbital interaction term, H_l is the contribution from the electronic angular momentum and H_d is the interaction between nuclear dipole and electronic angular momentum. In materials where two distinct magnetic lattices exist, two distinct sextets will be present in the spectrum, allowing calculation of individual lattice site magnetic moments. This information differs from that provided by magnetometry because magnetometry only provides the average magnetization of the sample. This has been particularly useful in verifying magnetic moments in true antiferromagnetic materials, which exhibit few magnetic properties but are still magnetically structured. The magnitude of the internal field is reflected in the spread of the sextet in the Mössbauer spectrum. Large internal fields may approach 600 kG, and will result in a spread of over 16 mm/s.

The magnetic moments in magnetic materials can change their orientation. Each magnetic moment in a paramagnet changes its direction continuously and quickly. In magnetic materials with strong magnetic order such as ferro- or ferrimagnets, all the net magnetic moments in a single domain point to the same magnetization direction. This magnetization direction can also be changed. By convention the reorientation of magnetic moments is termed relaxation. The average time between reorientations is termed the relaxation time τ. The measurement time in Mössbauer spectroscopy is equal to the Larmor precession period of the nucleus, which is about 10 ns for ^{57}Fe. If the magnetization direction of the sample is relaxing slower than 10 ns, Mössbauer spectrometer will be able to observe the magnetic hyperfine interaction and provide a spectrum with a sextet peak. If the relaxation time is shorter than 10 ns, the time average for the internal magnetic field becomes zero during Mössbauer measurement. The absorption peaks in the Mössbauer spectrum will only be a doublet.

French Nobel laureate Louis Néel (1949), while working with small magnetic particles, learned that the relaxation time is heavily dependeant on the domain size of the material being studied. This was achieved using small magnetic single domain particles, which were equal to or less than the volume of an average domain in the bulk material. Néel proposed an equation that modeled the relaxation time as a function of several variables:

$$\tau = \tau_0 \exp(KV/k_B T) \qquad (7.18)$$

where τ_0 is a constant, V is the volume of the particle, k_B is the Boltzmann constant, T is the temperature and K is the magnetocrystalline anisotropy constant. K is an intrinsic value for any magnetic material, and is of great importance for the applications of magnetic materials.

Néel's formula was developed using Mössbauer spectroscopy. In the superparamagnetic nanoparticles, the magnetization direction will relax at the temperature above the blocking temperature of the nanoparticles. It fluctuates among the easy axes of magnetization when there is no external magnetic field. The superparamagnetic relaxation of magnetic nanoparticles and its correlation with the particle volume and temperature can be studied by using Mössbauer spectroscopy and Néel's formula. Figure 7.12 shows the Mössbauer spectra of $MgFe_2O_4$ nanoparticles with a size of 6 nm at various temperatures. At 300 K, there is only a doublet pattern due to the nuclear quadruple splitting of ^{57}Fe, and there are no magnetic hyperfine interactions. When the temperature is reduced to 95 K, a sextet pattern emerges in addition to the reduction of the intensity of the doublet pattern. The sextet component increases as the temperature decreases to 70 K. At 55 K, the doublet component disappears and only the sextet pattern remains. The Mössbauer spectra of 12 nm $MgFe_2O_4$ nanoparticles are displayed in Fig. 7.13. There is also no magnetic hyperfine interaction in these nanoparticles at 300 K, and hence, there is only a quadruple interaction-induced doublet pattern. When the temperature decreases to 95 K, the spectrum shows a significant sextet component in comparison to the one in the spectrum obtained from 6 nm $MgFe_2O_4$ nanoparticles at the same temperature. For 12 nm nanoparticles even at 70 K, the Mössbauer spectrum shows a single sextet pattern. The mathematical fit of Mössbauer spectra has given a magnetic hyperfine field of 44 T for these nanoparticles at 55 K.

The changes of Mössbauer spectra in Fig. 7.12 and Fig. 7.13 clearly show the correlation between the relaxation time τ and the particle volume and temperature as (7.18) has expressed. ^{57}Fe Mössbauer spectroscopy has an experimental measurement time τ_L about 10 ns. When the relaxation time τ is smaller than τ_L, the fluctuation of the particle magnetization direction is so rapid that the average of internal magnetic hyperfine field is zero to the ^{57}Fe

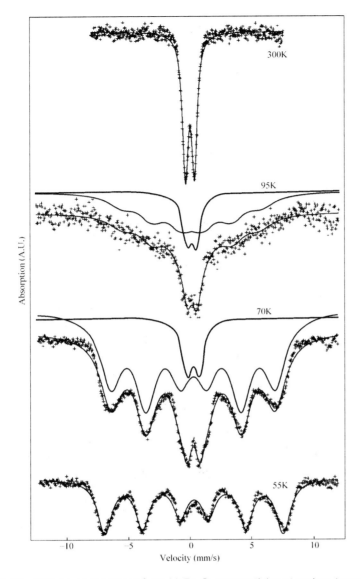

Figure 7.12 Mössbauer spectra of 6 nm $MgFe_2O_4$ nanoparticles at various temperatures

nucleus. Consequently, there is only a quadrupole doublet splitting in the spectrum. τ increases as temperature decreases. Eventually, τ approaches 10 ns, and a sextet pattern starts to appear. However, some ^{57}Fe nuclei still experience a rapid superparamagnetic relaxation and the doublet pattern remains visible. When the temperature is low enough, the relaxation in the nanoparticles are much slower and the spectrum becomes a pure sextet pattern. From (7.18), it is clear that the increase of τ is faster in the

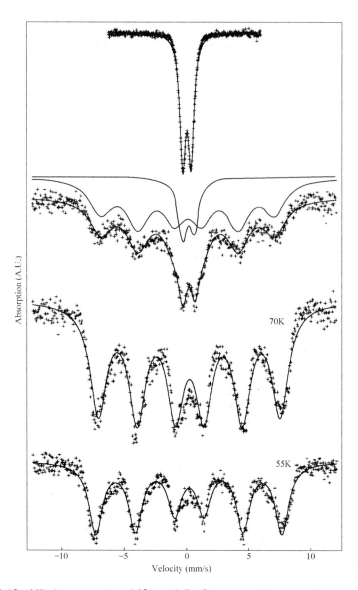

Figure 7.13 Mössbauer spectra of 12 nm MgFe$_2$O$_4$ nanoparticles at various temperatures

nanoparticles with a larger size as temperature decreases. Therefore, the sextet component in the Mössbauer spectrum of 12 nm MgFe$_2$O$_4$ nanoparticles at 95 K is stronger than the same component in the spectra of 6 nm nanoparticles at the same temperatures as shown in Fig. 7.12 and Fig. 7.13. At 70 K, the spectrum of 12 nm nanoparticles has already shown a pure sextet pattern. However, there still is a substantial doublet component in the spectrum of 6 nm nanoparticles at 70 K. The large temperature range for the superparamagnetic transition is considered as the result of the magnetic anisotropy barrier distribution in $MgFe_2O_4$ nanoparticles.

Unfortunately, the value of K may not be determined by Mössbauer spectroscopy studies alone because of the pre-exponential constant τ_0. In order to determine the two unknowns K and τ_0 one must solve the equation twice, at two different temperatures with two different relaxation times. A possible approach to determine the value of K for magnetic materials is using Mössbauer spectroscopy in combination with other magnetic techniques such as SQUID magnetometry since different techniques have different measurement times.

7.4 Neutron Powder Diffraction

Neutron diffraction is one of the most powerful tools for studying magnetic materials. For the characterization of the magnetic nanoparticles, neutron powder diffraction can provide the magnetic structural information in addition to the crystalline phases. Neutrons may interact with both the nucleus and the nuclear magnetic moment in the lattice. This gives rise to diffraction reflections from both the crystal structure and the magnetic order of the sample.

Neutron powder diffraction of crystalline materials follows Bragg's law and behaves similarly to X-ray powder diffraction (Cheetham and Wilkinson, 1992). The neutron travels with a wavelength given by the de Broglie relation:

$$\lambda = h/mv \qquad (7.19)$$

where h is Planck's constant, m is the neutron mass, and v is the neutron velocity. The wavelength is comparable with X-ray wavelength and the lattice constant in crystalline materials. Therefore, the neutron is well suitable to determine the atomic arrangement of materials.

The neutron beam for diffraction studies is produced by either a nuclear reactor source or a spallation source (Von Dreele, 1994). In a reactor source, neutrons are generated through the fission of ^{235}U or ^{239}Pu. The spallation source is an attractive alternative to a reactor source particularly with currently increasing concerns about the environment. In the spallation source, a beam of protons is accelerated to an energy above 500 MeV and collided with a heavy metal target such as W and ^{238}U. The bombardment of high-energy protons on the metal target shatters the heavy nuclei in the target to produce bursts of neutrons. Since the proton beam is in a pulsed mode in the accelerator, the generated neutron pulses are used to collect diffraction data in a time-of-flight fashion. In addition, the pulsed neutron beam is conveniently employed to conduct kinetic studies for crystal and/or magnetic phase transitions.

In both types of sources, neutrons are produced in high-energy environments and will encompass a broad range of energies. Since a high yield of one specific range of neutron energies is usually more useful for diffraction, the energy is regulated by passing the neutrons through a moderator. The moderator interacts with the neutrons and thermalizes the neutrons to lower the beam energy. The common moderator for a diffraction beam is light or heavy water. The thermalized neutron then passes through a monochromator, which, similarly to a photon monochromator, selects a narrow range of energies by using a single crystal and slit.

Neutrons are guided toward the sample and diffracted in the same manner as X-rays. However, the probability of interaction between the neutron and the sample is far lower than the case in X-ray diffraction. Consequently, it is common that up to 2—10 g of sample may be needed for achieving adequate diffraction.

Neutron diffraction of magnetic materials consists of nuclear and magnetic scattering. Compared to the wavelength of the neutrons, the size of a nucleus is too small to have a finite size. Therefore, the nuclear scattering of neutrons is treated as a point scattering. The intensity of the scattering shows no angular dependence. In contrast, X-ray scattering has a very strong angular dependence since scattering comes from the electrons surrounding the nucleus. X-ray scattering also shows a strong dependence on the atomic numbers. For neutron diffraction, the nuclear scattering only weakly depends upon the atomic number. The neutron scattering intensity is usually described by the neutron scattering length, b, which is closely related with the neutron scattering cross section of a nucleus. The scattering length can vary drastically even for adjacent elements.

The great variation of the neutron scattering length provides a powerful approach to distinguish certain adjacent elements on the periodic table and to determine the lattice site composition with a high precision. For instance, spinel ferrites, MFe_2O_4 (M = Co, Fe, or Mn) have a face-centered cubic structure with a large unit cell containing eight formula units (Fig. 7.14).

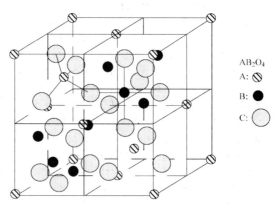

Figure 7.14 Spinel unit cell with half the A and B sites shown

There are two kinds of lattice sites for cation occupation. The A sites have a tetrahedral coordination formed by four oxygen anions, and the B sites have an octahedral coordination formed by six oxygen anions. Each unit cell contains 8 A sites and 16 B sites. When the A sites are occupied by M^{2+} cations and B sites are occupied by Fe^{3+} cations, the spinel is so called normal spinel. If A sites are completely occupied by Fe^{3+} cations and B sites are randomly occupied by M^{2+} and Fe^{3+} cations, the spinel is an inverse one. In most cases, the cation occupancy is intermediate with different degree of inversion, in which both sites are occupied by M^{2+} and Fe^{3+}. The change of the cation distribution in the spinel varies the magnetic properties of these oxides even though the chemical composition does not change. This is due to the fact that the magnetic couplings are varied as the cation occupancy changes. Therefore, the determination of cation distribution at A and B sites is essential for understanding the magnetic structure and magnetic properties of spinel ferrites. Since the difference of the electron density between Co^{2+} and Fe^{3+} and between Mn^{2+} and Fe^{3+} is too small, X-ray diffraction can not distinguish the elements in these two pairs. However, the neutron scattering length is very different for these elements. Fe has a scattering length of 1.01×10^{-12} cm, and the scattering lengths are 0.253×10^{-12} and -0.373×10^{-12} cm for Co and Mn, respectively. Consequently, the cation distribution in these spinel ferrites can be easily determined by neutron diffraction.

The power of neutron diffraction in the research of magnetic materials lies in the magnetic scattering of neutrons. The neutron carries a net magnetic spin of $1/2$, thereby allowing it to interact with magnetic nuclei. Magnetic scattering is quite similar to nuclear scattering, except one must also take into account the added vector quantity of the individual magnetic moments. The theoretical treatment for magnetic scattering is rather complex. However, analysis can be performed readily with some specialized computer software for analyzing diffraction patterns.

Magnetic scattering elucidates the magnetic structure a material and yields important information on the magnetic order. The added vector quantity of the magnetic moment will tend to lower the symmetry of the system, adding magnetic reflections to the diffraction pattern where nuclear reflections are forbidden by the crystal symmetry. There may also be overlapping reflections from the nuclear and magnetic scattering which strengthens the common peak intensity. The position and magnitude of magnetic reflections may be fitted by computer to yield the ferromagnetic or antiferromagnetic structures as well as the magnetic moments at each individual sublattice.

The Reitveld refinement (Cheetham, 1997; Young, 1993) is probably the most common method for the computer simulation of neutron diffraction pattern. In the Reitveld method, the experimental diffraction intensity I_o from the neutron data collection is directly compared to the calculated diffraction intensity I_c. I_c is obtained through the expression:

$$I_c = I_b + \sum P_i \qquad (7.20)$$

The calculated intensity, I_c, combines the contributions from the background scattering, I_b, and each of the Bragg diffraction peaks, P_i. A set of parameters is used to reflect the specific feature of the sample such as crystal space group and chemical composition. These adjustable parameters are refined following the least-square minimization of the weighted difference between the observed and calculated intensities.

$$P = \sum [w(I_o - I_c)^2] \qquad (7.21)$$

The quality of data fitting is indicated by the "goodness of fit", χ^2:

$$\chi^2 = P/(N_{obs} - N_{var}) \qquad (7.22)$$

where N is the number of total points. This complex analysis can be performed with the General Structure Analysis Software (GSAS), which is a set of programs developed at Los Alamos National Laboratory (Fig. 7.15).

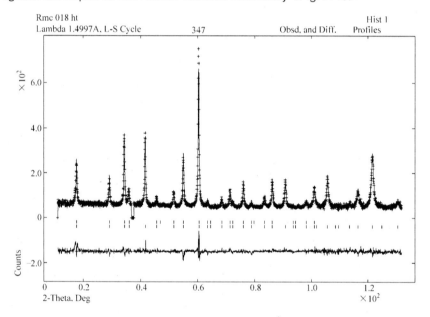

Figure 7.15 Neutron diffraction pattern ($\lambda = 1.4991$ Å) of $CoFe_2O_4$ nanoparticles at 523 K. Below the pattern, the first row of sticks marks the peaks from the magnetic scattering. The second row of sticks marks the peaks from nuclear scattering

Figure 7.15 shows the neutron diffraction pattern at 250℃ of the $CoFe_2O_4$ spinel ferrite nanoparticles with a size of 12 nm in diameter. The Rietveld refinement by the GSAS program clearly shows a magnetic structure in these nanoparticles even at such a temperature. The magnetic moments in the nanoparticles are aligned following an antiferromagnetic order, which is consistent with the ferrimagnetic nature of spinel ferrites. Cation distribution between the tetrahedral and octahedral sites shows that 72% tetrahedral sites are occupied by Fe and Fe occupancy is 63% at octahedral sites. The data refinement suggests a formula for these nanoparticles as $Co_{0.28}Fe_{0.72}(Co_{0.37}Fe_{0.63})_2O_4$ where cations in the brackets occupy the octahedral sites. The unit cell is cubic, and the lattice constant is 8.4131(7) Å. The magnetic moment is $2.42\mu_B$ at the tetrahedral lattice site and $-1.60\mu_B$ at the octahedral site.

7.5 Lorentz Microscopy

A detailed observation of micromagnetic structures of nanostructured magnetic materials can facilitate the understanding of magnetic behavior at the nanometer scale. Transmission electron microscopy (TEM) has proven to be an essential technique for studying nanostructured materials. The high spatial resolution of a transmission electron microscope may elucidate the mechanisms of magnetic transitions in magnetic microstructure. Lorentz microscopy is an excellent approach of utilizing the high spatial resolution of TEM to directly image the magnetic structure.

Lorentz forces are forces exerted on a moving electron by a magnetic field. The Lorentz force may be described by:

$$F = -e(v \times B) \quad (7.23)$$

where e is the charge of the electron, B is the magnetic induction acting on the electron and v is the velocity of the electron (McVitie and Chapman, 1995). For a uniformly magnetized sample of thickness t, the deflection caused by the magnetic field of a sample may be described by:

$$\beta_L = e\lambda Bt/h \quad (7.24)$$

where β_L is the angle of deflection, λ is the electron wavelength, h is Planck's constant, and B is the magnetic induction of the sample. Observation of the deflection caused by Lorentz forces in a magnetic sample can yield information about the positions of magnetic domains in the sample, as well as orientation of each individual domain.

Transmission electron microscopy is well suited for observing the Lorentz forces produced by a sample. The TEM produces images based on elastic scattering of the electrons by the sample. However the elastic scattering caused by the sample is generally two orders of magnitude greater than the deflection caused by Lorentz forces. Moreover, in an ordinary TEM the sample is placed within the objective lens, which may operate at a magnetic field up to 2 Tesla, high enough to magnetically saturate any sample and therefore mask the information one desires to obtain.

In order to perform Lorentz microscopy in a TEM, special modifications must be made. These modifications include placing a magnetic pole piece within the objective lens to shield the sample, and then adding corrective lenses to the objective lens to adjust for the pole piece. These modifications are easily performed, as Lorentz microscopy has become a popular technique for the studies of magnetic materials.

Two methods of Lorentz microscopy are available (Majetich and Jin, 1999). The first is the Foucault imaging mode, which makes use of a focused image and a moving objective aperture. The second is the Fresnel imaging mode, in which slight under - or overfocusing is used to elucidate the domain boundaries. Both methods of Lorentz microscopy have their own advantages and disadvantages. In general, Foucault imaging mode provides better information for most studies.

Foucault imaging is performed with the specimen in focus, which provides higher resolution then the defocused Frensel imaging. A specimen is photographed many times with the objective aperture at various angles from the transmitted beam, so that at some point the aperture will pass through an angle either favoring or excluding the beams diffracted due to Lorentz imaging. At that point the magnetic domains will appear either bright (favored) or dark (excluded). The disadvantage of using Focault imaging is that the user does not know at what angles (and in what direction) the Lorentz beams will be. An effective but labor-intensive way to overcome this shortcoming is to record a series of pictures with the objective aperture starting at zero angle and moving in one direction away from the center beam. The images from this series are digitally added together. A second series is taken as the objective aperture starts at zero angle and moves to higher angle in the other direction. The digital sums of the two series are then subtracted to yield the overall difference due to Lorentz deflection.

For Foucault imaging the magnetic domains will show up as either bright or dark bands in the specimen. The color is irrelevant, being only due to the direction of aperture movement. One only knows that a dark band is ordered antiparallel to the bright band, and the line making up the bright/dark boundary is the magnetic domain wall. Schematically, the information yielded by Foucault imaging may be described as Fig. 7.16.

Fresnel imaging is performed much differently. Like Foucault imaging, the goal is to image the deflections caused by Lorentz forces. Consider two adjoining domains ordered antiparallel to each other. If they are ordered antiparallel to each other, then they should deflect electrons either towards or away from each other. Therefore the domain boundary between the domains should appear either brighter (towards) or darker (away). As these lighter and darker bands will not appear when the specimen is fully focused, Fresnel imaging must be performed under slight defocus. However, such operation limits the spatial resolution of Fresnel imaging, which is why the more labor intensive Foucault imaging is often chosen in imaging magnetic domains. The schematic representation of Fresnel imaging is shown in Fig. 7.17.

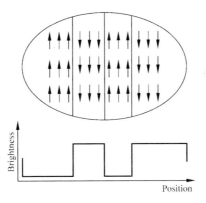

Figure 7.16 Intensity vs. position of Foucault mode TEM. Opposing domains appear alternately bright and dark

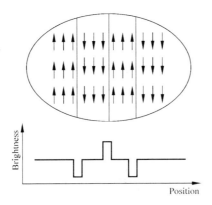

Figure 7.17 Intensity vs. position of Fresnel mode TEM. Domain walls appear alternately bright and dark

Lorentz microscopy also has the advantage of allowing in situ and real time studies of the magnetic order within a sample. These features combined with the ability to directly image magnetic domains make Lorentz microscopy an extremely useful tool in the magnetic characterization of nanostructured materials. Although the studies on magnetic nanoparticles by using Lorentz microscopy are limited, recent studies on $SmCo_5$, Fe_3O_4, and carbon-coated $Fe_{50}Co_{50}$ nanoparticles with a size as small as 5 nm in diameter demonstrate the power of Lorentz microscopy. Figure 7.18 shows the Foucault Lorentz images of a $SmCo_5$ nanoparticle. The time-dependent magnetization reversals in these superparamagnetic Fe_3O_4 nanoparticles have been clearly recorded by using the Foucault imaging mode (Fig. 7.18).

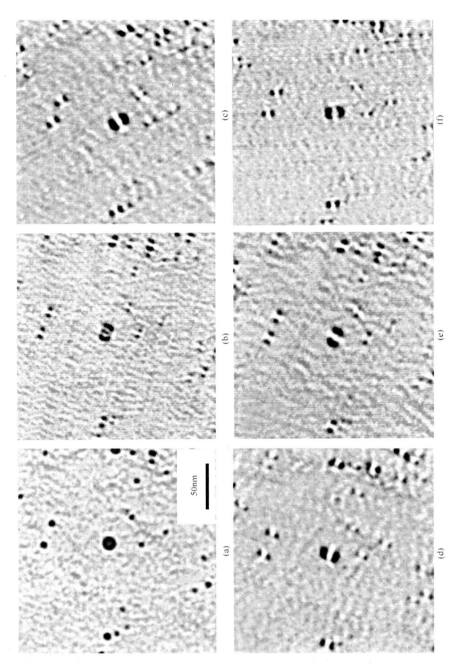

Figure 7.18 (a) Bright-field image of a SmCo5 nanoparticle. Foucault Lorentz images are shown for different shift angles: (b) 55°, (c) 81°, (d) 145°, (e) 207°, (f) 252°. (Courtesy of Professor S. A. Majetich. Reproduced with permission from S. A. Majetich and Y. Jin, Science 1999, 284, 471)

7.6 Summary

Studying the magnetic properties of nanoparticles is of great interest in understanding some fundamental issues of magnetism such as macroscopic quantum tunneling (MQT). It also is extremely important to the advancement of modern technologies. Magnetic nanoparticles have enormous potentials in applications including magnetic drug carriers, high-density data storage, ferrofluid technology, and magneto-refrigeration. The characterization methods that are discussed in this chapter are very effective in the magnetic studies of nanoparticulate materials. In addition, there are several very useful techniques for magnetic studies on a microscopic scale including scanning electron microscopy with polarization analysis (SEMPA), surface magneto-optic Kerr-effect (SMOKE) spectroscopy, spin-polarized low-energy electron microscopy (SPLEEM), and magnetic force microscopy (MFM) (Krishnan, 1995; Hartmann, 1999). These techniques are excellent to the study of nano-structured magnetic materials in a film format or magnetic nanoparticles assembled as a film. As more researchers are attracted to the research field of magnetic nanostructured materials, there will surely be tremendous advancements in existing magnetic characterization techniques. Moreover, some new magnetic methods will be invented and developed in near future.

References

Aharoni. A. in Magnetic Properties of Fine Particles. ed. by J. L. Dormann and D. Fiorani. North-Holland, Amsterdam pp. 3 (1992)
Baibich, M. N., J. M. Broto, A. Fert, F. Nguyen Van Dau, F. Petroff, P. Etienne, G. Creuzet, A. Friederich and J. Chazeles. Phys. Rev. Lett., **61**, 2472 (1988)
Bancroft, G. M., Mössbauer Spectroscopy: an Introduction for Inorganic Chemists and Geochemists. Wiley, New York (1973)

Acknowledgments We are grateful to our co-workers for their contributions to our studies on magnetic spinel ferrite nanoparticles, particularly to Dr. Qi Chen and Dr. Bryan B. Chakoumakos. Our research on spinel ferrite nanoparticles is supported by U. S. National Science Foundation. We gratefully acknowledge the helpful comments from Professor S. A. Majetich of Carnegie Mellon University.

Berkowitz, A. E., J. R. Mitchell, M. J. Carey, A. P. Young, S. Zhang, F. E. Spada, F. T. Parker, A. Hatten and G. Thomas. Phys. Rev. Lett., **68**, 3745 (1992)

Cheetham, A. K. and A. P. Wilkinson. Angew. Chem. Int. Ed. Engl., **31**, 1557(1992)

Cheetham, A. K., in New Trends in Materials Chemistry. ed. by R. Catlow and A. Cheetham, Klewer Academic Publishers, Dordrecht pp.35(1997)

Chien, C. L., Annu. Rev. Mater. Sci., **25**, 129(1995)

Clarke, J., Scientific American. August 1994, pp.46

Gallop, J. C., SQUIDs, the Josephson Effect and Superconducting Electronics. Adam Hilger, New York (1990)

Greenwood, N. N. and T. C. Gibb. Mössbauer Spectroscopy. Chapman & Hall, London (1971)

Hartmann, U., Annu. Rev. Mater. Sci., **29**, 53(1999)

Krishnan, K. M., MRS Bulletin. **20**(**10**), 24(1995)

Leslie-Pelecky D. L. and Reike, R. D., Chem. Mater., **8**, 1770(1996)

Long, G. J. and F. Grandjean. eds. Mössbauer Spectroscopy Applied to Magnetism and Materials Science. Plenum Press, New York (1993)

Majetich, S. A. and Y. Jin. Science. **284**, 470(1999)

McVitie, S. and J. N. Chapman. MRS Bulletin. **20**(**10**), 55(1995)

Parkin, S. S. P., Annu. Rev. Mater. Sci., **25**, 357(1995)

Rao, C. N. R., A. K. Cheetham and R. Mahesh. Chem. Mater., **8**, 2421 (1996)

Stoner, E. C. and I. P. Wohlfarth. Philos. Trans. R., Soc., A **240**, 599 (1948); reprinted in IEEE Trans. Magn. **27**, 3475(1991)

Tejada J., R. F. Ziolo and X. X. Zhang. Chem. Mater. **8**, 1784(1996)

Von Dreele, R. B., in Materials Science and Technology, eds. R. W. Cahn, P. Haasen and E. J. Kramer, (VCH Publishers, New York) pp.561(1994)

Von Helmolt, R., B. Holzapfel, L. Schultz and K. Samwer. Phys. Rev. Lett., **71**, 2331(1993)

Weinstock, H., ed. SQUID Sensors; Fundamentals, Fabrication and Applications. Klewer Academic Publishers, Dordrecht (1996)

Xiao, J. Q., J. S. Jiang and C. L. Chien. Phys. Rev. Lett., **68**, 3749(1992)

Young, R. A., ed. The Rietveld Method. Oxford University Press, Oxford (1993)

8 Electrochemical Characterization

Zhong Shi and Meilin Liu

8.1 Introduction

Nanostructured materials exhibit a host of interesting new phenomena directly related to their reduced dimensionality. Not only the electronic, magnetic, and optical properties but also chemical, electrochemical, and catalytic properties of nanostructured materials are very different from those of the bulk form and depend sensitively on size, shape, and composition (Kirk and Reed, 1992; Berry, et al., 1996). The large surface-to-volume ratio and the variations in geometry and electronic structure have a dramatic effect on transport and catalytic properties. For example, the reactivity of small clusters has been found to vary by orders of magnitude when the cluster size is changed by only a few atoms (Whetten, et al., 1985). Another example is hydrogen storage in metals. It is well known that most metals do not absorb hydrogen, and even among those that do, hydrogen is typically adsorbed dissociatively on surfaces with a hydrogen-to-metal ratio of 1. This limit is not applicable small clusters. It is shown that small positively charged clusters of Ni, Pd, and Pt generated in molecular beams and containing between 2 and 60 atoms can absorb up to 8 hydrogen atoms per metal atom (Cox, et al., 1990). The number of absorbed hydrogen atoms decreases with increasing cluster size and approaches 1 for clusters having more than 60 atoms, simplying that small clusters may be very useful for hydrogen storage. Yet another case of the effect of size on properties is observed in nanophase powders of lithium magnesium oxide, which display remarkable catalytic activity at temperatures much below those needed for non-nanophase magnesium oxide (Bowen, 1992).

In a nanostructured electrode, a large portion of atoms is located at surfaces or interfaces, dramatically influencing the transport of ionic and electronic defects through solids. In addition, short-range rearrangement or redistribution of ionic and electronic defects may cause electrical or chemical polarization; this effect is more pronounced or even becomes dominant in nanophase materials. Each interface will polarize in its unique way when the system is subjected to an applied stimulus. Nanocomposites consisting of two or more phases, each having different electrical properties, may exhibit significant space charge or interfacial polarization. Chemical polarization results from changes in composition or stoichiometry of a material while electrical polarization is a consequence of acquiring electric dipoles due to redistribution of charges or of aligning existing dipoles (e.g., complex defects or defect associates) in the material. A nanoporous electrode has enormous surface or interface area for chemical or catalytic reactions. For example, 1 cm^2 of a 5 μm thick film, comprising primarily of 20 nm particles of TiO_2, partially sintered together to form a microstructurally stable porous film, would have an actual surface area on the order of 750 cm^2. The high surface-to-volume ratio as well as the very small particle size make nanostructured electrodes exhibit unique electrocatalytic and photocatalytic properties (Boschloo, et al., 1997).

In porous nanocrystalline electrodes, no significant potential drop seems to be present between the center and the surface of an individual semiconductor particle, due to the very small particle size in combination with a low donor density (O'Regan, et al., 1990). Nevertheless, it has been demonstrated that efficient photocurrent generation can still take place in this type of electrode, which is explained by the efficient capture of the photoholes by redox species in the solution (Hagfeldt, et al., 1992). In contrast to dense macroscopic semiconductor electrodes, most efficient photoinduced charge separation in porous nanocrystalline electrodes takes place close to the conductive support (Hagfeldt, et al., 1992; 1994a; Björkstén, et al., 1994; Södergren, et al., 1994). Södergren, et al. (Södergren, et al., 1994) have derived a model in which transport of photo-generated electrons in the nanoporous film is driven by diffusion only. Despite the simplicity of this model, simulations for the spectral photocurrent response of nanoporous TiO_2 electrodes are in good agreement with experimental data. The influence of the applied voltage has been introduced into this model by taking into account that the electron concentration in the conductive support is determined by the applied voltage (Södergren, et al., 1994; Hagfeldt, et al., 1994a). Porous nanocrystalline metal oxide electrodes have a great potential for many applications, such as dye-sensitised photoelectrochemical solar cells (O'Regan, et al., 1990; O'Regan and Grätzel, 1991; Nazeeruddin, et al., 1993), electrochromic displays (Marguerettaz, et al., 1994; Hagfeldt, et al., 1994b; Liu and Kamat, 1995), fuel cells, chemical sensors, and lithium-ion batteries (Huang, et al., 1995). The electrochemical properties of nanostructured materials are critical to the performance of these devices.

In an electrochemical measurement, electrical potential or current can be readily controlled or measured. To date, various electrochemical methods have been successfully used to characterize nanophase materials. In this chapter, we will first introduce several methods for preparation of nanoelctrodes, then outline a number of electrochemical techniques commonly used to characterize the properties of nanophase materials, and finally present a few examples to illustrate how to characterize electrical and electrochemical properties of nanostructured materials using these techniques. For further details on electrochemical techniques and methods, see the excellent review articles and monographs by Bard and Faulkner (2000), Kissinger and Heineman(1984), Varma and Selman(1991), Abruna(1991) Scully, et al. (1993), Vanysek(1996) and Bard(1966).

8.2 Preparation of Nanostructured Electrode

One of the key challenges in characterizing the electrical and electrochemical properties of a nanosturctured material is to make an electrical contact to the material to be studied. In this section, several techniques will be introduced for preparation of nanostructured electrodes or nanophase materials on well-defined electrodes. The advantages, together with limitations, of each technique will also be discussed.

8.2.1 Electrodeposition and Electrophoretic Deposition

Recent advances have made electrodeposition an attractive approach for preparation of nanostructured materials (Searson, 1992; Searson and Moffat, 1994). Electrodeposition has a number of advantages over other physical deposition techniques. The rate of deposition is fast; the process is not equipment-intensive; the low processing temperature (usually room temperature) minimizes interdiffusion or chemical reaction; the film thickness can be precisely controlled by monitoring the delivered charge; composition and defect chemistry can be controlled; films can be deposited onto complex shapes; nonequilibrium phases can be deposited; and the driving force can be precisely controlled by adjusting the applied potential. One of the major challenges is to ensure that not only are the electrodeposited materials small in dimension, but also their size distribution is relatively narrow. To date, a wide range of nanophase materials have been successfully deposited, including metals, alloys, semiconductors, polymers, and ceramics.

Further, nanostructured metal-ceramic composites have been produced from suspensions of nanoparticles of alumina and metals such as gold (Buelens, et al., 1985), copper (Lee and Wan, 1988), and Ni (Oberle, et al., 1995) using electrophoretic deposition (EPD). EPD is a process in which powder particles are charged and suspended in a colloidal system and subsequently deposited on to an electrode under the influence of an applied electrical field (Ishihara, et al., 1996; Sarkar, et al., 1996). EPD consists of two steps: electrophoresis, the motion of particles in a stable suspension under the influence of an electric field, and deposition, the coagulation of a dense layer of particles on the electrode. The advantages of EPD include fast deposition, little restriction on the shape of substrate, simple deposition apparatus, low cost, and suitable for mass production (Ishihara, et al., 1996; Sarkar, et al., 1997). In this case, it is necessary to form a stable and well-dispersed suspension of nanoparticles to be deposited.

The successful use of traditional direct-current electrochemical methods for producing nanostuctures is limited in two ways: ① the substrate must be conductive and ② the controllable growth occurs only in the direction normal to the substrate. In order to make 0-, 2-, and 3-dimensional nanostructures the growth must be restricted in some way, for example, by applying a non-conductive mask. Another approach used to make electrochemical deposition spatially selective is to create regular (using lithography) or random arrays of metal nano-electrodes on a substrate, and deposition will be favored on the conductive surface. Alternatively, one might deposit mass-selected clusters on a substrate. Here are two examples to show the advantages of an electrodeposition process.

(1) Electrodeposition into nanopores

Nanophase materials have been grown in the pores of nanoporous membranes such as anodized aluminum or track-etch polymers (Martin, 1994), which have cylindrical pores of uniform diameter. The pores are used as templates to prepare nanoparticles of desired material. When a polymer, metal, semiconductor, or carbon is synthesized electrochemically within one of these pores, a nanocylinder of the desired material is obtained. Depending on the material and the chemistry of the pore wall, this nanocylinder may be hollow (a tubule) or solid (a fibril). Martin and co-workers (Nishizawa, et al., 1995) have shown that metal nanotubes can also serve as ion-selective membranes. The nanotube diameter can be as small as 0.8 nm, and the length of the nanotube can span the complete thickness of the membrane. These membranes show selective transport analogous to ion-exchange polymers. The ion permselectivity is thought to occur because of excess charge density that is present on the inner walls of the nanotubes. Since the sign of the excess charge can be changed potentiostatically, a metal nanotube can be either cation- or anion-selective, depending on the applied potential (Nishizawa, et al., 1995).

(2) Electrodeposition of nanocomposites

Electrochemical deposition is a very attractive processing route for the synthesis of composite materials. The low processing temperatures allow codeposition of materials (ceramics, metals, polymers, semiconductors) that would not tolerate each other at high temperatures, which are necessary for traditional processing, such as sintering, vapor-phase processing, or solidification used to fabricate composites. A simple approach is to suspend particulate material in a plating electrolyte and codeposit this with a desired metallic matrix. This can be accomplished both by electroless deposition and by electroplating. Commercial applications of this approach include codeposition of alumina, silicon carbide, or diamond with a metal such as nickel (Nastasi, et al., 1993). The key challenge in this approch is to prevent agglomeration of the particles prior to codeposition.

Another electrochemical scheme for preparation of composites is electrochemical infiltration. Recently, Lee, et al., have successfully used an electrochemical infiltration process to fill the pores (about 5 nm) of a silica xerogel film with nickel (Lee, et al., 1995). In this manner, room temperature processing was used to synthesize three-dimensionally interconnected nanoscale networks of metal and ceramic.

8.2.2 Formation of Nanoparticles in Polymers

Inorganic nanoparticles have often been prepared with an auxiliary medium, such as surfactant, the interface of monolayers and bilayer lipid membranes, polar headgroups of Langmuir-Blodgett films, clays and zeolites. One of the primary objectives of the various synthesis techniques is to control the particle size either by spatial restrictions, such as size of pores and entities in the media, or by reaction kinetic. Stabilizing nanoparticles is critical. The use of conducting polymers as matrices for nanoparticles not only enhances the stability, but also result in anomalous physical and chemical properties because of the interactions of nanoparticles with the macromolecules. A number of approaches used for preparing nanoparticles in a polymer matrix are briefly described below.

A popular method is a two-step, in situ generation of nanoparticles in a polymer matrix. The first step is the incorporation of a metal ion in a polymer matrix by immersion of the polymer in an aqueous solution containing the metal ions. The second step is the formation of the particles in the polymer matrix by reducing the metal ions with a proper reducing agent. An example is the formation of CuS particles in poly(vinyl alcohol)-poly(acrylic acid) matrix (Godovski, et al., 1993). The polymer mixture is first immersed into a $CuSO_4$ aqueous solution, where the acidic groups of poly(acrylic acid) serve as complexation sites for Cu^{2+} ions. Subsequently, the Cu^{2+} ions in the polymer matrix are reduced using Na_2S to form CuS particles (about 10 nm). Another similar approach is the preparation of nanoparticle particulate film. In this method, a polymer monolayer (instead of surfactant) is spread on an aqueous solution of metal salt, e.g., Cd, Zn, and Pb ions. Injection of a reactant gas (e.g., H_2S) into the gas phase of the enclosed system (Yuan, et al., 1990) initiates particle growth. The polymer monolayer provides a matrix for the size-controlled growth of semiconductor particles that can be transferred, essentially intact, to a solid substrate. CdS particles were formed in the poly (styrenephosphonate diethyl ester) (PSP) particulate monolayer using this approach.

Another widely used method is via polymerization of colloidal solutions containing metal ions and monomers. The particle size can be controlled by the reaction temperature and properties of the colloidal solution, thermal coagulation, and Ostwald ripening. As an example, polymerization of Pb(MA)$_2$ (lead methylacrylic acid) with styrene (Gao, et al., 1994) has been used to form PbS nanoparticles in the polymer. Since there are two C=C bonds in each Pb(MA)$_2$ molecule, it is easier to copolymerize with styrene to form Pb-polymer microgel. This was subsequently treated with H$_2$S gas to obtain PbS nanoparticles in the polymer matrix.

Ion exchange is another useful method to prepare organic polymer and inorganic soild nanocomposite. Polymers such as Nafion or Surlyn have cation-exchange sites where metal ions, such as Cd^{2+} or Pb^{2+}, can be introduced into the polymer matrix. Treatment of such ion-exchanged films with chalcogenide sources results in precipitation of nanoparticles of compound semiconductors within the hydrophilic regions of the polymer. This method has been extended to synthesize layered semiconductor clusters such as CdS (Spanhel, et al., 1992) as well as magnetic particles such as Fe$_2$O$_3$ (Ziolo, et al., 1992).

A somewhat different approach has been explored by Schrock and co-workers (Sankaran, et al., 1990) using direct incorporation of the metal ion Pb^{2+} into monomer units that eventually become part of a norbornene-derived copolymer following ring-opening metathesis polymerization (ROMP). Again, the polymer is designed to have carefully phase-separated and closely size controlled regions of high and low hydophobicity, and the semiconductor PbS is precipitated in the latter regions by exposure to H$_2$S from the gas phase.

Yet another approach is a hybrid of the polymer isolation and the surface-capped cluster or colloid approaches. In this method, a well-defined semiconductor cluster or colloid is prepared by the conventional capping techniques and dissolved in a solvent along with a soluble polymer. This mixed solution may then be deposited (e.g., by spin coating) onto a substrate and dried to produce a polymer film doped with the semiconductor clusters. This simple approach has provided some new examples of interesting photoconductive composites by using a photoconductive polymer such as polyvinylcarbazole (PVK) doped with clusters such as CdS (Wang, et al., 1993).

8.2.3 Electrochemical Self-Assembly

Among various approaches used to prepare nanostructured electrodes or devices, such as crystal growth, ion implantation, and vapor-phase molecular epitaxy, self-assembly is considered to be a very attractive method.

Molecular self-assembly uses the interactive forces of solid state lattice structures, chemical bonds, and van der Waals forces to form larger aggregates of atomic or molecular units with specific geometries, potentially leading to the design of a wide variety of nanostructures. This approach can be used to make nanostructures that are identical to one another (a truly monodisperse sample size).

Self-assembly of nanoparticles to the oppositely charged substrate surface is governed by a delicate balance of the adsorption and desorption equilibria. For example, the objective of immersion is to achieve efficient adsorption of one (and only one) monoparticulate layer of nanoparticles onto the oppositely charged substrate surface. It is equally important to prevent the desorption of the nanoparticles during the rinsing process. The optimization of the self-assembly in terms of maximizing the adsorption of nanoparticles from their dispersions and minimizing their desorption on rinsing requires the judicious selection of stabilizers and the careful control of the kinetics of the process. The self-assembly of CdS and PbS nanoparticles was found to be most efficient, for example, if the semiconductor particles were coated by a 1 : 3 mixture of thiolactic acid and ethyl mecaptane (Kotov, et al., 1995).

Let us take the self-assembly of alternating layers of polyelectrolytes and inorganic nanoparticles as an example to illustrate a layer-by-layer self-assembly process. First, a well-cleaned substrate is primed by adsorbing a layer of surfactant or polyelectrolyte onto its surface. The primed substrate is then immersed into a dilute aqueous solution of a cationic polyelectrolyte, for a time optimized for adsorption of a monolayer, rinsed, and dried. The next step is the immersion of the polyelectrolyte monolayer covered substrate into a dilute dispersion of surfactant-coated negatively charged semiconductor nanoparticles, also for a time optimized for adsorption of a monoparticulate layer, rinsed and dried. These operations complete the self-assembly of a polyelectrolyte monolayer-monoparticulate layer of semiconductor nanoparticle sandwich onto the primed substrate. Subsequent sandwich units are deposited analogously.

8.2.4 Mesoporous Electrodes

Fabrication of mesoporous electrodes or nanostructured interfaces is critical to the creation of a new generation of solid-state ionic devices and reactor systems for gas (e.g., O_2 and H_2) separation, electrosynthesis (e.g., methane conversion), and pollutant (e.g., NO_x, O_x, and H_2S) removal. Conventional mesoporous (or microporous) materials are referred usually to aluminosilicates (e.g., zeolites) and phosphates with pore diameters less than 1.5 nm and with long-range crystalline order. Since the discovery (Kresge, et al., 1992; Beck, et al., 1992) of a new class of mesoporous aluminosilicates (M41S), particularly MCM-41 having pores controllable from 2 to 10 nm and surface area greater than 700 cm^2/g, various mesoporous materials have been prepared using a liquid-crystal templating (LCT) mechanism, including oxides of Ti (Tanev, et al., 1994; Antonelli and Ying, 1995), V (Abe, et al., 1995), Mn (Tian, et al., 1997), Ta (Antonelli and Ying, 1996a), and Nb (Antonelli and Ying, 1996b), as well as doped aluminosilicates (Arean, et al., 1993), Nb-doped silica (Zhang and Ying,

1997), and Zr-Cr framework (Kim, et al., 1997) for molecular sieves. In an LCT synthesis, the structure is determined by the self-assembly of the surfactant molecules into micellar liquid crystals that serve as templates for the formation of a continuous solid framework of inorganic precursors. Subsequently, the surfactant molecules are removed using a thermal or a chemical process while maintaining the ordered structure of the inorganic framework. Amphiphilic molecules, such as surfactants, lipids, copolymers, and proteins, can self-assemble into a wide range of ordered structures, including spherical micelles, rod-like micelles, hexagonally ordered crystal, cubic crystals, lamellar phases, and inverse micellar liquid crystals (Israelachvili, 1992; Vinson, et al., 1991). These structures can also transform from one form to another with changes in solution conditions such as surfactant concentration, counterion polarizability and charge, pH, temperature, and the addition of co-surfactants or additives (Huo, et al., 1996). The equilibrium structures are determined by the thermodynamics of the self-assembly process and the inter- and intra-aggregate forces. The major driving forces for the amphiphiles to form well-defined ordered arrays are attractions between hydrophobic trails and the repulsion between the hydrophilic headgroups. The characteristics of the ordered structure are also influenced by steric effect and characterized by a dimensionless effective surfactant packing parameter (Hyde, 1992), $g = V/a_0 l$, where V is the total volume of the surfactant chain plus any auxiliary organic molecules between the chains, a_0 is the effective headgroup area of a surfactant molecule at the micelle surface, and l is the kinetic length of the surfactant tail. Typically, a small packing factor ($g<0.5$) favors the formation of a highly curved interface (spherical micelles and rod-like micelles), whereas a large packing factor ($g>0.5$) favors the formation of flat interfaces (flexible bilayers and planar bilayers) (Huo, et al., 1994).

In particular, structures for chemically selective catalysis or for fast transport can be tailored by adjusting the electrostatic and steric properties of the framework and the compensating ions as well as other processing parameters. For instance, the pore size can be readily tailored from 2 to greater than 12 nm by varying the chain lengths of the surfactant and addition of auxiliary organics (or swelling agents), making it possible to alter the rate of molecular gas transport through the pores. This provides an unique way to prepare novel mesostructured materials that are difficult to prepare using other methods. Recently, mesoporous SnO_2 (Chen and Liu, 1999), TiO_2, and Sn-TiO_2 (Peng, et al., 2000) have been successfully prepared using LCT mechanism for electrochemical applications.

In addition, tubules of spinel $LiMn_2O_4$ (200 nm outer diameter) have been prepared using thermal decomposition of an aqueous solution containing lithium nitrate and manganese nitrate at 1 : 2 molar ratio using a nanoporous alumina membranes as a template, which entails synthesis of a material within the pores of a nanoporous membrane (Nishizawa, et al., 1997).

8.2.5 Composite Electrodes Consisting of Nanoparticles

To evaluate nanophase materials as active electrode materials for lithium batteries, electrochemically functional composite electrodes consisting of the nanoparticles must be prepared. Lithium intercalated materials, such as $LiNiO_2$, $LiCoO_2$, $LiMnO_2$, and various forms of carbon, have been extensively studied for lithium-ion batteries. The defect structures and microstructures of these nanostructured materials have been tailored through ionic substitution and novel synthesis approaches in order to improve the stability, capacity, reversibility, and rate capability. Because these materials are typically prepared in powder form and have inadequate electronic conductivity, it is often necessary to add some conductive additives (such as carbon black) to improve the electrical conductivity. Traditionally, nanoparticles of active electrode materials are mixed with an organic binder (e.g., PTFE) and a conductive additive (e.g., carbon black) to form a composite electrode, which is then subjected to various electrochemical testing to characterize the electrochemical behavior. Detailed procedures can be found elsewhere (Natarajan, et al., 1998; Exnar, et al., 1997; Tsang and Manthiram, 1997; Striebel, et al., 1997).

The difficulty associated with this traditional approach, however, is that it is not clear how the binder and the additive may influence the electrochemical properties of the active nanoparticles in the composite. The performance of this composite electrode is sensitively influenced by many factors. For example, if the amount of carbon black is not sufficient or not well dispersed, the cycling behavior of the electrode would be poor even though the active electrode materials might be excellent. The cycling of the composite electrode is very time-consuming because of low diffusion coefficients of Li^+ ions inside the solid phase. However, simply because there are no other alternatives, this method is still widely used in study of electrode materials for battery applications.

8.2.6 Powder Microelectrode

Recently, microelectrode techniques have been used to directly study the electrochemical properties of small particles without the addition of any binder or other additives. Bursell and Björnbom (Bursell and Björnbom, 1990) applied a microelectrode technique to characterize a single-carbon agglomerate particles of size ranging form 50 to 120 μm in an alkaline solution. This technique has been successfully used to study single particles of electroactive materials rather than a mixture of electroactive particles, binder, and conductive additive. Uchida, et al., (Uchida, et al., 1997;

Nishizawa, et al., 1998) also used the microelectrode technique to study electrochemical behavior of a single electroactive particle ($LiCoO_2$, $LiMn_2O_4$, and graphite) and determined the chemical diffusion coefficient of lithium ion in the particles. The difficulty in this approach, however, is to make electrical contact to a single particle (especially for a nanoparticle) and the behavior of one individual particle may not be representative. Another more convenient and reliable approach is to use a powder microelectrode (Cha, et al., 1994) to study the direct electrochemical behavior of a number of particles trapped in a microcavity. Schematically shown in Fig. 8.1(a) is a powder microelectrode, which can be prepared by etching the tip of a traditional platinum microdisk electrode in aqua regia to form a microcavity at the tip. Then the sample nanoparticles (or agglomerates of nanoparticles) can be trapped into the microcavity to form a packed microelectrode. The diameter of the platinum wire is typically in the range of 10 to 100 microns. Shown in Fig. 8.1(b) is a SEM micrograph of a powder microelectrode packed with graphite particles (Shi and Liu, submitted). The depth of the microcavity can be controlled by the etching conditions. Microcavities can usually be packed easily and the powders within the microcavity can be directly studied using various electrochemical techniques.

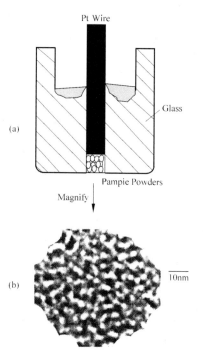

Figure 8.1 (a) A schematic diagram of a powder microelectrode (Cha, et al., 1994); and (b) a SEM micrograph of a powder microelectrode ($r_o = 12.5$ μm) packed with graphite particles (Shi and M. Liu, submitted)

The use of powder microelectrode offers the following advantages:

(1) It is an easy and convenient approach to rapid screening of new electrode materials.

(2) It provides valuable information directly from the sample particles without the need of adding additives such as a binder or carbon black. While it has been mostly used in studying macrosize powders, it can be extended to the study of nanosize powder electrodes.

(3) It offers an averaged behavior of a new material over a large number of particles, not the sole property of a single particle.

In addition to trapping of nanoparticles into a microcavaty by mechanical press, nanophase materials (or nanostructured electrodes) can also be prepared directly in a microcavity using electrodeposition, electrophoretic deposition, sol-gel, or self-assembly processes as discussed earlier.

8.3 Principles of Electrochemical Techniques

Both steady-state and transient techniques have been used to determine the electrical and electrochemical properties of nanophase materials. In a steady-state experiment, a constant current or potential is applied while the steady-state response of the material is monitored. In a transient experiment, the system is perturbed from an equilibrium or a steady state and the relaxation of the response of the system, such as current, potential, charge, impedance, optical reflectance, or other measurable properties of the system, is monitored as a function of time. In many cases, the experimental conditions are created under which the rate of the system response is controlled only by the process to be studied. In a potential step method, such as chronoamperometry or chronocoulometry, the experimental conditions are set up so that the system response is controlled completely by diffusion of electroactive species in order to determine the diffusivity of the electroactive species. In this section, several techniques for determination of electrical and electrochemical properties are discussed.

8.3.1 Impedance Spectroscopy

Impedance spectroscopy (IS) has emerged as a powerful experimental tool in the study of electrical, electrochemical, and catalytic properties of materials, including transport of ionic and electronic defects, dielectric polarization, electrode kinetics, and other bulk or interfacial phenomena in various material systems (Macdonald, 1987).

8.3.1.1 Definition of Impedance Functions

In an impedance measurement, a linear system is perturbed by an applied small-amplitude alternating stimulus (e.g., a voltage) of angular frequency ω,

$$\tilde{V}(\omega) = V_o \exp(j\omega t) \tag{8.1}$$

In responding to this periodic disturbance, the system will relax to a new pseudo-steady state. The response of the system (e.g., the current) will be at the same frequency of the perturbation, but may lead or leg behind the perturbation by a temporal phase angle θ,

$$\tilde{I}(\omega) = I_o \exp[j(\omega t + \theta)] \tag{8.2}$$

The ratio of the alternating voltage to the alternating current is defined as the impedance transfer function (in ohms) of the linear system,

$$Z(\omega) = \frac{\tilde{V}(\omega)}{\tilde{I}(\omega)} = \text{Re}\{Z(\omega)\} - j\text{Im}\{Z(\omega)\} = Z'(\omega) - jZ''(\omega) \tag{8.3}$$

Impedance function is, in general, a complex function; the real part of impedance, $\text{Re}\{Z\} = Z'$, characterizes the long-range transport (irreversible, energy dissipation processes) while the imaginary part of impedance, $\text{Im}\{Z\} = Z''$, characterizes polarization or magnetization (reversible, energy storage processes) in materials.

In addition, several other transfer functions related to impedance are often used. For instance, the reciprocal of the impedance function is called admittance function, $Y(\omega) = [Z(\omega)]^{-1}$. The ratio of the alternating electric induction, $\tilde{D}(\omega)$, to the alternating electric field, $\tilde{E}(\omega)$, is defined as the dielectric permittivity transfer function of a linear system,

$$\varepsilon_r(\omega) = \frac{\tilde{D}(\omega)}{\varepsilon_o \tilde{E}(\omega)} = \text{Re}\{\varepsilon_r(\omega)\} - j\text{Im}\{\varepsilon_r(\omega)\} = \varepsilon_r'(\omega) - j\varepsilon_r''(\omega) \tag{8.4}$$

where ε_0 is the absolute permittivity (or the permittivity of free space) and ε_r is the relative permittivity of the material under study.

The real part of the relative permittivity, $\text{Re}\{\varepsilon_r\} = \varepsilon_r'$, also called dielectric constant, characterizes the energy storage property of the material and depends in general on frequency. In fact, it is possible to determine the contributions of each polarization mechanism (electronic, ionic, dipole, and space charge polarization) from the frequency dependence of dielectric constant. In contrast, the imaginary part of the relative permitivity, $\text{Im}\{\varepsilon_r\} = \varepsilon_r''$, characterizes the energy dissipation in the material due to relaxation of dipole and space charge polarization and to resonance of ionic and electronic polarization in the material. In fact, the AC conductivity of a dielectric material can be expressed as

$$\tilde{\sigma}(\omega) = \omega\varepsilon''(\omega) \tag{8.5}$$

and the loss tangent is defined as

$$\tan\delta = \frac{\varepsilon''}{\varepsilon'} = \frac{\varepsilon_r''}{\varepsilon_r'} \tag{8.6}$$

where δ is the temporal phase angle between the charging current and the total current. Both $\tan\delta$ and ε_r'' reach a local maximum at each characteristic frequency for relaxation or resonance of dipoles in the material. Further, the reciprocal of the permittivity function is known as the modulus function, $M_r(\omega) = [\varepsilon_r(\omega)]^{-1}$.

These functional correlations completely characterize the dynamic electrical characteristics of a linear system. The measurement and analysis of each of these transfer functions in a wide frequency range is called impedance, admittance, permittivity, and modulus spectroscopy, respectively. All of them are also generally referred to as immittance spectroscopy. These transfer functions are interrelated as follows,

$$Z(\omega) = \frac{1}{j\omega\, C_o \varepsilon_r(\omega)} = \frac{1}{Y(\omega)} = \frac{M_r(\omega)}{j\omega\, C_o} \tag{8.7}$$

where C_o, the capacitance of the empty cell used for transfer function measurement, is given by

$$C_o = \varepsilon_o \frac{A}{d} \tag{8.8}$$

for a cell with two parallel electrodes of area A separated by a distance d. Thus, once one transfer function is determined, the other transfer functions can be calculated. All of them are important because of their different dependence on and dispersion with frequency. The analysis of one particular transfer function may offer better resolution in determining certain material properties. In general, analysis of impedance functions is convenient for conductors (dominated by transport) while analysis of permittivity functions is convenient for dielectrics (dominated by polarization). However, we shall take the general term, impedance spectroscopy (IS), to stand for the measurement and analysis of some or all of the four transfer functions, i.e., immittance spectroscopy.

8.3.1.2 Basic Assumptions

The fundamental assumption for impedance spectroscopy is that the response of the system is linear, time-invariant, and finite in the entire frequency domain. The assumption of linearity requires that the response of the system be independent of the amplitude of the applied alternating perturbation. Experimentally, the amplitude of the applied alternating perturbation should be sufficiently small to ensure the linearity of the system, particularly for systems involving interfacial reactions, which are highly non-linear in nature. The assumption of time stability implies that the system is unchanged during impedance measurement. For dynamic systems or non-stationary systems, such as corrosion, impedance data should be acquired as fast as possible to minimize the effect of time instability on the validity of impedance data. The response of the system is assumed to be finite and contains no singularities in the entire frequency domain, including the points where $\omega \to \infty$ and $\omega \to 0$.

While it is a basic assumption that the response is caused only by the alternating perturbation, the impedance of a system can be acquired not only in an equilibrium state ($\bar{V}=0$ and $\bar{I}=0$) but also in a steady state under the influence of a DC polarization (i.e., $\bar{V} \neq 0$ and $\bar{I} \neq 0$). That is, the system is perturbed by a small-amplitude alternating perturbation ($\tilde{V}$ or $\tilde{I}$) superimposed on a steady-state DC polarization ($\bar{V}$ or $\bar{I}$). In other words, the current and voltage can be described as

$$I = \bar{I} + \tilde{I}$$
$$V = \bar{V} + \tilde{V} \tag{8.9}$$

However, the impedance functions measured in a steady state depend, in general, on the amplitude of the DC polarization, $|\bar{V}|$ or $|\bar{I}|$ (or the conditions of the steady state) although they are independent of the amplitude of the alternating perturbation, $|\tilde{V}|$ or $|\tilde{I}|$. The effect of a DC polarization on the transfer function is more pronounced for non-linear processes, such as electrochemical reactions at an interface (Liu and Wu, 1998). In fact, the impedance arising from an electrochemical reaction obtained under different degree of DC polarization can be used for construction of Tafel plots, from which anodic and cathodic transfer coefficients can be determined (Hu and Liu, 1998).

8.3.1.3 Validity of Impedance Data

The linearity, time stability, and causality of a system can be verified by applying the Kramers-Kronig (K-K) transformations to the impedance data. The real and the imaginary part, or the phase and the amplitude, of a valid transfer function are inherently related by the K-K relations (Hilbert integral transform) (Urquidi-Macdonald, 1988; Kendig and Mansfeld, 1983; Macdonald and Brachman, 1956),

$$Z''(\omega) = \frac{-2\omega}{\pi} \int_0^\infty \frac{Z'(x) - Z'(\omega)}{x^2 - \omega^2} dx$$

$$Z'(\omega) - Z''(\infty) = \frac{2}{\pi} \int_0^\infty \frac{xZ''(x) - \omega Z''(\omega)}{x^2 - \omega^2} dx \tag{8.10}$$

$$\theta(\omega) = \frac{2}{\pi} \int_0^\infty \frac{\ln|Z(x)|}{x^2 - \omega^2} dx$$

A set of impedance data is invalid unless the calculated imaginary part of the impedance data from the real part of the impedance data using the K-K transformations matches the experimental data.

8.3.1.4 Equivalent Circuit Approximation

Under the conditions valid for impedance spectroscopy, the electrical behavior of an electrochemical system can be approximated by an equivalent circuit. The analogy between an equivalent circuit and an electrochemical system rests on the assumptions that ① it is possible to model the electrochemical system with a series-parallel combination of linear passive elements, ② each circuit element is associated with an independent physicochemical process or property, and ③ it is possible to determine the unique value of the elements from the impedance response.

Consider a simple electrochemical system, a pure ionic conductor (electrolyte) of thickness, d sandwiched between two electronically conductive electrodes (of area A). Under the assumption that specific adsorption of reactants and products is absent and that there are no complications due to the ionic double layer and mass transfer, an equivalent circuit as shown in Fig. 8.2 can be used to approximate the electrical behavior of the simple electrochemical system, where R_b represents the resistance to the motion of ionic defects through the bulk phase of the electrolyte, C_{dl} the double-layer capacitance, R_{ct} the resistance to charge transfer at the electrolyte-electrode interfaces, and Z_w the impedance to mass transfer in the vicinity of the electrolyte-electrode interfaces. The impedance (in $\Omega\text{-cm}^2$) of each element can be expressed as (Liu, 1998)

$$R_b = \frac{d}{\sigma_{i,el}}$$

$$R_{ct} = \frac{RT}{F}\left(\frac{1}{\alpha_a + \alpha_c}\right)\frac{1}{J_o} \tag{8.11}$$

$$Z_w = R_{mt}\frac{\tanh\sqrt{j\omega(L_d^2/D)}}{\sqrt{j\omega(L_d^2/D)}}$$

The last expression represents the impedance arising from transport or diffusion of electroactive species through a Nernst diffusion layer of thickness L_d with a diffusion coefficient D (Nernst's hypothesis). The mass-transfer time constant is

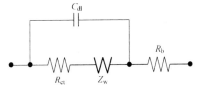

Figure 8.2 An equivalent circuit for a simple cell consisting of an electrolyte (a pure ionic conductor) and two identical electrodes

L_d^2/D and the steady-state mass-transfer resistance is $R_{mt} = \lim_{\omega \to 0} Z_W$.

In the simplest case, when the rate of mass transfer is much faster than the rate of other processes occurring in the system, Z_W may be neglected and the corresponding impedance function is shown in Fig. 8.3 in two different presentations: (a) Bode diagram and (b) Nyqist plot. When the rate of mass transfer in the vicinity of the electrolyte-electrode interfaces affects the rate of other processes or the overall process, Warburg impedance Z_W can no longer be neglected in impedance analysis. Shown in Fig. 8.4 are impedance functions of the system with different assigned values for each circuit element illustrating the effect of Warburg impedance Z_W on the overall impedance response of the system, Z. When $R_{ct} = R_{mt}$ as shown in Fig. 8.4(a) or $R_{mt} \gg R_{ct}$ as shown in Fig. 8.4(b), the overall impedance response of the system, $Z(\omega)$, is dramatically influenced by the mass-transfer time constant (L_d^2/D) or the Z_W. When $R_{mt} \ll R_{ct}$ as shown in Fig. 8.4(c), however, the mass transfer time constant (L_d^2/D) or Z_W has little effect on the overall impedance response of the system, $Z(\omega)$. In addition, Fig. 8.4 also demonstrates that the Warburg impedance is dramatically influenced by diffusivity of electroactive species and the thickness of a diffusion boundary layer (L_d), implying that impedance spectroscopy can also be used to determine the diffusion coefficients of electroactive specieces.

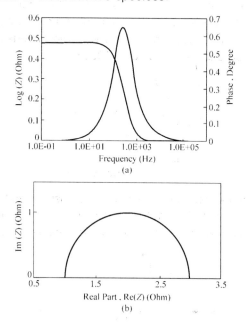

Figure 8.3 (a) Bode and (b) Nyqist presentation of an impedance response of the circuit shown in Fig. 8.2 when the rate of mass transfer is much faster than the rates of other processes occurring in the cell (so that the Z_W can be neglected)

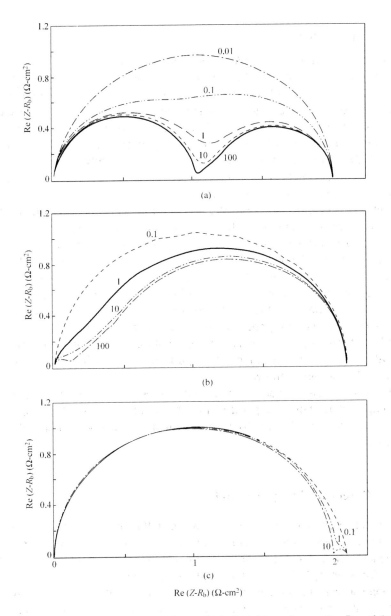

Figure 8.4 Predicted frequency dependence of interfacial impedance, $Z - R_b$, of the cell shown in Fig. 8.3. The number by each impedance spectrum represents the time constant (L_d^2/D) for mass transfer in seconds. The parameters used in the calculation where $R_{ct} = R_{mt} = 1.0$ Ω-cm², $C_{dl} = 0.01$ F/cm², and $L_d^2/D = 0.01, 0.1, 1, 10,$ and 100 s in (a); $R_{ct} = 0.1$ Ω-cm², $R_{mt} = 1.0$ Ω-cm², $C_{dl} = 0.1$ F/cm², and $L_d^2/D = 0.1, 1, 10,$ and 100 s in (b); $R_{ct} = 1$ Ω-cm², $R_{mt} = 0.1$ Ω-cm², $C_{dl} = 0.01$ F/cm², and $L_d^2/D = 0.1, 1, 10,$ and 100 s in (c)

8.3.1.5 Applications

With modern electronics, impedance spectra can be readily obtained in the frequency range from 10^{-4} to 10^7 Hz. As a non-destructive method with high accuracy and reproducibility, impedance spectroscopy has been widely used as an in situ technique to elucidate electrode kinetics and reaction mechanisms, to monitor electrical behavior of various materials, to separate the interfacial polarization from bulk response, to separate one process from another if their relaxation time constants are sufficiently different, and to separate ionic from electronic conduction in mixed ionic-electronic conductors (MIECs) (Liu, 1999; Liu and Hu, 1996).

While its principle of operation is simple, the interpretation of an impedance spectrum can be extremely difficult, particularly when multiple relaxation processes are present and their time constants are not sufficiently different to permit deconvolution of the spectrum. In this case, the best approach is to simplify the system under study as much as possible through careful cell design, including the insertion of reference electrodes. For example, a four-probe impedance measurement may make it possible to acquire the impedance response of only one component or one interface of a test cell, not the overall response of the whole cell, significantly simplifying the analysis and interpretation of impedance data while improving the reliability of impedance analysis.

In an impedance measurement with three or more electrodes, the alternating perturbation is applied to the entire cell, whereas the response of only a particular part of the cell, or a particular process occurring in the cell, is acquired.

8.3.2 Potential Sweep Method

Linear sweep voltammetry (LSV) or cyclic voltammetry (CV), is often the first method chosen by an electrochemist to study a new material for electrochemical and electrocatalytic applications. In LSV, the potential is ramped from an initial to a final value at a constant scan rate while the corresponding current is monitored. A linear sweep voltammogram is then just a plot of the measured current versus the applied potential. In cyclic voltammetry, the potential is swept back and forth between two chosen limits while the current is recorded continuously. In either case, the potential E is usually changed linearly with time, i.e.,

$$E = E_i \pm vt \qquad (8.12)$$

where E_i is the initial potential at $t=0$, which is chosen in a range where no faradaic process takes place, and v is the scan rate, dE/dt.

Shown in Fig. 8.5 is a typical experimental arrangement for LSV and CV measurements using a powder microelectrode. The electrochemical cell consists usually of a vessel that can be sealed to prevent air entering the solution, with inlet and outlet ports to allow the saturating of the solution with an inert gas, N_2 or Ar. The removal of O_2 is usually necessary to minimize or eliminate currents due to the reduction of O_2 which may interfere with the measurement. A standard cell configuration consists of three electrodes immersed in an electrolyte: the working electrode (WE), counter electrode (CE), and reference electrode (RE). The potential at the WE with respect to the RE is controlled by a potentiostat, which is usually interfaced with a computer. In addition to linear sweep voltammetry, various waveforms may be superimposed on the potential at the WE; the current flowing between the WE and CE is measured.

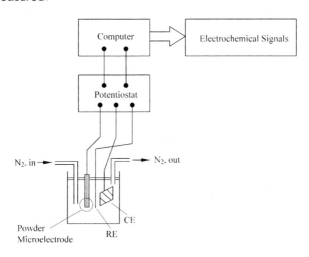

Figure 8.5 A schematic experimental arrangement for a potential sweep (e.g., linear sweep voltammetry and cyclic voltammetry) measurement using a powder microelectrode

When LSV or CV is used to study a system for the first time, it is usual to start with qualitative experiments in order to get a feel for the system, before proceeding to semiquantitative and finally quantitative ones from which thermodynamic and kinetic parameters may be calculated. In a typical qualitative study it is usual to record voltammograms over a wide range of sweep rates and for various ranges of potentials. Commonly, there will be several peaks, and by observing how these appear and disappear as the potential limits and sweep rate are varied, and also by noting the differences between the first and subsequent cycles, it is possible to determine how the processes represented by the peaks are related. At the same time, from the sweep rate dependence of the peak amplitudes, the role of adsorption, diffusion, and coupled homogeneous chemical reactions may be identified. The difference between the first and the subsequent cyclic voltammograms frequently provides useful mechanistic information.

Consider a redox reaction

$$O + ne^- \rightleftharpoons R$$

occurring at the surface of a nano-structured electrode, a typical cyclic voltammogram of this reaction is shown in Fig. 8.6. The anodic current can be expressed as (Bard and Faulkner, 1980; Christensen and Hamnett, 1994)

$$I = nFAc_R^* \left(\frac{\pi D_R F_v}{RT}\right)^{1/2} \chi(\sigma t) \tag{8.13}$$

where A is the area of the electrode, c_R^* is the bulk concentration of R, n is the number of electrons transferred in the reaction, D_R is the diffusion coefficient of R, v is the scan rate and $\chi(\sigma t)$ is a tabulated number that is a function of the electrode potential and contains the variation of the potential with time.

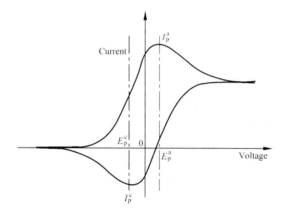

Figure 8.6 A schematic cyclic voltammogram expected form a reversible electrochemical reaction ($O + ne^- = R$) having a standard reduction potential E^0. E is the potential of the WE with respect to the RE and I is the current passing through the CE and the WE

The relationship between the peak current and the concentration of the electroactive species for the anodic and cathodic peak current, $I_{p.a}$ and $I_{p.c}$, is

$$I_{p.a} = 0.4463 nFAc_R^* (F/RT)^{1/2} v^{1/2} D_R^{1/2} \tag{8.14}$$
$$I_{p.c} = -0.4463 nFAc_O^* (F/RT)^{1/2} v^{1/2} D_O^{1/2} \tag{8.15}$$

the dependence of the peak current on $v^{1/2}$ is indicative of diffusion control.

A cyclic voltammogram is characterized by ① the amplitude of the separation of the potentials at which the anodic and cathodic peak currents occur, $\Delta E = E_{p.a} - E_{p.c}$, and ② the half-wave potential, $E_{1/2}$, the potential midway between the peak potentials. For a reversible system, the value of ΔE is about $(0.057/n)v$ at 25°C and independent of scan rate, though it is found in practice that ΔE increase slightly with v.

The half-wave potential is related to $E°$ by

$$E° = E_{1/2} + \left(\frac{RT}{nF}\right)\ln\left[\left(\frac{D_o}{D_R}\right)^{1/2}\right] \qquad (8.16)$$

At the other kinetic extreme, a voltammogram such as the one shown in Fig. 8.7 represents a completely irreversible reaction. For example, $O + ne^- \rightarrow R$ where R cannot be reoxidised to O (or anything else). In such a case, the (cathodic) peak current is given by

$$I_{p.c} = -2.99 \times 10^5 \beta^{1/2} AC_o^* D^{1/2} v^{1/2} \qquad (8.17)$$

and the potential at which this peak current occurs is

$$E_p = E_o - \frac{RT}{\beta F}\left[0.78 + \ln\left(\frac{D_o^{1/2}}{k°}\right) + \ln\left(\frac{\beta F v}{RT}\right)^{1/2}\right] \qquad (8.18)$$

where β is an asymmetry parameter for a electrochemical process, and $k°$ is standard rate constant.

For such an irreversible reaction, it can be shown that

$$|E_p - E_{p/2}| = 1.857\frac{RT}{\beta F} \qquad (8.19)$$

and β can thus be obtained from the separation of peak potential and half-peak potential. Thus, CV can be used to determine whether a reaction the potential at which it occurs, and the parameters n and β.

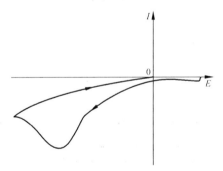

Figure 8.7 A schematic cyclic voltammogram expected from an irreversible process $(O + ne^- \rightarrow R)$

In the case of the reversible system discussed above, the electron transfer rates at all potentials are significantly greater than the rate of mass transport, and therefore Nernstian equilibrium is always maintained at the electrode surface. In fact, it is quite common for most of real electrochemical reactions that they are reversible at low sweep rates to become irreversible at higher ones after having passed through a region known as quasi-reversible at intermediate values. At low potential sweep rates the rate of electron transfer is greater than that of mass transfer, and a reversible cyclic voltammogram is recorded. As the sweep rate is increased, however, the rate of mass transport increases and becomes comparable to the rate of electron transfer. The most noticeable effect of this is an increase in the peak separation. Diagnostic tests for a system under the reversible, irreversible, and quasi-reversible conditions are given in Table 8.1 to 8.3 (Greef, et al., 1985).

Table 8.1 Diagnostic tests for cyclic voltammograms of reversible processes at 25°C (Greef, et al., 1985)

1. $\Delta E_P = E_P^A - E_P^C = \dfrac{59}{n}$ mV;
2. $|E_P - E_{P/2}| = \dfrac{59}{n}$ mV;
3. $\left|\dfrac{I_P^A}{I_P^C}\right| = 1$;
4. $I_P \propto v^{1/2}$;
5. E_P is independent of v;
6. At potentials beyond E_P, $I^{-2} \propto t$.

Table 8.2 Diagnostic tests for cyclic voltammograms of irreversible processes at 25°C (Greef, et al., 1985)

1. No reverse peak;
2. $I_P^C \propto v^{1/2}$;
3. E_P^C shifts $-30/\alpha_c n_\alpha$ mV for each decade increase in v;
4. $|E_P - E_{P/2}| = \dfrac{48}{\alpha_c n_\alpha}$ mV.

Table 8.3 Diagnostic tests for cyclic voltammograms of quasi-reversible systems (Greef, et al., 1985)

1. $|I_P|$ increases with $v^{1/2}$ but is not proportional to it;
2. $\left|\dfrac{I_P^A}{I_P^C}\right| = 1$ provided $\alpha_c = \alpha_A = 0.5$;
3. ΔE_P is greater than $59/n$ mV and increases with increasing v;
4. E_P^C shifts negatively with increasing v.

8.3.3 Potential Step Method

In a potential step experiment, the potential of the working electrode, with respect to a reference electrode, is stepped from a rest potential (at which there is no significant faradaic processes take place) to a potential at which the rate of the overall reaction is controlled completely by the mass transport of electroactive species while the current response or the cumulative charge is recorded as a function of time. Chronoamperometry and chronocoulometry are two examples of potential step techniques. The potential may be either positive or negative with respect to the reference electrode in order to cause, respectively, an oxidation or reduction reaction to take place. For a planar electrode the relaxation of diffusion-limited current is described by the Cottrell equation (Cottrell, 1902)

$$I_d = -nFAC_O^* \sqrt{\frac{D_O}{\pi t}} \quad (8.20)$$

where I_d is the diffusion-limited (cathodic) current for the reduction of O. Thus, a plot of I_d versus $t^{-1/2}$ will yield a straight line and slope of the line can be used to determine the diffusion coefficient of the electroactive species, D_O. It is important to note that the data should be collected over a wide time range to ensure the reliability of data. Usually, relaxations over 1 ms to 10 s are commonly recorded for analysis. In some cases, such as the study of solid-state diffusion in a nanophase material, current relaxation over a longer time should be acquired.

A somewhat more elaborate variation of the chronoamperometric technique is the symmetrical double potential step experiment where the potential is returned to its initial value after the potential was stepped at a potential for a period of time, τ. An example of the application of the double potential step technique to a mass transport controlled electrode reaction is shown in Fig. 8.8. If both the oxidized and reduced forms of the redox couple are stable and the potential to which the working electrode is returned on the

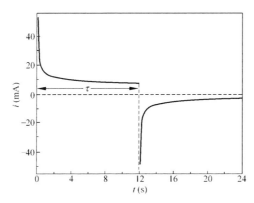

Figure 8.8 Double potential step chronoamperometric response for an one-electron oxidation and reduction of $O + e^- = R$

reverse step (which is the initial potential in the symmetrical case) is sufficient to result in a mass-transport-limited oxidation or reduction of the species produced by the first potential step, then the current obtained on the reverse step, i_r, is given by (Bard and Faulkner, 1980)

$$i_r = nFAD_o^{1/2} C_o^* \pi^{-1/2} [1/(t-\tau)^{1/2} - 1/t^{1/2}] \qquad (8.21)$$

Note that the current is negative if the second or reverse potential step results in an oxidation process and that D_o refers to the diffusion coefficient of the species initially present in solution (the oxidized species in this case). For an uncomplicated redox reaction, the ratio of the current that results from the forward step to the current produced by the reverse step is (Bard and Faulkner, 1980)

$$-i_r/i_f = [t_r/(t_r - \tau)] - [t_f/t_r] \qquad (8.22)$$

where t_f and t_r are the periods of time over which the potential was stepped at the forward and reverse direction, respectively. A plot of $-i_r/i_f$ versus t_r/τ constructed from the data in Fig. 8.8 is shown in Fig. 8.9. The beauty of the double potential step technique is similar to that of other reversal techniques like CV with current reversal; it can be used to probe the stability or reversibility of an electrogenerated redox couple. This is very important to electrode materials for rechargeable batteries. Variations of i_r/i_f from theoretical values can often be used to obtain kinetic data as well as diagnostic information about a redox process. This technique is better suited than CV for studying quasi-reversible electron transfer reactions with coupled homogeneous chemical reactions because the magnitude of the applied

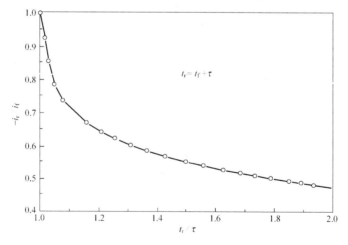

Figure 8.9 Plot of the chronoamperometric current ratio as a function of i_r/τ

potential steps can often be adjusted so as to render the electrode process mass transport limited.

A variant of double potential step chronoamperometry that is often used to study adsorption is double potential step chronocoulometry (Bard and Faulkner, 1980). Because cumulative charge (determined by the electronic integration of the current) is recorded as a function of time in chronoamperometry, and the influence of double-layer charging process at the beginning of the potential step on the total charge rapidly becomes negligible after a long time, the value of K can therefore be measured at quite long times. This permits rate constants perhaps an order of magnitude greater to be determined from charge, rather than current measurements (e.g., chromoamperometry) (Christie, et al., 1964). Under semi-infinite conditions, the chronocoulometric response in the forward direction can be expressed as

$$Q_f(t \leqslant \tau) = Q_{dl} + \frac{2nFAD_o^{1/2}C_o^* t^{1/2}}{\pi^{1/2}} \qquad (8.23)$$

and the charge removed in the reverse direction can be expressed as

$$Q_r(t > \tau) = Q_{dl} + \frac{2nFAD_o^{1/2}C_o^*}{\pi^{1/2}}[\tau^{1/2} + (t - \tau^{1/2}) - t^{1/2}] \qquad (8.24)$$

where Q_{dl} is the charge due to double-layer charging and τ is the forward step duration time.

The cumulative charge due to electrolysis of the electroactive species at a diffusion-controlled rate in the forward and reverse directions can be expressed as

$$Q(t \leqslant \tau) = \frac{2nFAD_o^{1/2}C_o^*}{\pi^{1/2}} t^{1/2} \qquad (8.25)$$

and

$$Q(t > \tau) = \frac{2nFAD_o^{1/2}C_o^*}{\pi^{1/2}}[t^{1/2} - (t - \tau^{1/2})] \qquad (8.26)$$

A typical Q versus t transient of double potential step chronocoulometry is shown in Fig. 8.10.

8.3.4 Controlled-Current Techniques

Most studies of the insertion (extraction) of lithium ions into (from) an intercalation compound have been focused on the diffusion of lithium ions within the compound. Since the rate of solid-state diffusion is relatively slow, the overall charge or discharge rate of a Li-ion battery is often determined by the

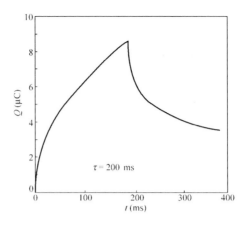

Figure 8.10 A typical response for a double-step chronocoulometric experiment

diffusion of Li^+ ions in the intercalation compounds. Thus, the chemical diffusion coefficient of Li^+ ion (D_{Li^+}) in an intercalation material (e.g., $Li_xMn_2O_4$ nanoparticles) is a critical parameter that determines the power density of batteries based on them. The values of D_{Li^+} can be determined using several methods, including galvanostatic intermittent titration technique (GITT) (Weppner and Huggins, 1977), current pulse relaxation (CPR) (Basu and Worrell, 1979), chronoamperometry (CA), impedance spectroscopy (Bruce, et al., 1992), and electrochemical permeation method (Ineba, et al., 1998). The reported values of D_{Li^+} in $Li_xMn_2O_4$ are rather scattered, varying from 10^{-6} to 10^{-9} cm^2 s^{-1}, depending on the way in which MnO_2 was prepared and on the characterization technique employed. To obtain D_{Li^+} using the first three methods discussed above, one has to know the real surface area of the sample (A) and, in some cases, the variation of the open-circuit potential with lithium concentration (dV_{oc}/dx) (Yamamoto, et al., 1985). Unfortunately, precise determination of A and dV_{oc}/dx is difficult, particularly in the case of porous electrodes; this is why there is a large discrepancy in the reported diffusivity data.

8.3.4.1 Galvanostatic Intermittent Titration Technique (GITT)

GITT (Weppner and Huggins, 1977) combines transient (current relaxation) and equilibrium (coulometric titration) measurements, as schematically illustrated in Fig. 8.11. In a typical GITT experiment, galvanostatic currents are applied to a test cell for a time interval τ to produce a change in stoichiometry of the electrode material. After such a titration pulse, a new equilibrium voltage will be established. The charge and discharge currents are selected so that a change in lithium content from $x=0$ to $x=1$ ($\Delta x=1$) for an insertion electrode material (e.g., $Li_xMn_2O_4$) would occur for a limited time

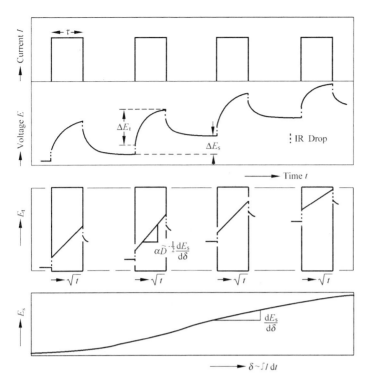

Figure 8.11 Principles of GITT for the evaluation of the thermodynamic and kinetic data of electrodes

(e.g., 40 h). Applying a constant current to the cell composed of the intercalated electrode during a short time (e.g., 1800 s) upon charging, the resulting cell potential transients are recorded. After interruption of the current pulse, the decay of the open-circuit potential is followed with time until the fluctuation of the open-circuit potential falls below 0.01 V versus. Li/Li$^+$. This potential value is just recorded as an electrode potential. The application and interruption of the constant current continue until the lithium content x reaches 0.4, after which the measurement is performed in the reverse direction, i.e., discharging, until x vanishes. Similar to the charge curve, the resulting cell potential transients and electrode potentials are obtained.

In a GITT measurement, there are no limits on the initial and boundary conditions for Fick's second law and hence a linear relationship between the potential and the square root of time is expected in a short time range of the galvanostatic potential transient. The chemical diffusion coefficient of lithium ions in the Li$_x$Mn$_2$O$_4$ electrode of planar symmetry can be calculated as a function of intercalated lithium content from GITT curve, as follows (Weppner and Huggins, 1977),

$$D_{Li^+} = \frac{4}{\pi}\left(\frac{V_m}{zFA}\right)^2\left[I_0\frac{dE/dx}{dE/d\sqrt{t}}\right]^2, \quad t \ll \frac{l_0^2}{D_{Li^+}} \quad (8.27)$$

where V_m is the molar volume of the electrode material ($cm^3\ mol^{-1}$), z the valence of lithium ion ($z=1$), F the Faraday constant, A the electrochemical active area of the electrode-electrolyte interface, I_0 the applied constant current, dE/dx the slope of the coulometric titration curve, and $dE/d\sqrt{t}$ the slope of the E versus square root of time curve.

Another advantage of this technique is that it can determine the partial conductivities due to the motion of a minority ionic defect in a predominantly electronically conducting electrode. The partial ionic conductivity of a mixed ionic-electronic conductor can be calculated from the concentration and the diffusion coefficient of ionic defect together with the variations of the steady-state and transient voltages (Bruce, 1995).

In addition, GITT provides the possibility to determine many other kinetic and thermodynamic parameters of the electrode as a function of stoichiometry (Weppner and Huggins, 1977), including partial ionic conductivity, ion mobility, Gibbs free energy of formation, etc.

8.3.4.2 Current Pulse Relaxation (CPR) Technique (Basu and Worrell, 1979)

The chemical diffusion coefficients of lithium ions in $Li_xMn_2O_4$ cathodes can also be measured using current pulse relaxation techniques in a two-electrode electrochemical cell:

$$Li\,|\,1M\ LiClO_4\ in\ PC\,|\,Li_xMn_2O_4$$

The cell is first galvanostatically charged to 4.5 V (versus. Li/Li^+) and then is discharged galvanostatically to a given lithium composition, x, at a constant current density (e.g., 0.1 mA cm^{-2}). At each composition, lithium is intercalated into the cathode by a constant-current pulse (e.g., 0.5 mA cm^{-2}) for a short period of time (e.g., 10 s). The diffusion coefficient of lithium ion can be calculated from the decay rate of the transient voltage using the following equation:

$$D_{Li^+} = \left[\frac{V_M(dE/dx)}{nFA}\frac{i\tau}{\Delta E/\Delta(t^{-1/2})}\right]^2 \quad (8.28)$$

where dE/dx is the slope of the equilibrium potential-composition curve, i the intensity of current pulse, τ pulse duration, and $\Delta E/\Delta(t^{-1/2})$ the slope of the plot of the relaxation potential versus $t^{-1/2}$ after the current pulse is interrupted. The quasi open-circuit voltage (OCV) curves of the cells can be measured using a galvanostatic intermittent charge-discharge method (Guohua, et al., 1996).

8.3.4.3 Electrochemical Permeation Method

Another method to study solid diffusion is electrochemical permeation method, which was originally developed for the determination of the diffusion coefficients of hydrogen in steel (Turnbull, et al., 1989; McBreen, et al., 1996) and gas molecules in Nafion (Ogumi, et al., 1985; 1984). Consider a MnO_2 nanostructured electrode with a thickness L as shown in Fig. 8.12(a). First, both surfaces are kept at the same potential so that Li^+ concentration is uniform throughout the electrode ($c=c_0$, $0<x<L$). At $t=0$, the potential of one surface ($x=0$) is stepped at a lower potential to increase Li^+ concentration at $x=0$ ($c_{x=0}=c_L$ at $t\geqslant 0$). Lithium ions diffuse through the electrode toward the other surface ($x=L$) due to a concentration gradient induced by the potential step. After lapse of time, Li^+ ions permeate to the other surface ($x=L$), where they are removed into the solution because Li^+ concentration at $x=L$ is electrochemically kept constant ($c_{x=L}=c_0$). The flux due to permeation of Li^+ ions can be monitored as a change in oxidation current at $x=L$. The current (I) initially increases with time and then reaches a steady-state value (I_∞). When the diffusion obeys Fick's law, the current transient is predicted by a "buildup" curve shown in Fig. 8.12(b).

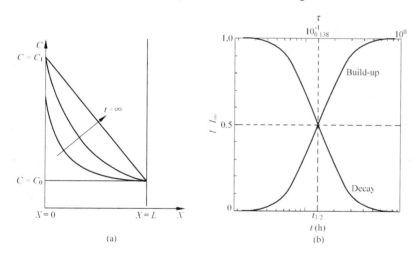

Figure 8.12 (a) Principle and (b) theoretical curves for current transient of an electrochemical permeation measurement (Ineba, et al., 1998)

After the steady state is attained, a potential step in the opposite direction gives a current transient shown by a "decay" curve in Fig. 8.12(b). The D_{Li^+} values can be determined from the current transients using the following relation (Ineba, et al., 1998):

$$D_{Li^+} = \frac{0.138L^2}{t_{1/2}} \tag{8.29}$$

where $t_{1/2}$ is the time when the current reaches half of the steady-state value ($I/I_\infty = 0.5$). It should be noted that both the real electrode area (A) and the variation of the open-circuit potential with lithium concentration (dV_{oc}/dx) are not needed for the calculation of D_{Li^+}, an advantage over other techniques.

8.3.5 Electrochemical Quartz Crystal Microbalance

The quartz crystal microbalance (QCM) or nanobalance (QCNB) is a piezoelectric device capable of extremely sensitive mass measurements. The quartz crystal typically oscillates in a mechanically resonant shear mode by application of an alternating, high-frequency electric field using electrodes which are usually deposited on both sides of the quartz disk. The mass sensitivity arises from a dependence of the oscillation frequency on the total mass of the (usually disk-shaped) crystal, its electrodes, and any materials attached to the electrode surface.

The in situ QCM was first applied to electrochemical problems by Nomura and co-workers (Nomura and Hattori, 1980; Nomura and Iijima, 1981) to determine Cu(II) and Ag(I) formed from electrodepositon. The applications of QCM to electrochemical systems will be distinguished from nonelectrochemical applications by referring to the former as EQCM (electrochemical QCM). EQCM has evolved as a powerful technique capable of detecting very small mass changes at an electrode surface where electrochemical reactions are taking place. This relatively simple technique only requires, in addition to conventional electrochemical equipment, an inexpensive radio-frequency oscillator, a frequency counter, and commercially available AT-cut quartz crystals. EQCM has been successfully used in the investigation of phenomena such as underpotential deposition, electrolyte adsorption, mass changes accompanying ion and solvent movement in redox polymer films, and electrochemically driven self-assembly. Recently, EQCM has been applied to the study of nanostructured materials (Hepel, 1998). In this section, we will provide the readers with a brief introduction to EQCM and with ways of obtaining useful information from in situ EQCM measurements. Further details about EQCM are referred to several review articles (Ward and Buttry, 1990; Buttry, 1991; Buttry and Ward, 1992).

8.3.5.1 Principle of Operation

Due to the converse piezoelectric effect, the application of an electric field across a piezoelectric material induces a mechanical strain (Curie and Curie,

1980). The vibrational motion of the quartz crystal results in a transverse acoustic wave that propagates back and forth across the thickness of the crystal between the crystal faces. Accordingly, a standing-wave condition can be established in the quartz resonator when the acoustic wavelength is equal to twice the combined thickness of the crystal and electrodes. The frequency f_0 of the acoustic wave fundamental mode is given by

$$f_0 = \frac{v_{tr}}{2d} \quad (8.30a)$$

where v_{tr} is the transverse velocity of sound in AT-cut quartz (3.34×10^4 m s^{-1}) and d is the resonator thickness.

Though mass changes at electrode surfaces are related to the change in the resonant frequency of the EQCM in an electrochemical process, it is not easy to get the quantitative correlation between frequency changes and mass changes. If one assumes that acoustic velocity and density of the foreign layer are identical to those of quartz, a change in thickness of the foreign layer is tantamount to a change in the thickness of the quartz crystal. Under these conditions, a fractional changes in mass results in a fractional change in the resonant frequency; appropriate substitutions yields the well-known Sauerbrey equation:

$$\Delta f = \frac{2f_0^2 \Delta m}{A(\mu_0 \rho_0)^{1/2}} \quad (8.30b)$$

where Δf is the measured frequency change, f_0 the frequency of the quartz resonator prior to a mass change, Δm the mass change, A the piezoelectrically active area, ρ_0 the density of quartz, and μ_0 the shear modulus of quartz. This equation is the primary basis of EQCM measurement wherein mass changes occurring at the electrode interface are evaluated directly from the frequency changes of the quartz resonator. It is generally considered adequate as long as the thickness of the film added to the QCM is less than 2% of the quartz crystal thickness. Typical operating frequencies of the EQCM vary from 5 MHz to 10 MHz, with the mass detection limits up to 1 ng cm^{-2}.

8.3.5.2 Electromechanical Model of the EQCM

In general, the mechanical vibrations in a piezoelectric crystal can be described in terms of electrical equivalent (Bottom, 1982). This also serves to enhance understanding of EQCM, particularly the conditions under which the Sauerbrey equation is valid. The equivalent circuit for a quartz resonator has an inductor (L_1), a capacitor (C_1), and a resistor (R_1) connected in series (Fig. 8.13). The relationship between this circuit and the quartz crystal is especially useful because the LCR branch is identical to a tank circuit, in which oscillations can be sustained by cycling of current between the capacitor and the inductor. The equivalent electrical parameters can be expressed in terms of crystal properties as follows, along with some typical experimental values for each parameter.

$$C_o = \frac{\varepsilon_Q \varepsilon_0 A}{d} \approx 10^{-12} \text{F}$$

$$C_1 = \frac{8Ah^2}{\pi^2 dc} \approx 10^{-14} \text{F}$$

$$R_1 = \frac{d^3 D_Q}{8Ah^2} \approx 100 \ \Omega \quad (8.31)$$

$$L_1 = \frac{d^3 \rho_Q}{8Ah^2} \approx 0.075 \text{ H}$$

where ε_Q is the dielectric constant of quartz, D_Q a dissipation coefficient corresponding to the energy losses during oscillation, h the piezoelectric stress constant, and c the elastic constant.

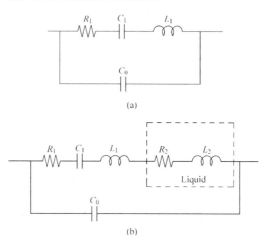

Figure 8.13 (a) Butterworth—van Dyke equivalent electrical circuit used to describe the mechanical properties of a quartz resonator. The components L_1, C_1 and R_1 in the circuit represent, respectively, the inertial mass, compliance, and energy dissipation in the crystal, and C_0 represents the static capacitance of the quartz crystal. (b) An equivalent circuit used to describe the mechanical properties of a quartz resonator immersed in a liquid. The inductance L_2 and resistance R_2 represent the mass and viscosity components of the liquid

8.3.5.3 Experimental Apparatus and Operation

Typically, a quartz crystal (diameters of 0.5 and 1.0 in.) with appropriately sized excitation electrode is mounted to the bottom of a glass cylinder, which assumes the role of the working electrode compartment of an otherwise conventional electrochemical cell. The two excitation electrodes are electrically connected to an oscillator circuit that contains a broadband RF amplifier, so that the electrode facing solution is at hard ground. The circuit is designed so that the crystal is in a feedback loop, therefore driving the crystal at a frequency at which the maximum current can be sustained in zero-phase-angle condition. Several oscillator designs are available, and a key requirement of the circuit is that it provide sufficient gain to allow for oscillation of the crystal in a viscous medium. The output of the oscillator is then connected to a conventional frequency meter for measurement. A critical feature of the EQCM is the potentiostat, which can be either a Wenking potentiostat or a more conventional potentiostat. The difference between these two potentiostats is in the working electrode: the Wenking potentiostat functions with the working electrode at hard ground, whereas the current commercial potentiostat functions with the working electrode at virtual ground. Finally, a computer is used to collect frequency and electrochemical data simultaneously, as well as control the waveform applied to the working electrode. This arrangement allows simultaneous measurement of the electrochemical charge, current, voltage, and EQCM frequency. Frequency counters are capable of sampling the frequency output of the oscillator at 100 ms intervals. This capability enables analysis of the kinetics of a wide range of electrochemical processes, including electrodeposition and dissolution, nucleation and growth, and ion-solvent insertion in redox polymer films.

8.3.5.4 Data Interpretation

Interpretation of EQCM data is rather straightforward. Since the electrochemical charge represents the total number of electrons transferred in a given electrochemical process, it corresponds to mass changes occurring at the electrode surface. Accordingly, under ideal conditions, the frequency change measured with the EQCM will be proportional to the electrochemical charge and will be related to the apparent molar mass by

$$\Delta f = MW \cdot \frac{C_f Q}{nF} \qquad (8.32)$$

where MW is the apparent molar mass (g mol^{-1}), Q the electrochemical charge, and C_f (Hz g^{-1}) the sensitivity constant derived from the Sauerbrey relationship. Thus, a plot of Δf versus Q is particularly useful in the determination of MW/n, the molar mass per electron transferred.

An alternative approach to data analysis involves the relationship between the electrochemical current and the first derivative of the frequency change with respect to time, as

$$i = \frac{d(\Delta f)}{dE} \cdot \frac{nvF}{MW \cdot C_f} \qquad (8.33)$$

where v is the potential scan rate in units of $V\ s^{-1}$. This format is particularly useful for cyclic voltammetry experiments, as $d(\Delta f)/dt$ should appear similar in form to the voltammograms if the electrochemical events are accompanied by corresponding mass changes.

8.4 Application to Nanostructured Electrodes

In this section, electrode materials for lithium-ion batteries, e.g., lithium transition metal oxides and carbon, will be used as examples to illustrate how to characterize the electrical and electrochemical properties of nanophase materials using the techniques discussed earlier. Further, the relationship between these fundamental properties (such as chemical diffusion coefficient of Li^+ ion, potential window, and reversibility) and the performance of battery electrodes (such as energy density, rate of charge/discharge, cell voltage, and rechargeability) will be elaborated.

8.4.1 Characterizing the Reversibility of Battery Electrode Materials

Once a nanostructured electrode is prepared, the first and the most important criterion in determining its viability as an electrode material for a secondary lithium battery is its reversibility, which is often studied using cyclic voltammetry and chronopotentiometry (Julien and Nazri, 1994). Cyclic voltammetry is a powerful tool for probing kinetic and thermodynamic properties of a new electrode material. The peak shape and the number of observable redox processes within a given potential range often provide additional criteria for the selection of viable electrode material. Electrode 'cycling' or multiple double-step chronopotentiometry is then applied to promising compounds in order to obtain information regarding the available capacity and possible cycle life of an electrode. In general, voltammetric techniques are valuable in initial screening or in discriminating 'good' from 'bad' battery electrode materials. However, the behavior of a composite or thin-film electrode depends not only on the active materials but also on the electrode porosity and additives (such as binders and conductivity enhancers). Required voltammetric scan rates are typically in the μV/s range in order to compensate for slow mass transfer of electroactive ions in solid electrode materials. When a powder microelectrode is used, cyclic voltammograms are typically characterized by symmetric, well-defined peaks resembling those observed at thin homogeneous films. Depending on the particle size and transport properties of the material with respect to incorporated ions, voltammetric scan rates in the mV/s range may be used.

This method should therefore be advantageous when studying poorly conductive battery electrode materials such as manganese oxides. However, effects due to a distribution of particle size are expected to result in broadened voltammetric signals.

In a typical cyclic voltammogram, the current is proportional to the area of the working electrode. Assuming that the nanoparticles to be studied are spheres, the current observed on a nanoparticle is proportional to the square of the particle diameter. Typically the currents observed on a microelectrode are about 3 to 5 orders of magnitude smaller than those observed on an ordinary electrode. This reduction in current greatly reduces the distortion of voltammograms due to iR drop. Because the particles trapped in a microcavity are very small, they can be readily oxidized or reduced entirely when the potential is cycled at a very slow rate (e.g., 1 mV/s to 10 mV/s). Thus, the multiscan cyclic voltammetry is just like an accelerated charge-discharge cycling test. The reversibility and the cycle stability of nanoparticles can be examined by continuous scanning reversed at different potentials.

Shown in Fig. 8.14(a) are the cyclic voltammograms of a packed graphite microelectrode (Pt diameter is 25 μm) in 1 M $LiN(SO_2CF_3)_2$/EC+DMC (1 : 2 by vol) for the first five cycles. Three reversible redox peaks are clearly observable, suggesting that there are three redox reactions. During the first five cycles, no significant changes in CV curves were observed, implying that the graphite powder has a very good reversibility. Shown in Fig. 8.14(b) are several charge/discharge curves for the same graphite sample. The capacity remains 95% of the secondary-cycle-capacity after 5 cycles. These results indicate that cyclic voltammetry using powder microelectrode can be used as a characterization tool for rapid screening of new electrode materials.

Another example based on powder microelectrode technique is shown in Fig. 8.15, which shows the cyclic voltammograms of a packed cobalt-doped lithium nickel oxide microelectrode (Pt diameter is 25 μm) in 1 M $LiN(SO_2CF_3)_2$/EC + DMC (1 : 2 by vol) at different sweep rates. The relationship between the peak currents and peak potentials and the square root of sweep rates remain linear when the sweep rate is changed in the range of 0.1 mV/s to 10 mV/s, indicating that reversible Li^+ intercalation/deintercalation process is a diffusion-controlled process. Often, it is the transport of lithium ions in the solid state that determines the current density or rate capability.

EQCM can also be used to study the reversibility of nanostructured materials. Since EQCM can directly monitor the mass change during an intercalation/deintercalation process of Li^+ ion within a nanophase material. The mass change in a nanostructured electrode material can be measured sensitively by EQCM technique, and the reversibility of the electrode material can be readily evaluated by the change in the mass after each cycle. Since the change in mass is determined from the change in resonance frequency which may also be influenced by many other factors (such as the viscosity of

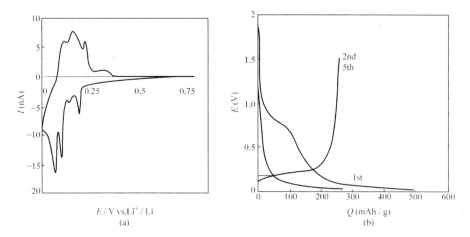

Figure 8.14 (a) Cyclic voltammograms of a graphite packed powder microelectrode in 1 M LiN(SO_2CF_3)$_2$/EC+DMC(1:2 by vol) for the first five cycles. Sweep rate is 0.1 mV/s; (b) Charge/discharge curves for a graphite/1 M LiN(SO_2CF_3)$_2$/EC+DMC(1:2 by vol)/Li battery during the first five cycles. $i = 0.1$ mA/cm^2

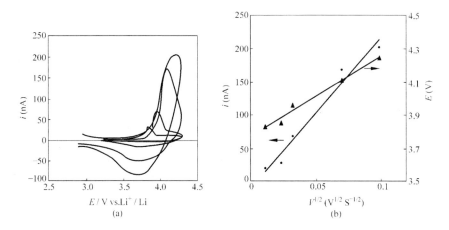

Figure 8.15 (a) Cyclic voltammograms of a packed LiNi$_{0.8}$Co$_{0.2}$O$_2$ microelectrode in 1 M LiN(SO_2CF_3)$_2$/EC+DMC(1:2 by vol) at different sweep rates. $v = 10, 5, 1, 0.5, 0.1$ mV/s (from outer to inner). (b) Relationship between $I_{p,a} \cdot E_{p,a}$ and $v^{1/2}$

solvent), only those nanostructured materials which can be deposited on the substrate of quartz can be studied using this method.

The simplest, yet important and widely used, technique in the study of nanomaterials as battery electrodes is galvanostatic charge-discharge cycling. In fact, each material must be tested using this method to determine the viability as an electrode material. VO$_2$ nanocrystalline (Tsang and Manthiram,

1997), spinel $Li_xMn_2O_4$ nanotubules (Natarajan, et al., 1998), and $Li_xMn_2O_4$/PEO nanocomposites (Striebel, et al., 1997) have been studied as positive electrode materials for rechargeable lithium batteries. However, the voltages of these nanophase materials in lithium batteries range from 2 to 4 V versus Li^+/Li. Limit of discharge voltage is even at 1 V for VO_2 whereas the same materials with larger particle size exhibit voltages ranging from 3 to 4.3 V versus Li^+/Li.

8.4.2 Characterizing the Transport Properties

Another critical property, which determines the rate capability or power density of nanostructured electrode materials in lithium-ion batteries, is the diffusion coefficient of the Li^+ ions in the solid framework of the host structure. The determination of ion diffusion coefficients in solid materials has always been a challenging task. In particular, experimental techniques that have been traditionally developed for diffusion processes occurring in liquid ionic media may not be applicable to study of nanophase solids.

Various experimental methods have been used to determine the chemical diffusion coefficient of Li^+ ions in intercalation electrodes. Most of the methods used in the early studies utilized modifications of the Cottrell transient techniques developed on the basis of the solution to Fick's second equation for an instantaneous planar source of diffusing species in a semi-infinite geometry (Bard and Faulkner, 1980). Basically, the technique consists of applying short (a few seconds) galvanostatic (or potentiostatic) pulses to promote excess concentration of the diffusion species at the electrode surface and then following the consequent voltage (or current) relaxation as a function of time after switching off the current (Bonino, et al., 1980; Bonino and Scrosati, 1985). Alternatively, the technique involves the application of a galvanostatic pulse of very short duration (< 1 s) to the cell and the analysis of the resulting overvoltage-time transient during the pulse. Here the overvoltage is a function of the concentration of the diffusing lithium ions at the electrode-electrolyte interface.

For example, GITT was used for the determination of the chemical diffusion coefficient of the Li^+ ions in Li_xC_6 carbon electrodes (Choi, et al., 1995). The technique consisted of applying to (the electrode) voltage steps and monitoring the charge increment $q(t)$ versus time t, thus allowing the evaluation of the value of D_{Li^+} for the whole composition range of the Li_xC_6 electrode.

The chemical diffusion coefficient of Li^+ in $Li_{1-x}NiO_2$ was determined using impedance spectroscopy (Bruce, et al., 1992). This alternating current technique is particularly suitable for kinetic studies in thin-film insertion electrodes (Ho, et al., 1980) where the semi-infinite diffusion boundary conditions, necessary for the direct current transient techniques, can no longer be assumed. The interpretation of the impedance results requires the construction of an equivalent circuit that is representative of the electrochemical system under study. The most commonly used circuit to interpret the impedance response of an electrode-electrolyte interface is shown in Fig. 8.2.

8.5 Summary

Nanophase materials are not only fascinating from a scientific point of view but also have significant technological implications. For example, nanostructured electrode-based batteries, fuel cells, and chemical sensors may dramatically improve energy efficiency, environmental quality, and a new generation of electrical vehicles. One of the key challenges in characterizing electrical and electrochemical properties of nanophase materials is to establish an electrical contact to the materials to be studied. A number of preparative approaches have been described for preparation of nanostructured electrodes. Among them, preparation of nanophase electrodes on a microelectrode or powder microelectrode makes it possible to directly investigate the properties of the nanophase materials, eliminating the effect of binders and other additives which would be otherwise added to form an electrochemically functional electrode. Electrodepostion or electrophoretic deposition is gaining popularity in fabrication of nanostructured electrodes because of its low deposition temperature, easy control, and fast deposition rate. Formation of nanoparticles in a polymer matrix provides a convenient way of preserving the microstructural stability of nanoparticles. Electrochemical self-assembly has also been successfully used for fabrication of nanoelectrodes and devices. Further, composite electrodes, consisting of nanoparticles, binders, and conductive additives, are also widely used in screening and evaluation of nanoparticles for battery applications. The use of composite electrode is effective for study of nanophase materials to which an electrical contact can not be readily established otherwise. Once a nanoelectrode is prepared, several electrochemical techniques can be used to study its kinetic and thermodynamic properties, including impedance spectroscopy, potential sweep methods, potential step methods, current-controlled techniques, and electrochemical quartz crystal microbalance. While each technique has its own unique advantages over other techniques, it is often found that the combination of several techniques is the best approach to improve accuracy and reliability. For instance, the combination of impedance spectroscopy, current-controlled techniques, and potential-step techniques is uniquely suited for investigations into solid-state diffusion of ionic defects in intercalation compounds. Similarly, the combination of impedance spectroscopy, open cell voltage measurements, and steady-state permeation measurements is a powerful approach to separating ionic from electronic transport in mixed ionic-electronic conductors.

References

Abe, T., A. Taguchi, and M. Iwamoto. Chem. Mater. **7**, 1429(1995)
Abruna, H. D., Electrochemical interfaces ; modern techniques for in-situ interface characterization. VCH Pub., New York (1991)
Antonelli, D. M., and J. Y. Ying. Angew. Chem. Int. Ed. Engl., **34**, 2014 (1995)
Antonelli, D. M., and J. Y. Ying. Chem. Mater. **8**, 874(1996a)
Antonelli, D. M., and J. Y. Ying. Inorg. Chem., **35**, 3126(1996b)
Arean, C. O., et al., Mater. Chem. Phys., **34**, 214(1993)
Bard, A. J., and L. R Faulkner. Electrochemical methods: fundamentals and applications. Wiley, New York (1980) the 2nd Ed., (2000)
Bard, A. J., Ed., from 1966, Electroanalytical Chemistry: a series of advances, Marcel Dekker, New York (20 volumes have published)
Basu, S., and W. L. Worrell. in Fast Ion Transport in Solids. Eds.; P. Vashishta, J. N. Mundy and G. K. Shenoy, Elsevier North Holland, 149(1979)
Beck, J. S., et al., J. Am. Chem. Soc., **114**, 10834(1992)
Berry, A. D., R. J. Tonucci, and P. P. Nguyen. MRS Symp. Proc., **431**, 387 (1996)
Björkstén, U., J. Moser and M. Grätzel. Chem. Mater. **6**, 858(1994)
Bonino, F., and B. Scrosati. in Solid State Batteries (Eds: C. A. C. Sequeira and A. Hooper), NATO ASI Series, Martin Nijhoff, Dordrecht. 15(1985)
Bonino, F., M. Lazzzri, C. A. Vincent, and A. R. Wandless. Solid State Ionics1, 311(1980)
Boschloo, G. K., A. Goossens, and J. Schoonman. J. Electroanal. Chem. **428**, 25(1997)
Bottom, V. G., Introduction to Quartz Crystal Unit Design. Van Nostrand Reinhold, New York (1982)
Bowen, K., Z. Phys., D **26**, 46(1992)
Bruce, P. G., L. Lisowaka-Oleksiak, M. Y. Saidi, and C. A. Vincent. Solid State Ionics. **57**, 353(1992)
Bruce, P. G., Solid State Electrochemistry. Cambridge University Press, (1995)
Buelens, C., J. P. Celis and J. R. Roos. Trans. Inst. Met. Fin., **63**, 6(1985)
Bursell, M., and P. Björnbom. J. Electrochem. Soc., **137**, 363(1990)
Buttry, D. A., in Electroanalytical Chemistry. Ed.; A. J. Bard. Vol. **17**, Marcel Dekker, New York (1991)
Buttry, D. A., and M. D. Ward. Chem. Rev., **92**, 1355(1992)

Cha, C. S., C. M. Li, H. X. Yang, and P. F. Liu. J. Electroanalytical Chem., **368**, 47(1994)

Chen, F., and M. Liu. Chemical Communications., 1829(1999), Chen F., Z. Shi, and M. Liu. Chem. Comm., 2095(2000)

Choi, Y.-M., S.-I. Pyun, J.-S. Bae and S.-I. Moon. J. Power Sources. **56**, 25(1995)

Christensen, P. A., and A. Hamnett. Techniques and mechanisms in electrochemistry. Blackie Acadmic & Professional, 172(1994)

Christie, J. H., G. Lauer, and R. A. Osteryoung. J. Electroanal. Chem., **7**, 60(1964)

Christie, J. H., G. Lauer, R. A. Osteryoung, and F. C. Anson. Anal. Chem., **35**(1979)

Cottrell, F. G., Z. Physik. Chem., **42**, 385(1902)

Cox, D. M., P. Fayet, R. Brickman, M. Y. Hahn, and A. Kaldor. Catal. Lett., **4**, 271(1990)

Curie, P., and J. Curie. C. R. Acad. Sci. **91**, 294(1980)

Exnar, I., L. Kavan, S. Y. Huang, and M. Gratzel. J Power Sources **68**, 720 (1997)

Gao, M., Y. Yang, B. Yang, F. Bian, and J. Sher. J. Chem. Soc. Chem. Commun., 2779(1994)

Godovski, D. Y., A. V. Vokkov, I. V. Karachevtser, M. A. Moskvino, A. L. Volynskii and N. F. Bakeev. Polymer Science USSR. A **35**, 1308(1993)

Greef, R., R. Peat, L. M. Peter, D. Pletcher, and J. Robinson. Instrumental methods in electrochemistry. Ellis Horwood Limited (1985)

Guohua, L., H. Ikuta, T. Uchida and M. Wakihara. J. Electrochem. Soc., **143**, 178(1996)

Hagfeldt, A., U. Björkstén and S.-E. Lindquist. Sol. Energy. Mater. Sol. Cells. **27**, 293(1992)

Hagfeldt, A., S.-E. Lindquist and M. Grätzel. Sol. Energy. Mater. Sol. Cells. **32**, 245(1994a)

Hagfeldt, A., N. Valchopoulos, and M. Grätzel. J. Electrochem. Soc., **141**, L83(1994b)

Hepel, M., J. Electrochem. Soc., **145**, 124(1998)

Ho, C., D. Raistrick, and R. A. Huggins. J. Electrochem. Soc., **127**, 343 (1980)

Hu, H., and M. Liu. Solid State Ionic. **109**, 259(1998)

Huang, S. Y., L. Kavan and M. Grätzel. J. Electrochem. Soc., **142**, L142 (1995)

Huo, Q., D. I. Margolese, and C. D. Stucky. Chem. Mater. **8**, 1147(1996)

Huo, Q., D. I. Margolese, U. Ciesia, P. Feng, T. E. Gier, P. Sieger, R. Leon, P. M. Petroff, F. Schuth, and G. D. Stucky. Nature. **368**, 317(1994)

Hyde, S. T., Pure Appl. Chem., **64**, 1617(1992)

Ineba, M., S. Nohmi, A. Funabiki, T. Abe, and Z. Ogumi. Materials for Electrochemical Energy Storage and Conversion II——Batteries, Mater. Res.

Soc. Symp. Proc., MRS Warrendale PA USA, v496, pp.493—498(1998)
Ishihara, T., K. Sato, and Y. Takita. J. Am. Ceram. Soc., **79**, 913(1996)
Israelachvili, J. N.. Intermolecular and Surface Forces. Academic Press, London (1992)
Julien,C. and G.-A. Nazri (Eds.). Solid State Batteries: Materials Design and Optimization. Kluwer, Dordrecht, The Netherlands. (1994)
Kendig and Mansfeld. Corrosion **39**, 466(1983)
Kim, J. M., C. Shin, and R. Ryoo. Catal. Today. **38**, 221(1997)
Kirk, W. P., and M. A. Reed. Nanostructures and Mesoscopic Systems. Academic Press, New York (1992)
Kissinger, P. T., and W. R. Heineman (editors). Laboratory techniques in electroanalytical chemistry, Dekker, New York (1984)
Kotov, N. A., I. Dékány, and J. H. Fendler. J. Phys. Chem., **99**, 13065 (1995)
Kresge, C. T., M. E. Leonowicz, W. J. Roth, J. C. Vartuli and J. S. Beck. Nature. **359**, 710(1992)
Lee, C.C., and C. C. Wan. J. Electrochem. Soc., **135**, 1930(1988)
Lee, T. J., K. G. Sheppard, A. Ganburg, and L. Klein. in Electrochemical Microfabrication II. Eds.: M. Datta, K. Sheppard, and J. Dukovic, Proceedings Volume 94-2, The Electrochemical Society, Pennington, NJ. (1995)
Liu, D., and P. V. Kamat. J.Electrochem.Soc., **142**, 835(1995)
Liu, M., J. Electrochem. Soc., **145**, 142(1998)
Liu, M., and H. Hu. J. Electrochem. Soc., **143**, L109(1996)
Liu, M., and Z. L Wu. Solid State Ionics. **107**, 105(1998)
Macdonald,J. R.. Impedance Spectroscopy-Emphasizing Solid Materials and Systems. John Wiley and Sons. (1987)
Macdonald, J. R., and M. K. Brachman. Rev. Mod. Phys., **28**, 393(1956)
Marguerettaz, X., R. O'Neill and D. Fitzmaurice. J. Am. Chem. Soc., **116**, 2629(1994)
Martin, C. R.. Science. **266**, 1961(1994)
McBreen, J., L. Namis, and W. Beck. J. Electrochem. Soc., **113**, 1218 (1996)
Nastasi, M., D. M. Parkin, and H. Geleiter (eds.). Mechanical Properties and Deformation Behavior of Materials Having Ultra Fine Microstructures. Kluwer, Boston (1993)
Natarajan, C., K. Setoguchi, and G. Nogami. Electrochim Acta. **43**, 3371 (1998)
Nazeeruddin, M., A. Kay, I. Rodicio, R. Humphry-Baker, E. Müller, P. Liska, N. Vlachopoulos and M. Grätzel. J. Am. Chem. Soc., **115**, 6382 (1993)
Nishizawa, M., K. Mukai, S. Kuwabata, C. R. Martin, and H. Yoneyama. J. Electrochem. Soc., **144**, 1923(1997)
Nishizawa, M., R. Hashitani, T. Itoh, T. Matsue, and I. Uchida. Electrochemical and Solid-State Letters. **1**, 10(1998)

Nishizawa, M., V. P. Menon, and C. R. Martin. Science. **268**, 700(1995)
Nomura, T., and M. Iijima. Anal. Chim. Acta. **131**, 97(1981)
Nomura, T., and O. Hattori. Anal. Chim. Acta. **115**, 323(1980)
Oberle, P. R., M. R. Scanlon. R. C. Cammarata, and P. C. Searson. Appl. Phys. Lett.. **66**, 19(1995)
Ogumi, Z., T. Kuroe, and Z. Takehara. J. Electrochem. Soc.. **132**, 2601 (1985)
Ogumi, Z., Z. Takehara, and S. Yoshizawa. J. Electrochem. Soc.. **131**, 769(1984)
O'Regan, B., and M. Grätzel. Nature. **353**, 737(1991)
O'Regan, B., J. Moser, M. Anderson and M. Grätzel. J. Phys. Chem.. **94**, 8720(1990)
Peng, Z., Z. Shi, and M. Liu. Cnem. com..2125(2000)
Sankaran, V., C. C. Cummines, R. R. Schrock, R. E. Cohen, and R. J. Silbey. J. Am. Chem. Soc.. **112**, 6858(1990)
Sarkar, P., and P. S. Nicholson. J. Am. Ceram. Soc.. **79**, 1987(1996)
Sarkar, P., S. Datta, and P. S. Nicholson. Composites. 28B, 49(1997)
Scully, J. R., D. C. Silverman, and M. W. Kendig (editors). Electrochemical impedance : analysis and interpretation. Philadelphia : ASTM. (1993)
Searson, P. C.. Solar Energy Materials and Solar Cells. **27**, 377(1992)
Searson, P. C., and T. P. Moffat. Critical Reviews in Surface Chemistry. **3**, 171(1994)
Shi, Z., and M. Liu. J. American Ceramic Soc.. Submitted.
Södergren, S., A. Hagfeldt, J. Olssen and S.-E. Lindquist. J. Phys. Chem.. **98**, 5552(1994)
Spanhel, L., E. Arpac, and H. Schmidt. J Non Cryst Solids. 147—48, 657 (1992)
Striebel, K. A., S.-J. Wen, D. I. Ghantous, and E. J. Cairns. J. Electrochem. Soc.. **144**, 1680(1997)
Tanev, P. T., M. Chibwe, and T. J. Pinnavaia. Nature. **368**, 321(1994)
Tian, Z. R., W. Tong, J. Y. Wang, N. G. Duan, V. V. Krishnan, and S. L. Suib. Science. **276**, 926(1997)
Tsang, C., and A. Manthiram. J. Electrochem. Soc.. **144**, 520(1997)
Turnbull, A., M. Saenz de Santa Maria, and N. D. Thomas. Corros. Sci.. **28**, 89(1989)
Uchida, I., H. Fujuyoshi, and S. Waki. J. Power Source. **68**, 139(1997)
Urquidi-Macdonald, M., S. Real, and D. D. Macdonald. J. Electrochem. Soc.. **133**, 2018(1988)
Vanysek, P.. Modern techniques in electroanalysis. John Wiley & Sons, New York (1996)
Varma, R., and J. R. Selman (edited). Techniques for characterization of electrodes and electrochemical processes. Wiley, New York (1991)
Vinson, P. K., J. R. Bellare, H. T. Davis, W. G. Millar and L. E. Scriven. J. Coll. Interface Sci.. **142**, 74(1991)

Wang, Y., N. Herron, and J. Caspar. Physics and Chemistry of Nanometer Scale Materials Mater Sci Eng B Solid State Adv Technol Elsevier Sequoia SA Lausanne 1 Switz. B **19**, 61(1993)
Ward, M. D., and D. A. Buttry. Science. **249**, 1000(1990)
Weppner, W., and R. A. Huggins. J. Electrochem. Soc.. **124**, 1569(1977)
Whetten, R. L., D. M. Cox, D. J. Trevor, and A. Kaldor. Phys. Rev. Lett.. **54**, 1494(1985)
Yamamoto, T., S. Kikkawa, and M. Koizumi. Solid State Ionics. **17**, 63 (1985)
Yuan, Y., I. Cabasso, and J. H. Fendler. Chem. Mater.. **2**, 226(1990)
Zhang, L., and J. Y. Ying. J. AIChE. **43**, 2793(1997)
Ziolo, R. F., E. P. Giannelis, B. A. Weinstein, M. P. O'Horo, B. N. Ganguly, V. Mehrotra, M. W. Russell, and D. R. Huffman. Science. **257**, 219 (1992)

9 Mechanical Property Characterization

H. Mizubayashi, H. Tanimoto, M. Suganuma, A. Shimatani and H. Saka

9.1 Elasticity Study of Metal Nanometer Films

The ULSI technology has brought about a great demand for understanding and controlling of the mechanical properties as well as electromigration in thin films, e. g., see Kraft and Arzt (1998) Nix (1997) Koch (1994) and references therein. Some new materials are synthesized as thin films, where the elasticity study may give an insight into their properties. For example, although cementite, Fe_3C, is a classic material, its Young's modulus has not been measured because no bulk specimens of Fe_3C have been available. Very recently, a single phase Fe_3C can be synthesized as films several hundred nm thick (Yumoto, et al., 1996a, 1996b) and its Young's modulus is determined by means of the recently developed elasticity methods (Li, et al., 1998; Mizubayashi, et al., 1999a). The present section is devoted to the vibrating reed method oriented to thin film measurements and its application to Ag nm films, Al nm films and Ag/Pd multilayer films.

9.1.1 Vibrating Reed Method

Figure 9.1(a) is a schematic of a silicon reed substrate which has been cut out from a FZ-Si single crystal and polished into a vibrating reed with a thick end for clamping (Mizubayashi, et al., 1992, 1999b). The surface 1 scheduled for deposition of a thin film specimen is a mirror surface, and the surface 2 for deposition of an electrode is a smooth surface with small undulations. Homogeneity of the reed thickness, b, along the long axis is within 2%. A metal film electrode is deposited on its surface 2 by means of, e. g., RF-sputtering of Al, where the surface 2 is covered by a thin plate with a window to form the film electrode on it minimizing contamination of the surface1. Figure 9.1(b) is a schematic drawing of a measurement setup, where the resonant flexural vibration of a Si reed substrate with strain amplitude of 10^{-6} is electrostatically excited and detected through one electrode. The internal friction of the Si reed substrate is, typically, 1×10^{-7} at 80 K and 1×10^{-6} at 300 K. When the thermoelastic damping comes in, the internal frequency can be strongly increased from the above values (see p. 494 in Nowick and Berry,

1972). When the thickness of a specimen film may be in the nm range, a change in the resonant frequency of the Si reed substrate may be very small. On the other hand, for the measurement setup shown in Fig. 9.1(b), a steady polarized potential U between the reed and the electrode is modified by a dynamic voltage to excite the resonant vibration of the reed. For such a case, an attained resonant frequency, f_a, under U shows a deviation from the intrinsic resonant frequency, f, where the relationship $(f_a - f)/f = -\gamma \zeta^{-3} U^2$ is expected; ζ denotes a spacing between the reed and the electrode and γ a constant (Barmatz and Chen, 1974; Mizubayashi, et al., 1992). Figure 9.2 shows examples of the f_a versus. U^2 data found before deposition and after deposition of a Ag/Pd multilayer films several hundred nm thick, where f_o and f_1 found at $U^2 = 0$ are the intrinsic resonant frequency before and after deposition, respectively. For the Young's modulus measurements of thinner films, an amount $\Delta f = f_1 - f_o$ may be smaller than $f_a - f$; therefore, the calibration to find the intrinsic resonant frequency f is inevitably required for this method in addition to the temperature control within, e.g., ± 0.005 K during the measurements (see Mizubayashi, et al., 1992 for details).

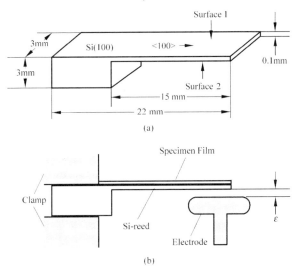

Figure 9.1 (a) Schematic drawing of a silicon reed substrate. (b) Schematic of a measurement setup (Mizubayashi, et al., 1992, 1999b)

Young's modulus, E_f, of a specimen film can be determined using the equation

$$E_f = \frac{E_s}{3}\left[\frac{2(\Delta f/f)}{d/b} + \frac{\rho_f}{\rho_s}\right]\left[\frac{1-v_f^2}{1-v_f v_s}\right] \tag{9.1}$$

where E_s denotes the Young's modulus of the silicon reed substrate, ρ_f and ρ_s are the density of a specimen film and that of the substrate, v_f and v_s are the

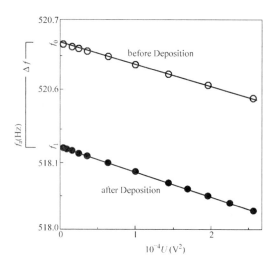

Figure 9.2 Examples of the bias voltage dependence of the resonant frequency (Mizubayashi, 1992)

Poisson ratio of the film and that of the substrate, respectively (Uozumi, et al., 1976; Berry and Pritchet, 1981; Mizubayashi, et al., 1992).

9.1.2 Elasticity Measurements on Ag and Al Films

In Mizubayashi, et al. (1999b) a Ag film is deposited onto the surface 1 (see Fig. 9.1(b)) after the hydrogen termination by means of vapor deposition in 10^{-5} Pa at room temperature with a deposition rate below 0.3 nm/s. The thickness, d, of the Ag film is estimated from an areal weight of the film. Figure 9.3(a) shows the mean grain size of Ag films along the film thickness, GS_t, which is estimated from the (111)—X-ray diffraction (XRD) profiles and Scherrer's method. The GS_t shows good agreement with the film thickness up to $d \approx 30$ nm and a tendency of saturation for the thicker films, suggesting that nucleation and growth of crystallites simultaneously take place during deposition. The grain size, GS_s, along the film surface is estimated from the scanning tunnel microscope (STM) surface topograph (not shown here). For the films 17 nm thick, the GS_s is comparable with film thickness. For thicker films, grains are composed of two groups, one with GS_s of 20 nm to 30 nm and the other with GS_s of 50 nm to 70 nm, again suggesting that nucleation of crystallites and growth of existing crystallites simultaneously take place during deposition.

Figure 9.3(b) shows the Young's modulus E_f versus the film thickness d data of Ag films, where the open symbols denote the data after Mizubayashi, et al. (1999b) and the filled symbols, those after Uozumi, et al. (1976). In

Fig. 9.3(b), the dashed lines <hkl> denote the Young's modulus along the <hkl> direction, $E^{<hkl>}$, reported for Ag single crystal. The arrows a, b and c denote the Young's modulus, $E_{<111>-bf}$, calculated for the <111>-oriented polycrystalline film, $E_{<110>-bf}$ for that of the <110>-oriented film and $E_{<100>-bf}$ for that of the <100>-oriented film, respectively. As seen in Fig. 9.3(b), Young's modulus E_f of the Ag films thicker than 50 nm is found nearby $E_{<111>-bf}$, where the data after Mizubayashi, et al., 1999b and those after Uozumi, et al. (1976) show good agreement with each other. On the other hand, the Young's modulus E_f appears to show a decrease with decreasing film thickness below about 50 nm.

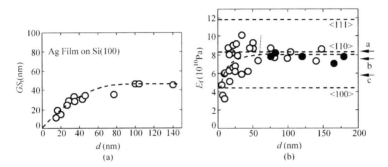

Figure 9.3 (a) The mean grain size, GS_t along the Ag film thickness for the open symbol specimens in (b). (b) Young's modulus E_f vs. film thickness d data found in the Ag films. The open symbols are after Mizubayashi, et al. (1999b) and the filled symbols after Uozumi, et al. (1976). See Fig. 9.4 and text for the dashed curve 1 (Mizubayashi, 1999b)

Figure 9.4 is an enlarged drawing of Fig. 9.3(b) for Ag films less than 80 nm thick. The decrease in the lattice parameter along the film thickness by less than 0.2% is detected for the thinner Ag films, indicating that the lattice spacing along the film surface increases by less than 0.5% after a simple calculation assuming the bulk Poisson ratio (Mizubayashi, et al., 1996). Tensile stress calculated from the postulated stretcher strain is about 400 MPa as a maximum. Computer simulation work on copper and nickel reports a decrease in Young's modulus with increasing lattice spacing (Jones, et al., 1992). The combination of the postulated stretcher strain and the computer simulation work predicts a decrease in E_f by about 10% as a maximum, which is much smaller than the observed decrease in E_f. On the other hand, as mentioned for Fig. 9.3(a), both GS_s and GS_t show a decrease with decreasing film thickness below about 40 nm, presumably suggesting that the effects of the surface, interface and grain boundary regions increase for the thinner films. For Ag nanocrystalline materials with mean grain size of about 30 nm, it is reported that Young's modulus found after extrapolation to the bulk density is about 85% that of the Ag polycrystalline bulk specimen (Qin, et al., 1998). In Fig. 9.4, the filled triangle is plotted at $E_f = 0.85E_{<111>-bf}$ and $d = 30$ nm after the data on the Ag nanocrystalline materials and appears to show good agreement with the data found for Ag films. In Mizubayashi, et al. (1999b), it is supposed that a Ag film is composed of the surface, interface and grain boundary regions with an effective thickness t_x and a remaining region with a mean diameter $(d - t_x)$, where the volume fractions of the regions are counted as $(1 - R_r)$ and R_r with $R_r = [(d - t_x)/d]^3$, respectively. It is further assumed that Young's modulus E_r in the remaining region is the same as the bulk value E_b, and the Young's modulus E_x in the surface, interface and grain boundary regions can be defined and is the same in these regions. Under the rule of thumb, the dependence of the Young's modulus of the Ag film on the film thickness is evaluated as a composite material under the condition that the strains are equal in a constrained film. In Fig. 9.4, the dashed curve 1 is a calculated result with $E_x = 0.36E_b$, $t_x = 1.5$ nm and $E_b = E_{<111>-bf}$ and appears to follow the outline of the observed results including Young's modulus reported for Ag nanocrystalline materials (Qin, et al., 1998). The above value of E_x corresponds to $0.69E^{<100>}$.

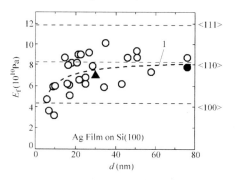

Figure 9.4 An enlarged drawing of Fig. 9.3(b) for the thickness range below 80 nm. The filled triangle denotes the data reported for nanocrystalline Ag (Qin, et al., 1998). See text for the dashed curve 1

Figure 9.5(a) shows the resonant frequency, f_a, and internal friction, Q^{-1}, of a Si reed substrate with a Ag film 10 nm thick. The internal friction Q^{-1} shows a steep increase above about 200 K and the shoulder around 215 K (the 215 K shoulder below). (The 155 K sharp internal friction peak reported in Tsakalakos and Hilliard(1983) is not observed in Fig. 9.5(a).) Figure 9.5 (b) shows the internal friction data observed after heating to 370 K, where the 215 K shoulder disappears. Figure 9.6(a) shows effects of heating on the 215 K shoulder observed for a Ag film 16 nm thick on a Si reed substrate. The 215 K shoulder disappears after heating to 370 K. Figure 9.6(b) shows the film thickness dependence of an amount of the internal friction in Ag films associated with the 215 K shoulder. The 215 K shoulder in Ag films shows a steep increase with decreasing film thickness d. A bulk Ag specimen does not show the 215 K shoulder and the steep increase around 200 K, suggesting that they are associated with anelastic relaxation processes in the surface, interface and/or grain boundary regions. The steep increase in the internal friction around 200 K has been also found in Al nm films (Mizubayashi, 1992; Berry and Pritchet, 1981) and in nanocrystalline Au (see Section 9.2), suggesting that it may reflect the common anelastic relaxation process in nanostructured metals.

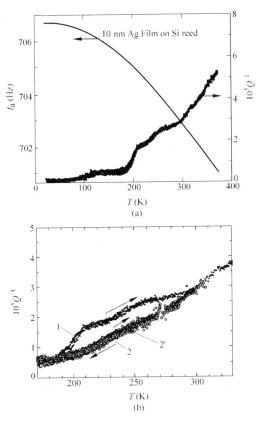

Figure 9.5 (a) The resonant frequency, f_a, and internal friction, Q^{-1}, of a Si reed substrate with a Ag film 10 nm thick. (b) 1 = enlarged drawing of Q^{-1} shown in (a), 2 and 2' = Q^{-1} after heating to 370 K (Mizubayashi; et al., 1999b)

Figure 9.7 shows the Young's modulus E_f versus. film thickness d data found in the Al films (Mizubayashi, et al., 1996). E_f nearby $E^{<100>}$ is found for Al films thicker than 15 nm and shows a decrease with decreasing film thickness below 15 nm. The dashed curve 2 is calculated for Al films assuming $E_x = 0.28 E_b$, $t_x = 0.5$ nm and $E_b = E_{<111>-bf}$ for d below 15 nm and E_f for d above 15 nm being independent of d. This value of E_x for Al films corresponds

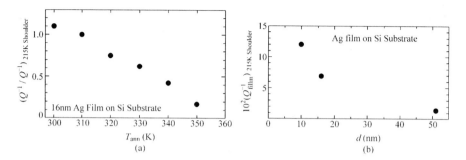

Figure 9.6 (a) Effects of heating up on the 215 K shoulder observed for a Ag film 16 nm thick on a Si reed substrate. (b) The film thickness d dependence of an amount of the internal friction in Ag films associated with the 215 K shoulder (Mizubayashi, 1992)

to $0.31E^{<100>}$. The values of E_x and t_x found above for Ag films and Al films are preliminary, and further work is required to clarify the origin for the decrease in Young's modulus observed for nm films. However, the outline of dependence of Young's modulus on film thickness suggests that Ag mono-metal and Al mono-metal may not be safe for wiring less than 50 nm wide and for wiring less than 15 nm wide, respectively.

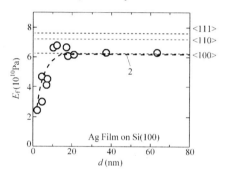

Figure 9.7 Young's modulus E_f vs. film thickness d data found in the Al films (Mizubayashi, et al., 1996). See text for the dashed curve 2

9.1.3 Supermodulus Effect in Ag/Pd Multilayers

The supermodulus effect (SME) in metal multilayers (MMLs) was firstly reported in Au/Ni, Cu/Ni, Cu/Pd and Ag/Pd MMLs (Tsakalakos and Hilliard, 1983; Henein and Hilliard, 1983). In spite of much effort, the data reported on the SME showed scattering among laboratories or specimens, giving rise to a suspicion on the appearance of the strong SME (Baker, et al., 1993). In Mizubayashi, et al. (1993, 1994a, 1999c) and Yamaguchi, et al. (1996) the composite reed method (see Fig. 9.1) oriented to MML film measurements is equipped to clarify this issue, and the SME in Ag/Pd MMLs and its disappearance after diffusion annealing is found, suggesting that the SME in Ag/Pd MMLs is of intrinsic singular behavior.

In Mizubayashi, et al. (1993, 1994a, 1999c) and Yamaguchi, et al. (1996) Ag/Pd MML films are deposited on the Si(100) surface of Si reed substrates at room temperature by an RF-sputtering method, where the deposition starts with a Ag layer. The modulation wavelength, λ, means the repetition of Pd layer $\lambda/2$ thick on Ag layer $\lambda/2$ thick below. The total thickness of the Ag/Pd MML films is about 800 nm (see Mizubayashi, et al., 1993 for details). The two types of specimens are obtained and will be referred to as A- and B-specimen. For A-specimens, the deposition rates of Ag and Pd are about 0.70 and 0.35 nm/s, respectively, and the adhesion between the Ag/Pd MML film and the Si reed substrate is strong enough to determine Young's modulus E_f of the films by means of the composite reed method. For B-specimens, the deposition rates of Ag and Pd are about 0.35 and 0.35 nm/s, respectively, and adhesion between the Ag/Pd MML film and the Si reed substrate is weaker than for the A-specimens. To be on the safe side, for the B-specimens the Ag/Pd MML film is stripped from the Si reed substrate, and then E_f of Ag/Pd MMLs is measured using the free-film reed with gage length of 300 μm. Figure 9.8 shows the E_f versus λ data found for Ag/Pd MML films. The data found for the A- and B-specimens show good agreement with each other, suggesting that constraint by a Si substrate does

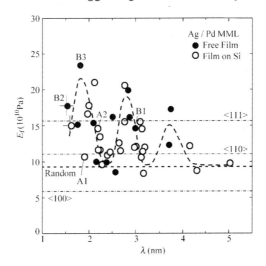

Figure 9.8 The dependence of Young's modulus E_f of Ag/Pd MML films on the wavelength λ. The open and the filled symbols are the data for the A- and B-specimens, respectively. The horizontal dashed lines $<100>$, random, $<110>$ and $<111>$ denote E_f calculated from the data of the bulk crystals (Mizubayashi, et al., 1993, 1994a, 1999c; Yamaguchi, et al., 1996)

not modify the SME. The SME beyond 100% is found for λ between 1.5 nm and 5 nm, where E_f shows local maximums at around $\lambda = 1.9$, 2.8 and 3.7 nm or oscillations with increasing λ.

Figure 9.9 shows the Young's modulus E_f of the A-specimens on Si reed substrates after heating to T_H, where the heating rate is about 0.04 K/s. The SME found in Ag/Pd MML films in the as-deposited state disappears after alloying at elevated temperatures, suggesting that the SME found in Ag/Pd MML films is of intrinsic. Further, the Ag/Pd MML film showing the stronger SME is more stable against annealing at elevated temperature than the Ag/Pd

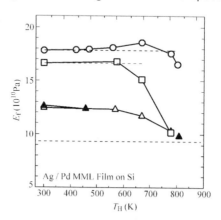

Figure 9.9 Young's modulus E_f of the A-specimens on Si reed substrates after heating to T_H (Mizubayashi, et al., 1993)

MML films showing the weaker SME, suggesting that the interatomic potential may be higher for the Ag/Pd MML films showing the stronger SME. Since the chemical reaction between Ag/Pd MML and Si comes in at elevated temperatures, free-film specimen is desired to study the effect of annealing at higher temperatures. Free-films stripped from the A-specimens have a curled shape with radius of 1 to 2 mm due to plastic deformation during stripping. The radius of the curled films decreases during annealing due to the stress or strain relaxation and then increases during annealing for interdiffusion at elevated temperatures. Such a change in the radius of the curled films is monitored as a change in the flexural vibration frequency during the step isothermal annealing, and the activation enthalpy, H, for interdiffusion is determined using a change of slopes method (Damask and Dienes, 1963). Figure 9.10 shows H determined for the A1- and A2-specimens which are indicated in Fig. 9.8. In Fig. 9.10, the open symbols denote the data for three free films of the A1-specimen, and the vertical arrow 1 denotes the temperature bounds above which the XRD indicate the progress in interdiffusion. The filled symbols and the vertical arrow 2 denote the results found for the free film of the A2-specimen, where arrow 2 denotes the temperature bounds above which

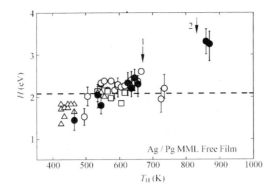

Figure 9.10 The activation enthalpy H found from step annealing. The open and filled symbols denote the data for the A1- and A2-specimens shown in Fig. 9.8 (Mizubayashi, et al., 1999c)

interdiffusion proceeds. The horizontal dashed line denotes H_{SD} reported for interdiffusion of Pd in Ag or Ag in Pd. H for interdiffusion found in the A1-specimen which shows no SME is similar to H_{SD}. The annealing behavior of Ag/Pd MML reported in Henein and Hilliard (1984) is similar to that found in the A1-specimen. In contrast, H for interdiffusion found in the A2-specimen shows a strong SME about 1.5 times as large as H_{SD}.

The change in the Young's modulus due to annealing is studied using flat free films of the B-specimens, B1 to B3 (see Fig. 9.8). For the B-specimens, a flat free film is sandwiched by two plates of fused quartz and then annealed as a whole for a half hour in 10^{-4} Pa, and then E_f is determined at room temperature by means of the free-film reed setup. Figure 9.11 shows the E_f

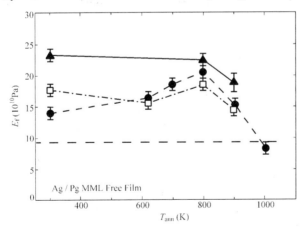

Figure 9.11 Young's modulus E_f of the B-specimens after annealing for 1.8 ks at temperature T_{ann}. ●: B1, □: B2, ▲: B3. See Fig. 9.8 for the B1 to B3 specimens (Mizubayashi, et al., 1999c)

versus annealing temperature, T_{ann}, data. The XRD indicates that interdiffusion starts above 800 K. In Fig. 9.11, all the specimens, B1 to B3, show a decrease in E_f for T_{ann} beyond 800 K, indicating that H for interdiffusion is about 1.5 times as large as H_{SD}. These results indicate that the atomic potential in the high-SME Ag/Pd MMLs is increased as a whole, as depicted schematically in Fig. 9.12. Since the stress or strain relaxation below 800 K causes a minor change in E_f, the lattice strain plays a minor role on the SME in Ag/Pd MML. We surmise that a certain modification in the electron system (Wu, 1982; Jankowski, 1992) is mainly responsible for the SME in Ag/Pd MML.

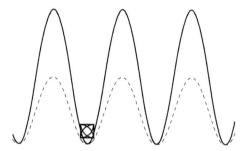

Figure 9.12 The solid curve depicts a model of the interatomic potential in Ag/Pd MML, and the dashed curve that in Ag/Pd homogeneous alloy (Mizubayashi, et al., 1999c)

9.2 Mechanical Behavior of High-Density Nanocrystalline Gold

High-density nanocrystalline gold (n-Au) specimens with almost the theoretical density can be prepared by the gas deposition method, and allow us to study their intrinsic mechanical properties. In the elastic range, Young's modulus found below 80 K for high-density n-Au specimens with mean grain size of 26 nm to 60 nm is higher than 92% of the bulk value, suggesting that Young's modulus in the grain boundary region is higher than 70% of the bulk value. Above 200 K, anelastic strain due to a certain relaxation in the grain boundary region increases with increasing temperature, giving a decrease in Young's modulus. Plasticity tests were made near room temperature for n-Au specimens with mean grain size of 15 nm to 60 nm. The Hall-Petch relationship is observed in Vickers microhardness tests with holding time of 10 s, and creep deformation proceeds in quasi-static tensile tests for 10^3 to 10^5 s, indicating that the preferable plastic deformation process near room temperature in n-Au is governed by not only the stress level as well as the strain rate.

A polycrystalline material with mean grain size d_g less than 100 nm is generally called a nanocrystalline (n-) material. Nowadays, n-metals can be prepared by several methods such as gas condensation and compaction (Birringer, et al., 1984; Gleiter, 1989), gas deposition (Sasaki, et al., 1996; Sasaki, 1998; Tanimoto, et al., 1996a), equal-channel angular pressing (Segel, et al., 1981; Valiev, et al., 1991; Nemoto, et al., 1999), high-energy ball milling (Koch and Cho, 1992) and crystallization of amorphous alloys. In the former two, n-metals are synthesized by consolidation or deposition of nanometer size gas-condensed ultrafine particles where atomic structure in grain boundary (GB) regions may, in general, differ much from those found in conventional polycrystalline materials (bulk materials, hereafter). Such n-metals have a large fraction of atoms in their GBs, and their properties may reflect those of the GB regions. In this section, we focus our attention on n-metals synthesized by gas condensation or gas deposition.

Since the early days of n-metal research, much attention has been devoted to their mechanical properties (Gleiter, 1993 and see the other papers in the same book), where the Hall-Petch relation predicts an increase in strength in proportion to $(1/d_g)^{0.5}$. On the other hand, a decrease in elastic moduli is expected because of an increased fraction of atoms in GBs. Because of small specimen sizes, strength of n-metals has mostly been investigated by Vickers microhardness tests, where mostly reported is an increase in the Vickers microhardness with increasing $(1/d_g)^{0.5}$ in the grain size range from about 10 nm to 100 nm and saturation or a decrease of the Vickers microhardness in the range of smaller grain sizes (Nieman, et al., 1989, 1992; Weertman, 1993; Weertman and Sanders, 1994; Kobelev, et al., 1993; Kumpmann, et al., 1993; Siegel, 1994; Siegel and Fougere, 1995; Qin, et al., 1995; Bonetti, et al., 1997). For the elastic property, the pioneering works report a drastic decrease in the elastic moduli of n-metals (see Table 9.1 and Nieman, et al., 1989, 1992; Weertman, 1993; Kobelev, et al., 1993; Korn, et al., 1988; Weller, et al., 1991; Sanders, et al., 1995), tentatively suggesting that the elastic moduli in the grain boundary region are much lower than in the crystalline region. Then the combination of the increase in the Vickers microhardness with increasing $\left(\frac{1}{d_g}\right)^{0.5}$ and the drastic decrease in the elastic moduli of n-metals reported in the literature predicts that dislocations under a high stress can be trapped by the grain boundaries with very low elastic moduli. However, recent elasticity works suggest that the high-density and low-porosity n-metal specimens do not show such a strong decrease in the elastic moduli (see Table 9.1 and Shen, et al., 1995; Sanders, et al., 1997a; Sakai, et al., 1999). In the following, the intrinsic elastic and plastic properties of n-metals found using high-density and low-porosity n-Au specimens will be given.

Figure 9.13 shows a schematic of a gas deposition apparatus (Sasaki, et al., 1996; Sasaki, 1998; Tanimoto, et al., 1996a, 1999; Sakai, et al., 1999)

Table 9.1 Elastic moduli of nanocrystalline metals and coarse-grained counterparts (Sakai. et al., 1999)

material	nanocrystalline. specimen			coarse-grained counterpart		measurement method	reference
	grain size (nm)	E (GPa)	G (GPa)	E (GPa)	G (GPa)		
Au	60	78.5±2 (80 K)		82.9 (80 K)		tensile test	Sakai. et al., 1999
	26—40	76.5±2 (80 K)				tensile test	Sakai. et al., 1999
	60	79.0±4 (20 K)		84 (20 K)		vibrating reed	Sakai. et al., 1999
Ag	60	~0.8E_b		(E_b = 82.7)		ultrasonic	Kobelev. et al., 1993
Cu	10—22	106±2 (RT)*		124 (RT)		tensile test	Sanders. et al., 1997a
	10—22	112±4 (RT)*	41.2±1.5 (RT)	131 (RT)	48.5 (RT)	ultrasonic	Sanders. et al., 1997a
	26	107				nanoindentation	Shen. et al., 1995
	15—61	45±9 (RT)		130 (RT)		tensile test	Nieman. et al., 1989
Pd	36. 47	129. 119 (RT)*		132 (RT)		tensile test	Sanders. et al., 1997a
	16—54	123±6 (RT)*	44.7±2 (RT)	132 (RT)	47.5 (RT)	ultrasonic	Sanders. et al., 1997a
	12	82±4 (RT)				ultrasonic	Sanders. et al., 1995
	5—15	44±22 (RT)		121 (RT)		tensile test	Nieman. et al., 1992
	—	88 (RT)	32 (RT)	123 (RT)	43 (RT)	ultrasonic	Korn. et al., 1988
	6		35 (RT)		43 (RT)	torsional vib.	Weller. et al., 1991

* E = 121±2 GPa for n-Cu and E = 130±1 GPa for n-Pd are reported after extrapolation to the porosity-free state.

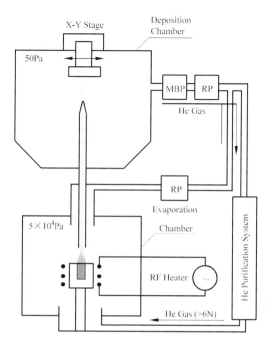

Figure 9.13 Schematic of a gas deposition apparatus, where a mechanical booster pump is abbreviated as MBP, and a rotary pump as RP (Sakai, et al., 1999)

composed of an evaporation chamber, a deposition chamber and a helium circulation system with purification columns, where the purity of circulating helium gas is about 99.9999%. The helium gas pressure during operation is about 5×10^4 Pa in the evaporation chamber and about 50 Pa in the deposition chamber, respectively. In the evaporation chamber, 99.99+% gold in a graphite crucible is heated to 1700—1900 K by the RF-induction heating system and ultrafine gold particles are formed just above the crucible. In more detail, the observation of an n-Au specimen by the atomic force microscope (AFM) suggests that small clusters of 10 to 20 ultrafine particles are simultaneously formed during the formation of the ultrafine particles above the Au melt (Tanimoto, et al., 1996a). Such clusters and ultrafine gold particles are immediately transferred to the deposition chamber through a transfer pipe with a jet nozzle and deposited on a glass substrate faced to the jet nozzle. A position of the glass substrate is controlled by a computer in order to prepare an n-Au specimen of a desired shape. A size of the n-Au specimens is 0.01—0.06 mm thick, 1 mm wide and 13 mm long. The ultimate velocity of ultrafine gold particles at the jet nozzle is estimated to be several tens of percent of the sound velocity, where a high-density specimen is expected because of high-speed bombardment of particles onto the substrate. Estimated time of flight before the bombardment is the order of 10^{-3} s, where, remarkably, reduction of contamination on particle surfaces is expected (Sasaki, 1998). After deposition, a n-Au specimen on a glass substrate is slowly immersed into water, where the n-Au specimen peels off the glass substrate and floats on the surface of water.

In Sakai, et al. (1999) and Tanimoto, et al. (1999) evaluation of a probable contamination of n-Au specimens during preparation is made by thermal desorption in a high vacuum for gaseous elements and by electron probe micro-analysis (EPMA) in SEM for heavy elements, where the following is found: the gaseous molecule most often observed is H_2O and a small amount of H_2 is detected. An amount of desorbed H_2O is independent of the specimen mass and shows an increase with increasing area of the specimen surface. The observed amount of desorbed H_2O is explained when we assume that the specimen surface was covered by monolayer of H_2O. In contrast, EPMA measurements indicate no detectable contamination by heavy elements.

In Sakai, et al. (1999) and Tanimoto, et al. (1999) the density of n-Au specimens is evaluated by the Archimedes' method and/or from measurements of mass and shape, with the results compiled in Table 9.2. The density of the n-Au specimens is very near to that reported for bulk Au. TEM observations made for some specimens indicate no detectable pore in the n-Au specimens. The mean grain size of the n-Au specimens is estimated from X-ray diffraction profiles and Scherrer's method using the (111), (200) and (220) reflections, and the results are listed in Table 9.1. The mean lattice spacing observed for n-Au specimens shows good agreement with that reported for bulk Au.

Table 9.2 Density of nanocrystalline Au specimens (Sakai, et al., 1999)

measurement method	sample (mean grain size)	density (g/cm^3)
Archimedes' method	n-Au (60 nm)	19.5±0.4
shape and weight	n-Au (20—60 nm)	19.3±0.8
(literature)	bulk-Au	19.32

Figure 9.14(a) is a schematic drawing for a tensile test apparatus with load cell capacity of 50 N (Sakai, et al., 1999). A tensile testing bed is handmade, where supporting pipes and a pulling rod are made from fused quartz in order to minimize the effect of thermal expansion. Figure 9.14(b) is a schematic for specimen clamping, where a n-Au specimen is sandwiched at its ends by conventional Au thin plates and then clamped by specimen holders. For clamping, conventional Au thin plates are plastically deformed to grip the n-Au specimen, where a clamping force is adjusted to avoid plastic deformation in the n-Au specimen. Figure 9.15 shows an example of calibration of the tensile test apparatus, where displacement of a fused quartz thin wire is evaluated after the subtraction of displacement of the tensile test apparatus. For the fused quartz thin wire, linear elasticity is observed in the

strain range. Young's modulus evaluated from the observed linear elasticity is 73.6±0.6 GPa, which shows good agreement with 73.0 GPa reported in the literature. For the elasticity measurements, displacement of a n-Au specimen is evaluated after subtracting displacement of the apparatus. Figure 9.16 shows an example of tensile test at 81.7 K on a n-Au specimen. In contrast to a fused quartz thin wire shown in Fig. 9.15, the stress versus strain curve observed for n-Au specimens shows, in general, deflection in the low-strain range. Since an amount of the deflection varies among specimens, the deflection may be an artifact. Young's modulus is evaluated from the linear elasticity found in the higher-strain range being free from such an artifact.

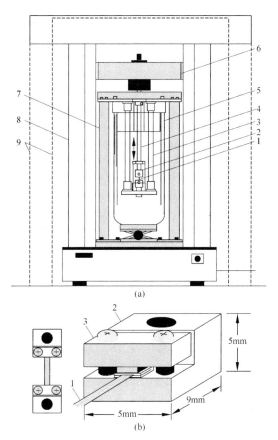

Figure 9.14 (a) Schematic drawing for a tensile test apparatus. 1: specimen, 2: specimen holder, 3: supporting pipes (fused quartz), 4: pulling rod (fused quartz), 5: thermos bottle, 6: cross head, 7: frame for testing bed, 8: main frame, 9: thermal shields. (b) Schematic drawing for specimen holder (Sakai, et al., 1999)

Figure 9.17 shows the dependence of Young's modulus on the measurement temperature observed for n-Au specimens. Below about 200 K,

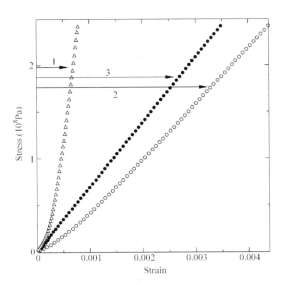

Figure 9.15 An example of calibration of the tensile test apparatus using a fused quartz thin wire 6.22×10^{-9} m^2 areal and 9.70 mm long, where strain rate is 3.7×10^{-4} s^{-1} and test temperature is 297.6 K. Arrow 1 denotes displacement of the apparatus alone, arrow 2 displacement of the apparatus with a fused quartz thin wire, and arrow 3 displacement of the fused quartz thin wire, respectively (Sakai, et al., 1999)

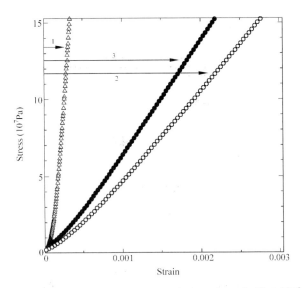

Figure 9.16 An example of tensile tests on a n-Au specimen 8.20×10^{-9} m^2 areal and 9.09 mm long, where strain rate is 3.7×10^{-4} s^{-1} and test temperature is 81.7 K (Sakai, et al., 1999)

the dependence of Young's modulus on the measurement temperature observed for n-Au specimens is very similar to that of coarse-grained Au, and the values of Young's modulus of n-Au specimens are lower by several percent than of coarse-grained Au. Among the data found in tensile tests, the values of Young's modulus of n-Au specimens tend to decrease with decreasing mean grain size. The values of Young's modulus found for n-Au specimens in Sakai, et al.(1999) are listed in Table 9.1 together with those reported for various n-metals.

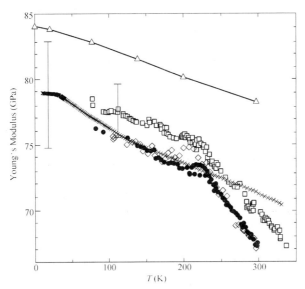

Figure 9.17 Young's modulus evaluated from tensile tests for three n-Au specimens with mean grain sizes of 60 (symbol □), 40 (symbol ◇) and 26 (symbol ●) nm and that found in vibrating reed measurements at about 700 Hz on an n-Au specimen with mean grain size of 60 nm (symbol ×). Young's modulus reported for coarse grained Au (symbol △) is also depicted (Sakai, et al., 1999)

As mentioned, a drastic decrease in the elastic moduli in n-metals has been reported in the pioneering works on the relatively-low-density n-metal specimens (Nieman, et al., 1989, 1992; Weertman, 1993; Kobelev, et al., 1993; Korn, et al., 1988; Weller, et al., 1991; Sanders, et al., 1995). In contrast, the values of Young's modulus in the high-density n-Au specimens (Sakai, et al., 1999), and the values of Young's modulus reported in the high-density n-Cu (Shen, et al., 1995; Sanders, et al., 1997a) and n-Pd (Sanders, et al., 1997a) specimens are very similar to those in the bulk counterpart. As the first step, an n-Au sample is supposed to be composed of the grain boundary region with the effective thickness t_{GB} and the remaining crystalline region with the mean diameter ($d_g - t_{GB}$), where volume fractions of the regions are counted as $(1 - R_{cry})$ and R_{cry} with $R_{cry} = [(d_g - t_{GB})/$

$d_g]^3$, respectively. It is further assumed that Young's modulus E_{cry} in the remaining crystalline region is the same as that in bulk Au, and Young's modulus E_{GB} in the grain boundary region can be defined. Then under rule of thumb, Young's modulus of a n-Au specimen is evaluated as a composite material under the condition that the stresses are equal while the strains are additive. The evaluation is made using Young's modulus data of the n-Au specimens with $d_g = 33$ (as the mean value for 26 to 40) nm and that of the n-Au specimens with $d_g = 60$ nm (see Fig. 9.17) and the results are listed in Table 9.1. Although a recent molecular dynamics simulation for n-Cu (Phillpot, et al., 1995) suggests t_{GB} in n-Cu being about 1 nm, the dependence of Young's modulus on d_g found in n-Au appears to be explained for t_{GB} being 2 nm to 3 nm, where E_{GB}/E_{crys} is evaluated as 0.7 to 0.8. It may be suggestive that the present ratio of E_{GB}/E_{crys} being 0.7 to 0.8 is similar to the ratio between the amorphous and crystalline states of alloys. On the other hand, although the density of the present n-Au specimens is very near to the bulk density (see Table 9.2), an effect of porosity on Young's modulus (Sakai, et al., 1999 and references therein) may be more or less expected. Therefore, the present value of t_{GB} is its upper bound, and that of E_{GB}/E_{crys} is its lower bound.

Above about 200 K the values of Young's modulus of n-Au specimens observed in tensile tests show a rapid decrease with increasing temperature. Figure 9.18 shows the stress relaxation behaviors of n-Au in the elastic range. The stress relaxation is not detected at 103 K, but is observed in the temperature range above about 200 K. The stress relaxation is not explained as a single relaxation; however, it is accelerated as a whole with increasing temperature. Figure 9.19 shows an example of internal friction measurements of n-Au at about 150 Hz, where a steep increase in the internal friction above about 200 K can be seen. (See Tanimoto, et al., 1996b for the low-

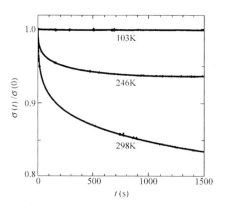

Figure 9.18 Examples of stress relaxation measurements on an n-Au specimen with mean grain size of about 26 nm. $\sigma(0)$ is about 0.14 GPa (Sakai, et al., 1999)

temperature internal friction peaks at around 90 K, 60 K and 25 K.) It is noted that for Al polycrystalline nm-films (Mizubayashi, et al., 1992) and Ag polycrystalline nm-films (Mizubayashi, et al., 1994b), a steep increase in the internal friction above about 200 K is also observed. We surmise that in n-metals, the anelastic relaxation processes in GBs are thermally activated above about 200 K. It is also noted that the hardness of n-Cu specimens shows a rapid decrease above about 200 K (Huang, et al., 1997).

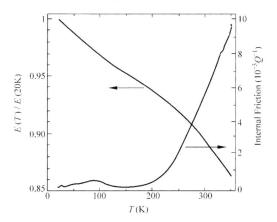

Figure 9.19 The internal friction (Q^{-1}) and Young's modulus in an n-Au specimen with mean grain size of 26 nm, where the measurement frequency is about 150 Hz (Sakai, et al., 1999)

Figure 9.20 shows stress versus strain curves at 77 and 293 K found in tensile tests of n-Au specimens with the mean grain size of 36 nm. The stress versus strain data shows an almost linear elasticity at 77 K in a wide strain

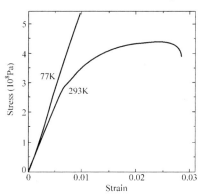

Figure 9.20 Tensile tests at 77 and 293 K on an n-Au specimen with mean grain size of 36 nm, where strain shown is calculated from the total displacement of a specimen and the apparatus. The nominal strain rate is about 1.5×10^{-4} s^{-1} (Tanimoto, et al., 1999)

range below 0.01. In contrast, at 293 K the plastic deformation starts at a strain of about 0.005, and a considerable amount of plastic deformation proceeds by creep when the strain rate is as low as 10^{-4} s^{-1}. Figure 9.21 shows creep strain under a given stress observed for n-Au specimens with mean grain size of 44 nm at RT. The plastic deformation starts at a threshold stress (see the arrow in Fig. 9.21), and plastic strain shows a linear increase with stress beyond the threshold stress. Recoverable strain (anelastic strain) shows a complicated change with stress. Figure 9.22 shows the dependence of the strain rate at the steady state on stress, where the linear relationship between the strain rate and stress is found in the stress range indicated by arrows 1 (threshold) and 2 (upper limit), suggesting Coble-type creep (Rai and Ashby, 1971).

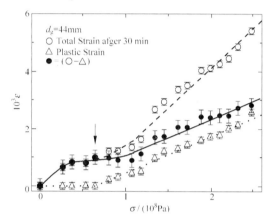

Figure 9.21 Creep tests under a given stress made at RT in n-Au specimens (Tanimoto, et al., 1999)

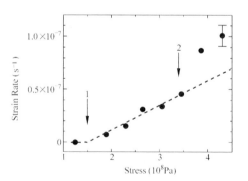

Figure 9.22 Strain rate found in the steady-state creep on the stress-control tests conducted at RT for n-Au specimens with $d_g \approx 36$ nm (Tanimoto, et al., 1999)

As short-term tests, Vickers microhardness measurements were made at room temperature (RT) in Sakai, et al. (1999) and Tanimoto, et al. (1999), where the as deposited n-Au specimens on the glass substrates were used. A benchmark test was made using a conventional Au foil of 20 μm thick on a glass substrate, where the data found with the ratio B/D of about 3 showed good agreement with the data observed with B/D being larger than 3 within an accuracy of about 10%, where B is the specimen thickness and D the indentation depth. Figure 9.23 shows the Hall-Petch plot for the results found in n-Au specimens with mean grain size of 15 to 60 nm with indentation load of 10 gf and loading time of 10 s, where B/D is kept to be larger than 3. Although the number of data points are limited, the results shown in Fig. 9.23 appear to be explained by the Hall-Petch plot, where the Hall-Petch slope is estimated as about 10 GPa $nm^{0.5}$. However, it may be premature to say that the Hall-Petch relation found in n-Au is indicative of the dislocation mechanism, because the preliminary STM observation does not appear to be explained by the dislocation mechanism (not shown here). The results shown in Fig. 9.20 to Fig. 9.23 indicate that the preferable deformation process varies with strain

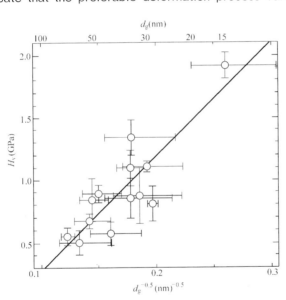

Figure 9.23 The H_V vs. $(d_g)^{0.5}$ data observed at RT for n-Au specimens (Tanimoto, et al., 1999)

rate and temperature. Figure 9.24 shows a schematic for the dependence of the yield or flow stress near room temperature on the inverse of the deformation rate and $(1/d_g)^{0.5}$ which is supposed for n-Au. It is noted that the role of creep on the plastic deformation behavior of n-materials is stirring up much interest (Valiev, 1997; Sanders, et al., 1997b; Wing, et al., 1997; Rittner, et al., 1997), because creep may govern plastic deformation of n-materials at a low strain rate above RT, giving superplasticity, dynamic control of the magnetic domain structure and so on. To clarify this issue in n-Au specimens, further work is required.

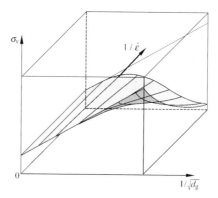

Figure 9.24 Schematic drawing for the dependence of the yield or flow stress near room temperature on the inverse of the deformation rate and $(1/d_g)^{0.5}$ supposed for n-Au (Sakai, et al., 1999, Tanimoto, et al., 1999)

9.3 FIB/TEM Observation of Defect Structure Underneath an Indentation

Defect structure beneath and near spherical indentations made with a load ranging from 20 mN to 120 mN in Si has been studied in detail by TEM. For preparation of foil specimens FIB milling was applied. Beneath the 20 mN indentation no defects were introduced, but for 45 mN and 60 mN indentations many dislocations were introduced.

9.3.1 Introduction

The indentation hardness test is widely used to determine the mechanical properties of a variety of materials. The continuous hardness tests where the load-displacement curves are recorded continuously have gathered much attention in determining mechanical properties of advanced materials (Pethica, et al., 1983; Newly, et al., 1982). An analysis of such a load-displacement curve provides information regarding near-surface plastic and elastic properties. However, in order to get any realistic picture on what happens inside the material under consideration, careful investigation by transmission electron microscopy (TEM) of microstructure underneath and near an indentation is indispensable (Page, 1992).

However, the problem with TEM is that preparation of TEM foil specimens is not easy, especially when the microstructure near the particular indentations is to be studied. Recently a new technique utilizing a focused-ion beam (FIB) system has been applied to the preparation of TEM specimens (Young, et al., 1990), and it is now possible to prepare TEM specimens from a pre-selected region with pinpoint accuracy (Ishitani, et al., 1996). This is an extremely powerful way to characterize the deformation structure beneath and near an indentation.

In this section, this novel technique will be described in some detail and a few examples will be given.

9.3.2 FIB Milling

Figure 9.25(a) shows the principle of a FIB system. Ga ions emitted from an ion gun are accelerated by 30 kV and deflected to bombard the surface of a specimen. In doing so, the surface of the specimen is sputtered and two trenches were milled in such a way that a thin wall left behind between the two trenches contains an indentation under consideration, as is schematically shown in Fig. 9.25(b). The fabrication can be carried out by monitoring the surface of the specimen being fabricated using scanning ion microscopic observation, as shown in Fig. 9.26.

9.3.3 Experimental Procedures

As a model specimen a dislocation-free single crystal of Si was used. Spherical indentations were performed on the (100) surface in air, at room temperature, using a load ranging from 20 mN to 120 mN with a 5 s dwell time with a CSIRO UMIS 2000 nanoindenter. The indented surfaces were examined with an optical microscope and then in a scanning electron microscope (SEM). The thin foil specimens were examined in the high-voltage electron microscope of Nagoya University, Hitachi HU-1000D, operated at an accelerating voltage of 1000 kV.

9.3.4 Load-Displacement Curve

Figure 9.27(a)—(c) shows a few examples of load-displacement curve. When the force is less than 20 mN, the behavior of material under the indenter is just elastic; no anomaly was observed. When the force is 45 mN, a well-defined pop-in phenomenon was observed but a pop-out phenomenon was not

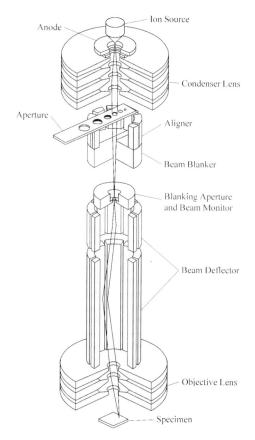

Figure 9.25 (a) The principle of FIB. (b) Geometry of the specimen prepared by FIB

Figure 9.26 SEM micrograph of a specimen prepared by FIB

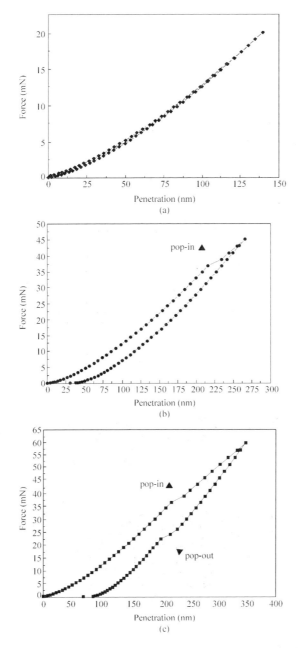

Figure 9.27 Examples of load-displacement curves of nanoindentation on Si surface. (a) 20 mN, (b) 45 mN, (c) 60 mN

observed. On the other hand when the force exceeded 60 mN, both pop-in and pop-out phenomena were observed.

9.3.5 TEM Observation

Figure 9.28 shows typical plan-view TEM micrographs of a 60 mN indentation where both pop-in and pop-out took place. Figures 9.28(a) and (b) were imaged in the bright-field and dark-field modes, respectively. Just underneath the indentation there exists a pyramid-like zone, which is deformed both elastically and plastically and contains a high density of dislocations and slip

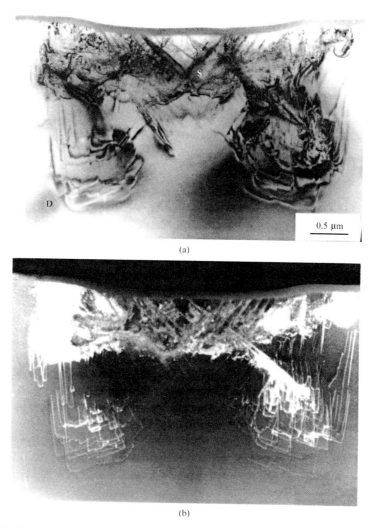

Figure 9.28 Low-magnification TEM micrograph of a 60 mN indentation: (a) bright-field image, (b) dark-field image (H. Saka, 2002)

bands. Underneath the periphery of the indentation, many L-shaped dislocations were nucleated and propagated into the interior of the crystal. Those dislocations running vertically are identified to be 30° dislocation with Burgers vector 110 (Hill and Rowcliff, 1974). Those running horizontally are identified screw dislocations.

Figure 9.29 shows typical plan-view TEM micrographs of a 45 mN indentation where only pop-in was observed. Figures 9.29(a) and (b) were imaged in the bright-field and dark-field modes, respectively. Comparison of Fig. 9.29 with Fig. 9.28 revealed that the major difference is that in Fig. 9.28 L-shaped dislocations are dominating, while in Fig. 9.29 only straight vertical dislocations exist underneath the periphery of the indentation. In other words,

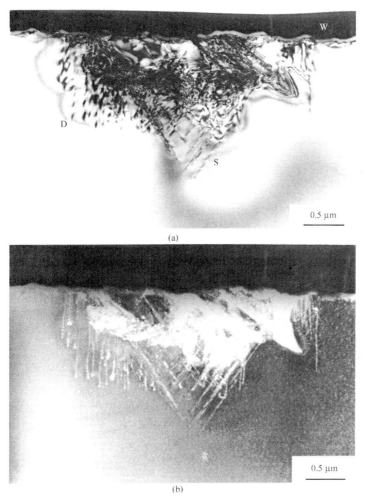

Figure 9.29 Low-magnification TEM micrograph of a 45 mN indentation: (a) bright-field image, (b) dark-field image (H. Saka, 2002)

in Fig. 9.29 only 30° dislocations are activated, while in Fig. 9.28 both 30° and screw dislocations are activated.

Figure 9.30 shows a SEM micrograph of indentations made with a load of 20 mN. In this case, as can be seen from the force-penetration curve, the specimen behaves in a purely elastic way. No image of the indenter was printed on the surface within the resolution SEM allows. Therefore, 100 mN indentation were made, as markers, in a grid array and 20 mN indentations were performed between two 100 mN indentations.

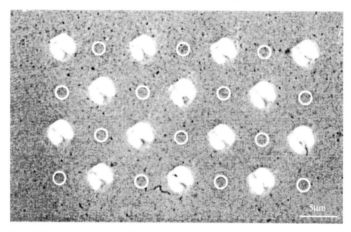

Figure 9.30 SEM micrograph of 10 mN indentations. 10 mN indentations were performed between a grid array of 100 mN indentations (H. Saka, 2002)

Figure 9.31 shows cross-sectional TEM micrographs of a 20 mN indentation imaged in the bright-field mode. I indicates the positions of the two neighboring 100 mN indentations. The region indicated by one arrow just in

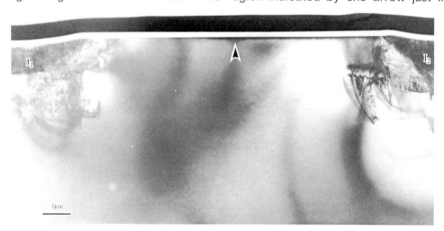

Figure 9.31 Low-magnification bright-field TEM micrograph of a 20 mN indentation. A 20 mN indentation was performed near the region indicated by arrows

between two l's is that region where 20 mN indentation was performed. Obviously nothing happened underneath the 20 mN indentation. This is in good agreement with the force-penetration curve in the sense that the material behaves elastically.

9.3.6 Conclusion

In order to understand what is happening in material under an indenter, cross-sectional TEM observation is very useful. For preparation of foil specimens which contain indentations under consideration, FIB milling is very powerful.

References

Baker, S.P., M.K. Small, J.J. Vlassak, B.J. Daniel, and W.D. Nix. Mechanical Properties and Deformation Behavior of Materials Having Ultra Fine Microstructures. edited by M. Nastasi, et al., Kluwer, Dordrecht (1993)
Barmatz, M., and H.S. Chen. Phys. Rev., B, 9, 4073(1974)
Berry, B.S., and A.C. Pritchet. J. Physique (Paris), 42, C5-1111(1981)
Birringer, R., U. Herr, and H. Gleiter. Phys. Lett., 102A, 365(1984)
Bonetti, E., L. Pasquini and E. Sampolesi. Nanostruct. Mater., 9, 611 (1997)
Damask, A.C., and G.J. Dienes. Point Defects in Metals. Gordon and Breach, New York (1963)
Gleiter, H., Prog. Mater. Sci., 33, 223(1989)
Gleiter, H., Mechanical Properties and Deformation Behavior of Materials Having Ultra-Fine Microstructures. Edited by M. Nastasi, et al., Kluwer, Amsterdam (1993)
Henein, G.E., and J.E. Hilliard. J. Appl. Phys., 54, 728(1983)
Henein, G.E., and J.E. Hilliard. J. Appl. Phys., 55, 2895(1984)
Hill, M.J., and Rowcliff, D.J., J. Materi. Sci., 9, 1569(1974)
Huang, Z., L.Y. Gu and J.R. Weertman. Scripta Mater., 37, 1071(1997)
Ishitani, R., and Yaguchi, T., Microscopy Res. and Tech., 35, 320(1996)
Jankowski, A.F., J. Appl. Phys., 71, 1782(1992)
Jones, R.S., J.A. Slotwinski and J.M. Jintmire. Phys. Rev., B 45, 13624 (1992)
Kobelev, N.P., Ya.M. Soifer, R.A. Andrievski and B. Gunther. Nanostruct. Mater., 2, 537(1993)
Koch, C.C., and Y.S. Cho. Nanostruct. Mater., 1, 207(1992)
Koch, R., J. Phys. Condens. Matter. 6, 9519(1994)
Korn, D., A. Morsch, R. Birringer, W. Arnold and H. Gleiter. J. Phys.

(Paris). **49**, C5-769(1988)

Kraft, O., and E. Arzt. Acta mater.. **46**, 3733(1998)

Kumpmann, A., B. Gunther and H.-D. Kunze. Mechanical Properties and Deformation Behavior of Materials Having Ultra-Fine Microstructures. Edited by M. Nastasi, et al.. Kluwer, Amsterdam (1993)

Li, S. J., M. Ishihara, H. Yumoto, T. Aizawa and M. Shimotomai. Thin Solid Films. **316**, 100(1998)

Mizubayashi, H., Y. Yoshihara and S. Okuda. Phys. Stat. Sol.. (a) **129**, 475(1992)

Mizubayashi, H., T. Yamaguchi, W. Song, A. Yamaguchi and R. Yamamoto. J. Magnetism and Magnetic Mater.. **126**, 197(1993)

Mizubayashi, H., T. Yamaguchi, W. Song, A. Yamaguchi and R. Yamamoto. J. Alloys and Compounds. **211/212**, 442(1994a)

Mizubayashi, H., T. Yamaguchi and Y. Yoshihara. J. Alloys and Compounds. **211/212**, 446(1994b)

Mizubayashi, H., S. Harada and T. Yamaguchi. J. Physique IV (Paris). **6**, C8-799(1996)

Mizubayashi, H., S.J. Li, H. Yumoto, and M. Shimotomai. Scripta Mater.. **40**, 773(1999a)

Mizubayashi, H., J. Matsuno and H. Tanimoto. Sciripta Mater.. **41**, 443—448(1999b)

Mizubayashi, H., S. Iwasaki, T. Yamaguchi, Y. Yoshihara, W. Soe and R. Yamamoto. J. Magnetism and Magnetic Mater.. **198/199**, 605—607(1999c)

Nemoto, M., Z. Horita, M. Furukawa and T. G. Langdon. Mater. Sci. Forum. **304—306**, 59(1999)

Newly,D., Wilkins,M.A., and Pollcok,H.M.. J.Phys.. **E15**,119(1982)

Nieman, G. W., J. R. Weertman and R. W. Siegel. Scripta Metall.. **23**, 2013 (1989)

Nieman, G. W., J. R. Weertman and R. W. Siegel. Nanostruct. Mater.. **1**, 185(1992)

Nix, W.D.. Mater. Sci. Engng.. **A234—236**, 37(1997)

Nowick, A. S., and B. S. Berry. Anelastic Relaxation in Crystalline Solids. Academic Press, New York (1972)

Page,T.F.. J.Materi. Res.. **7**,450(1992)

Pethica,J. B., Hutchings, R., and Oliver, W. C.. Phil. Mag.. **A 48**, 593 (1983)

Phillpot, S.R., D. Wolf and H. Gleiter. J. Appl. Phys.. **78**, 847(1995)

Qin, X. Y., X. J. Wu and L. D. Zhang. Nanostruct. Mater.. **5**, 101(1995)

Qin, X. Y., X. R. Zhang, G. S. Cheng and L. D. Zhang. NanoStructured Mater.. **10**, 661(1998)

Rai, R., and M.F. Ashby. Met. Trans.. **2**, 1113(1971)

Rittner, M. N., J. R. Weertman, J. A. Eastman, K. B. Yoder and D. S. Stone. Mater. Sci. Engng.. **A 234—236**, 185(1997)

Sanders, P.G., A.B. Witney, J.R. Weertman, R.Z. Valiev and R.W. Siegel. Mat. Sci. Engng., **A 204**, 7(1995)

Sanders, P.G., J.A. Eastman and J.R. Weertman. Acta Mater., **45**, 4019 (1997a)

Sanders, P.G., C.J. Youngdahl and J.R. Weertman. Mater. Sci. Engng., **A 234—236**, 77(1997b)

Sakai, S., H. Tanimoto and H. Mizubayashi. Acta Mater., **47**, 211(1999)

Sasaki, Y., K. Shiozawa, H. Tanimoto, Y. Iwamoto, E. Kita and A. Tasaki. Mater. Sci. Engng., **A 217/218**, 344(1996)

Sasaki, Y.. Thesis. University of Tsukuba, (1998)

Segal, V.M., V.I. Reznikov, A.E. Drobyshevskiy and V.I. Kopylov. Russian Metallurgy. Metally **1**, 99(1981)

Shen, T.D., C.C. Koch, T.T. Tsui and G.M. Pharr. J. Mater. Res., **10**, 2892(1995)

Siegel, R.W.. Nanostruct. Mater., **4**, 121(1994)

Siegel, R.W., and G.E. Fougere. Nanostruct. Mater., **6**, 205(1995)

Tanimoto, H., H. Fujita, H. Mizubayashi, Y. Sasaki, E. Kita, and S. Okuda. Mater. Sci. Engng., **A 217/218**, 108(1996a)

Tanimoto, H., H. Mizubayashi, H. Fujita and S. Okuda. J. Phys. IV (Paris) **6**, C8-199(1996b)

Tanimoto, H., S. Sakai and H. Mizubayashi. NanoStructured Mater., **12**, 751(1999)

Tsakalakos, T., and J.E. Hilliard. J. Appl. Phys., **54**, 734(1983)

Uozumi, K., H. Honda and A. Kinbara. Thin Solid Films. **37**, L49(1976)

Valiev, R.Z., and N.K. Tsenev. Hot Deformation of Alminum Alloys. Edited by T.G. Langdon, H.D. Merchant, J.G. Morris and M.A. Zaidi, (The Minerals, Metals and Materials Society, Warrendale, PA) 319(1991)

Valiev, R.Z.. Mater. Sci. Engng., **A 234—236**, 59(1997)

Weertman, J.R.. Mat. Sci. Engning., **A166**, 161(1993)

Weertman, J.R., and P.G. Sanders. Solid State Phenomena. **35—36**, 249 (1994)

Weller, M., J. Diehl and H.-E. Schaefer. Phil. Mag., **A 63**, 527(1991)

Wing, N., Z. Wang, K.T. Aust and U. Erb. Mater. Sci. Engng., **A 234—236**, 150(1997)

Wu, T.B.. J. Appl. Phys., **53**, 5265(1982)

Yamaguchi, T., H. Mizubayashi, Y. Yoshihara, W. Song, A. Yamaguchi, and R. Yamamoto. J. Magnetism and Magnetic Mater. **156**, 279(1996)

Young, R., J., Kirk, E.C.G., Williams, D.A., and Ahmed, H.. in Specimen Preparation for Transmission Electron Microscopy of Materials II. edited by R. Anderson. Mater. Res. Soc., Pittsburgh, 205(1990)

Yumoto, H., K. Kaneko, M. Ishihara, Y. Kato and K. Akashi. Thin Solid Films. **281/282**, 311(1996a)

Yumoto, H., S. Onozumi, Y. Kato, M. Ishihara and K. Kishi. Cryst. Res. Technol., **31**, 159(1996b)

10 Thermal Analysis

Qing Jiang and Ke Lu

10.1 Introduction

This chapter provides an introduction to thermal analysis technique such as different scanning calorimeter (DSC) and differential thermal analysis (DTA) to characterize the thermal properties of nanophases and nanostructured materials. The principles introduced here should be generally applicable to all materials.

Nanophases are usually single crystals with sizes in the nanometer region (typically less than 100 nm in at least one dimension). Nanostructured materials are single or multi-phase polycrystals of which the grain sizes are in the nanometer region. Owing to this extremely small size, nanocrystals or nanostructured materials are structurally characterized by a large proportion of surface or interface atoms (ions). Since the vibrational and positional enthalpy and entropy of surface or interface atoms of nanophases and nanostructured materials evidently differ from that of atoms within the crystals, which may significantly alter a variety of physical, mechanical, and chemical properties with respect to the conventional bulk materials (Gleiter, 1988; Lu, 1996; Suryanarayana, 1995). As the nanophases, such as nanoparticles, nanowires and thin films will and have been widely applied in semiconductor industry, these differences must be identified.

The thermal property of materials is one of the most important physical properties of materials. The study of the thermal property of nanophases and nanostructured materials at present essentially concentrates on melting temperatures, thermodynamics and kinetics for nanocrystal-liquid transition and nanocrystal-glass transition, grain growth kinetics of nanophases and nanostructured materials, the specific heat of nanostructured materials, surface (interface) enthalpy. Since these thermal properties are strongly related to the size of nanocrystals especially when the radius of nanocrystals is smaller than 10 nm, an important task of these thermal analyses is to find the size-dependent functions of these thermodynamic amounts of nanostructured materials. All of these thermal properties of nanophases and nanostructured materials can, or only can, be obtained by a thermal analysis technique. Thus, thermal analysis is widely utilized in the study of nanophases and nanostructured materials. These works will enrich thermodynamic and kinetic theories of materials science and supply evidence of thermal stability of nanophases and nanostructured materials due to the metastable nature of nanophases and nanostructured materials.

10.2 Fundamental Techniques

Although there are many different calorimeters to measure thermodynamic parameter of materials, such as the adiabatic calorimeter, the drop calorimeter, the levitation calorimeter, the modulation calorimeter, the most popular commercial calorimeter is the differential scanning calorimeter (DSC) for the study of nanophases and nanostructured materials. This is because the DSC has a higher measuring accuracy and its measuring temperature range among 100—1000 K is suitable for most cases. When a higher temperature (> 1000 K) is required, the thermal differential analyzer (DTA) can be utilized. However, the latter has a lower measuring accuracy and cannot measure the heat capacity of materials with a satisfied accuracy. Thus, in this section, we will essentially introduce measuring techniques of DTA/DSC in detail and indicate the characteristics of the above instruments (Cezairliyan, 1988; Jiang, et al., 1998a; Speyer, 1994).

To begin, it is helpful to read the first law of thermodynamics,

$$dU = dQ - dW \qquad (10.1)$$

where U is the internal energy, Q is heat, and W is work. This equation states that the change in energy of a system is dependent on the heat that flows in or out of the system and how much work the system does or has done on it. Under conditions where no work is done on/by the system, the change in internal energy of the system is equal to the heat flowing in or out of it. For condensed phases, which are effectively incompressible, the volume dependence on the change in internal energy is negligible. Thus, the internal energies of liquids and solids are considered a function of temperature alone. For this reason, the internal energy of a system may loosely be referred to as the "thermal energy" (Speyer, 1994). Through measuring the thermal energy and temperature by DTA/DSC with different heating rates (B), the thermodynamics and kinetic functions of phases are obtained, which essentially include:
(1) The heat released or absorbed in a phase transition at constant pressure (latent heat, H_t). A DSC plot of dQ/dt versus temperature may be translated to a plot of dQ/dt versus time (t), using the heating rate. The heat released/absorbed in a reaction is simply the area under the peak.

$$Q = \int_{-\infty}^{\infty} (dQ/dt)dt \qquad (10.2)$$

$Q \approx H_t$ when the pressure is constant. For nanostructured materials, through measuring the difference between the transition enthalpy of glass-nanocrystal and that of glass-coarse-grained crystals, the message of interface enthalpy can be obtained.

(2) Heat capacity (C_p) at constant pressure (the ability of a substance to hold thermal energy) is defined as

$$C_p = (\partial H/\partial T)_p = dQ/dT \qquad (10.3)$$

(3) Information with respect to the kinetics of a reaction may be established from DTA/DSC output by determining the fraction of reactant transformed and then fitting these data to some reaction kinetics models (see Section 10.3.2). Also the grain growth of a reactant can be considered in a similar way.

To better finish the above purposes, the measuring principle of DTA/DSC must be fully understood. Figure 10.1 (Speyer, 1994) is a schematic of the DTA design. The device measures the difference in temperature between a sample and a reference that are exposed to the same heating schedule via symmetric placement with respect to the furnace. The reference material is any substance, with about the same thermal mass as the sample, which undergoes no transitions in the temperature range of interest. The temperature difference between sample and reference is measured by a "differential" thermocouple in which one junction is in contact with the underside of the sample crucible, and the other is in contact with the underside of the reference crucible. The sample temperature is measured via the voltage across the appropriate screw terminals and similarly for the reference temperature. Sample and reference temperature measurements require cold junction temperature correction.

When the sample undergoes a transition, it will either absorb (endothermic) or release (exothermic) heat. If the sample and the reference are exposed to a constant heating rate, the X-axis is often denoted as temperature since temperature is proportional to time. Plotting temperature on the X-axis thus implies an experiment in which a constant heating rate was used.

If there is good "communication" between the heating elements and the sample or the reference thermocouple junctions, then the control system can make its power adjustments based on one of those temperatures. If there is substantial insulation between these locations, which may be necessary for heat flow uniformity to both sample and reference or to permit the introduction of special gases (usually via argon or nitrogen with purity of 99.99%), then a separate control thermocouple placed near the furnace windings is used.

Although the output traces of a DSC are visually similar to a DTA, the operating principles of these devices are entirely different. Figure 10.2

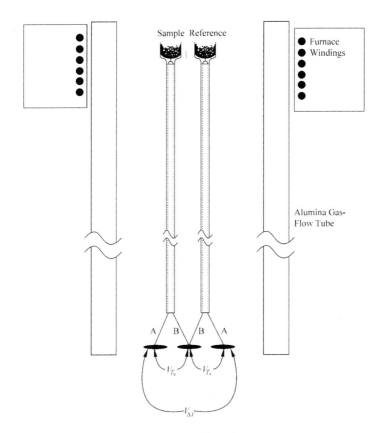

Figure 10.1 Schematic of a differential thermal analyzer

(Speyer, 1994) shows a schematic of the DSC design. There are separate containers for both the sample and the reference, and associated with each are individual heating elements as well as temperature measuring devices. Both cells are surrounded by a refrigerated medium (usually via flowing water or liquid nitrogen) which permits rapid cooling.

The sample and reference chambers are heated equally into a temperature regime where a transition takes place within the sample. As the sample temperature infinitesimally deviates from the reference temperature, the device detects it and reduces the heat input to one cell while adding heat to the other, so as to maintain a zero temperature difference between the sample and reference, establishing a "null balance". To maintain this null balance, the quantity of electrical energy per unit time that must be supplied to the heating elements (over and above the normal thermal schedule) is assumed to be proportional to the heat released per unit time by the sample. Hence, the y-axis is expressed in terms of energy per unit time (watts). This DSC has two control cycle portions. One portion strives to maintain the null balance

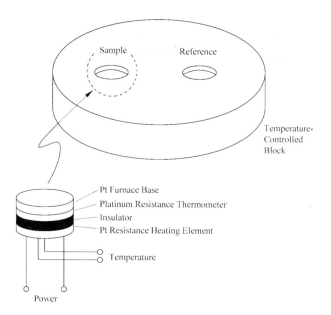

Figure 10.2 Schematic of a power-compensated differential scanning calorimeter

between sample and reference, while the other strives to keep the average of the sample and reference temperature. These processes switch back and forth quickly so as to maintain both simultaneously.

The above description of a DSC is called a "power-compensated" DSC marketed by Perkin-Elmer Corporation. Another type of DSC, "heat flux" DSC, actually operates based on a DTA principle, where a single heating chamber is used. A calibration constant within the computer software (determined using standard materials) converts the amplified differential thermocouple voltage to energy per unit time, which is plotted on the y-axis of the instrument output.

Nanophases with smaller mass of samples can be measured by some self-made DTA (the so-called scanning microcalorimeter) with a high sensitivity of $1J/m^2$ (Lai, et al., 1995; 1996). The system is designed using ultrathin SiN membranes that serve as a low thermal mass mechanical support structure for the calorimeter. Calorimetry measurements of the system are accomplished via resistive heating techniques applied to a thin film Ni heating element that also serves as a thermometer. A current pulse through the Ni heater generates heat in the sample via Joule heating. The voltage and current characteristics of the heater were measured to obtain real-time values of the temperature and the heat delivered to the system. This instrument has directly measured melting temperature and melting enthalpy of Sn nanophases with radii among 2.5—25 nm. Figure 10.3 shows the view of this microcalorimeter (Lai, et al., 1995).

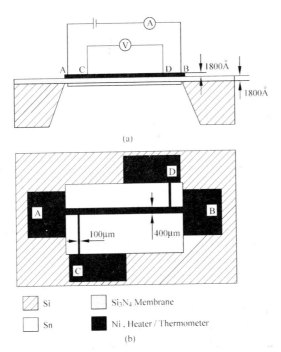

Figure 10.3 Cross-sectional (a) and planar (b) view of microcalorimeter. System is based on a reduced thermal mass sample holder made using a silicon nitride membrane in a thickness of 180 nm and supported at the perimeter by the Si substrate. A thin Ni film layer is deposited on the top side of the membrane and serves as both heater and thermometer. The sample of interest is deposited on the bottom side of the membrane

Figure 10.4 (Jiang, et al. 1998b) shows an example of a DSC trace of the heating of a crystalline Indium. The material underwent an endothermic transition representing the melting of the crystalline phase. The melting temperature is represented by the onset of the peak (M_s). The convention for determining the melting temperature is to extend the straight-line portions of the baseline and the linear portion of the upward slope, marking their intersection. The M_s is actually the melting temperature of bulk crystals. The point (M_d) at which the first indication of a deviation from the baseline is observed is the onset of the transition on the crystalline defects, such as surfaces, interfaces, dislocations, etc. However, it may be difficult to reach universal agreement as to where that location is. At the peak point of the curve (M_p), the transition is finished.

Any thermal analysis instruments must be calibrated using well-characterized pure metals, such as In, Sn, Pb and Zn. The software permits input of peak area values and onset temperatures determined by a test run, as well as values from the literature, into a program. It then automatically applies abscissa and ordinate corrections to all future data collected by the instrument.

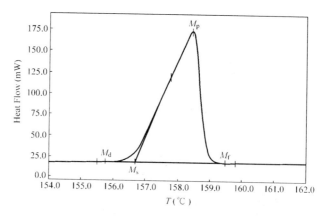

Figure 10.4 Melting curve of indium on a DSC. M_s is the onset temperature of the melting or the beginning melting temperature of the bulk crystals. M_d is the onset of the transition on the crystalline defects. M_p is the peak temperature where the transition is finished. M_f is the apparent finishing temperature of the melting

If the instrument is continuously utilized, this calibration can be carried out once every month. If the instrument has not been used for some time, the instrument must be calibrated before the measurement. Note also that the calibrated value is heating rate dependent.

10.3 Experimental Approach

There are abundant methods (chemical, physical and mechanical) to produce nanophases and nanostructured materials (see Handbook of Nanophase and Nanostructured Materials-Synthesis, Chapter 1 to Chapter 11). If the masses of nanostructured materials are more than several milligrams, a commercial DSC can directly measure these materials (Jiang, et al., 1998b). However, the measurement may be carried out only below a specific temperature at which grain growth occurs. If the materials are nanophases, which are easy to grow and aggregate, some special methods must be used to separate nanocrystals in order to avoid these growths and aggregations. A typical method is the use of controlled pore glass materials (CPG) with narrow pore-size distributions (Jackson and McKenna, 1990). This kind of material is suitable for low melting temperature nanophases that can be filled in the CPG when they melt. In addition, when the mass of the nanophases is small in nanograms, sometimes the so-called microcalorimeter (DTA) with a very high sensitivity is necessary (see Section 10.2). In the following, measuring methods detail will be introduced for some special research targets.

10.3.1 Melting of Nanophases and Nanostructured Materials

DTA/DSC can determine the melting temperature and latent heat of transition. In a DTA scan at a constant heating rate where a solid sample fuses, the reference material increases in temperature at the designated heating rate, while the sample temperature remains at the melting temperature until the transition is complete. Thus, when temperature versus reference temperature is plotted, a linear deviation from the baseline value is seen on the leading edge of the endotherm. The peak of the endotherm represents the temperature at which melting terminates (M_p). When the transition is complete, the temperature of sample is lower than its surroundings; it thus heats at an initially accelerated rate, returning to the temperature of the surroundings. The corresponding DTA trace returns to the baseline. Although the melting transition is complete at the peak of the endotherm, it is still the entire area under the peak that represents the latent heat of fusion. Recall that the enthalpy of the reference increases during the time of the transition since its temperature increases, as dictated by the heating rate. The time rate of change of reference enthalpy was subtracted from the time rate of change in sample enthalpy, leaving only the enthalpy change due to the latent heat of transition in the sample. However, during sample melting (linear portion of the endotherm between onset and maximum of the endotherm), there is no enthalpy change due to temperature increase, simply because there is no temperature increase. Only after the completion of the transition does the sample temperature catch up to that of the reference; this corresponds to the exponential decay portion of the endotherm. This portion of the peak thus needs to be included in the integral, so that both sample and reference have shown the same contribution to enthalpy change due to increased temperature.

The linear drop and exponential recovery shape of these transitions also appear in power-compensated DSC traces, but for different reasons. The temperature-measuring device measures its own temperature. The device adds power to the sample side as needed to compensate for the cooling effect on the chamber due to sample melting. This energy requirement increases linearly since the sample temperature increases linearly. When melting is over, the need for extra heat flow to the sample chamber side drops exponentially as the chamber temperature quickly catches up to the setpoint.

Heating rate is an important consideration in the investigations (Jiang, et al., 1997b). Slower heating rates will more accurately depict the onset temperature of a transition. They will also diminish the accelerating effects of self-feeding reactions. Further, two transitions which are very close in temperature range may be more distinctly seen as separate peaks, whereas they may be mistaken for a single transition under a rapid heating rate. On the other hand, slower heating rates make the peaks shorter and broader (in time). Thus, a transition with a minute thermal effect may result in a peak height no more intense than the thickness of a line under slow heating rates and therefore will not be observed. In such a case, a faster heating rate will increase the peak intensity, making it visible.

10.3.2 Kinetics of Glass-Nanocrystal Transition and Grain Growth of Nanostructured Materials

DSC can measure the kinetics of transition from metastable phases (Lu, 1996; Speyer, 1994). In these types of reactions, the sample temperature will not remain at a constant value during the transition. During the glass-nanocrystal transition, for example, heat released at the glass-crystal interface raises temperature of the sample. The rates of such transitions generally have an exponential temperature dependence, causing them to proceed more quickly, which in turn causes a more rapid temperature rise, and so on. As a result, these "self-feeding" reactions will show irregular temperature/time profiles when externally heated at a constant rate. To maintain the sample at the setpoint temperature during a self-feeding reaction, small sample mass (e.g. <10 mg) and excellent thermal contact between the sample and its container, as well as the container and the chamber, are required.

Since metallic glass and nanophases are usually powders or thin ribbons, the packing of the particles or the thin ribbons will have an effect on the transition temperature of the sample. Good mechanical contact between the sample and the bottom of the crucible will improve instrument sensitivity to transitions. Using samples shaped to match the crucible, or finely crushed granules, as opposed to more spherical or odd-shaped chunks may optimize the surface contact. However, because there exists air among the particles, which has poor thermal contact, its lag time is much larger than a disk-like sample (the best shape of the sample is that the diameter of the sample is 5.8 cm and the thickness of the sample is 2 mm). Optimum mechanical contact minimizes the lag time between the time when a reaction occurs and when heat propagates to/from the point of temperature measurement, and the reaction is recorded.

The properties of the sample must be known in advance, to some extent, in order to decide if granulating it will alter its behavior. For example, the added surface energy of crushing a glass may change the crystal nucleation and growth mechanism from a bulk to a surface effect. Grinding of metals will strain and harden them, which will in turn cause a recrystallization exotherm when the sample is heated. This exotherm will not appear when the sample is exposed to the same thermal schedule a second time.

For glass-nanocrystal transition, the particle size in the sample container will affect the rate of reaction. Heat flow from the surroundings to the interior of larger particles would be relatively suppressed because of diminished access. This, in turn, would cause a time delay in the onset of the transition in larger particles. The heat released from transitions within more massive particles would also be more insulated from escape, further stalling detection of the transition. The latent heat of transition would then alternatively tend to increase large particle temperature more than in smaller particles. The greater temperature rise in larger particles accelerates the reaction rate causing the reaction to terminate earlier.

Normally, the value of activation energy is related to the order of phase transition. A higher order of phase transitions may correspond to a smaller value of activation energy. The study of the activation energy for phase transition can help us to understand the kinetics of phase transition and the stability of metastable phase.

An isothermal experiment on DSC can be run in which the sample temperature is quickly brought up into a temperature regime of interest and held during the isothermal crystallization transition. Determining the fraction transformed as a function of time from partial areas under the DSC peak is made by means of a DSC software. The crystallization kinetics of a glass may be described by the Johnson-Mehl-Avrami (JMA) model,

$$x(t) = 1 - \exp[-k(t-t_0)^n] \quad (10.4)$$

where $x(t)$ as function of time (t) is the transition fraction, k is a time independent but temperature dependent rate constant, t_0 is incubation time, and n is referred to as the mechanism constant. If in the derivation of this equation, it was assumed that nucleation occurs homogeneously (e.g., without impurity surfaces to catalyze the process) and the crystalline regions grew as spheres, an n value of 4 would be established. For different assumptions about the process, the mechanism constant will vary from unity to 4. Therefore, if experimental data can be fit to this model, and if the model is appropriate to the phenomenon studied, then statements can be made about the mechanism of the experimentally measured transition. To utilize Eq. (10.4) in the experiment, We rearrange it as

$$\ln \ln[1 - x(t)] = -\ln(k) - n\ln(t) \quad (10.5)$$

From $x(t)$ and t_0 calculated by the kinetic software made by Perkin-Elmer corporation, the values of $x(t)$ and t can be inserted into this function to generate a plot that should take the form of a straight line. By crystallizing samples at various isothermal temperatures, a series of fraction transformed traces and corresponding JMA traces can be generated. The slopes of the lines in the JMA plot represent the mechanism constant. Each slope should be the same, presuming that the mechanism of the reaction has not changed for crystallization at various isothermal temperatures.

The rate constant is generally taken to have an Arrhenius temperature dependence,

$$\ln(k) = \ln(k_0) - E_x/(RT_x) \quad (10.6)$$

where k_0 is a pre-exponential constant, E_x is the activation energy of the glass-nanocrystal transition and T_x denotes the crystallization temperature. The activation energy is an important descriptive factor of the extent of the exponential temperature dependence of a transition and is the compared value for the kinetics of reactions. For each JMA trace, a value of $\ln(k)$ corresponding to each T_x can be determined from the y-intercept. These values can then be put into Eq. (10.6). The best fitline to these points would establish the activation energy of the reaction as the slope (after multiplying by $-R$). Note that experimental methods involving rapid heating followed by isothermal crystallization become difficult to realize at higher temperatures where the reaction proceeds rapidly.

Kissinger method is another method to determine E_x from DSC measurements in terms of different B values for anisothermal glass-nanocrystal transition (Lu, 1996). The Kissinger equation is expressed as

$$\ln(B/T_x^2) = -E_x/(RT_x) + C \qquad (10.7)$$

where C is a constant, and B, T_x, E_x and R have the same meaning as above. Through measuring T_x with different B values, a plot of $1/T_x$ versus $\ln(B/T_x^2)$ is carried out. The slope of the regression line on the measured points is equal to $-E_x/R$.

Since different B values are be utilized during the calculation of Eq. (10.7), some related measuring problems must be considered to obtain a correct T_x. B values of DSC can vary among 0.1 and 500 K min^{-1} (DSC-7) or 0.01 and 500 K min^{-1} (Pyris 1 DSC). However, if mass of samples and crucibles is large (more than 100 mg), thermal responsibility of sample is not enough when $B > 100$ K min^{-1}. At this heating condition, temperature of sample can not reach the given temperature shown on DSC. Thus usually samples for thermal analysis have a mass less than 10 mg. To realize a large heating rate, Al crucible (17 mg) is suggested to use (Jiang, et al., 1997b). Besides, to guarantee an identity between the given and the real temperature, it is best that the temperature for each B value is calibrated just before the actual measurement. However, when $B < 1$ K min^{-1}, temperature error induced by different heating rates is smaller than 0.5 K. Thus, when the heating rate is low, it is allowed that a moderate B, such as 0.2 K min^{-1}, is used to calibrate for heating rates being smaller than 1 K min^{-1}.

Similarly, the above consideration can be utilized in the investigation of the grain growth of nanostructured materials (Lu, 1996; Lu and Dong, 1997). The grain growth temperature can be determined by means of the DSC runs at constant heating rates. However, owing to the relatively small heat effect during grain growth, sample weight for DSC measurement is usually large in order to get evident exothermic peak. The grain sizes should be determined by using X-ray diffraction or electronic microscopic observations. When the characteristic temperatures for the grain growth process are determined at different heating rates, the activation energy for the grain growth is obtained by using Eq. (10.6) or Eq. (10.7).

10.3.3 Heat Capacity of Nanostructured Materials

As the name implies, heat capacity refers to the ability of a substance to contain thermal energy. The mechanism by which solids sustain thermal energy is the vibration of atoms about a mean lattice position. Metallic solids have the additional energy storage mechanism of electron motion. Liquids have more mechanisms of energy storage than solids, e.g., atoms can translate and rotate among themselves, as well as vibrate. Thus, liquids generally have a higher heat capacity than their crystalline counterparts.

C_p measurements can be made on DSC. A usual method is the scanning method. Firstly, an empty pan is heated at a heating rate of $B = 20 - 40$ K min^{-1} with a mass of 5—40 mg. C_p values are obtained by the difference of heat flows between measurement run with and without a sample. The measured heat flow values are calibrated by the comparison with a measurement of a standard sample (sapphire) with similar mass of the sample exposed to the same heating rate (Speyer, 1994). To increase the measuring accuracy, measuring error can be decreased by taking many measurements (according to statistical theory, with N measurements the measuring error is $N^{-1/2}$ of the original error) (Jiang, et al., 1995). This operation has special importance for comparison of heat capacities among coarse-grained polycrystals, nanostructured materials, and glasses since their difference is in the error range of DSC. In addition, sample and reference should be centered in the crucibles, the crucibles should be centered in the containers, and are of nearly identical total heat capacity. Note also that many times of putting in and taking out samples for C_p measurements lead to that the crucible and sample are not located at same position of container, which lead to an additional error. Therefore, a calibration with the standard sample for the heat capacity measurement may increase, but not decrease, error. (measuring error of C_p is about $\pm 1\%$).

During the measurement of C_p, although argon gas is fitted in the furnace, the oxidation must specially be noticed. When T is higher, a slight oxidation will result in a heat release and a depression of the temperature coefficient of C_p functions. Thus, when oxidization is easily present for some alloys, the sample must be sealed. In order to avoid oxidation-induced measuring error, the mass of the sample should be measured after the measurement. According to our experience, if the mass increases after the measurement is smaller than 1%, the measured result is acceptable (Jiang, et al., 1997b).

A particularity for C_p measurement of nanophases and nanostructured materials is that transitions from metastable phases to stable phases related to T and t must be avoided. On DSC with a usual B, there exists an evident temperature where this kind of transition arises. For nanostructured materials, grains evidently begin to grow at $0.4—0.6 T_m$.

An essential error source for C_p measurements is that there is a distinct shift in the baseline at the beginning of the measuring curve. The cause of this can be visualized by assuming an empty reference container and a sample container with a significant mass of sample. Starting at room temperature, the sample and reference temperatures would be the same, thus ΔT is zero. Since more thermal energy is required to raise the temperature of the large mass in the sample container as compared to the empty reference container, the sample lags the reference in temperature. This lag initially increases until the difference in temperature reaches a steady-state value, at which point it may remain more or less constant as long as the heating rate does not change.

Since the transition enthalpy is obtained by an integration of the measured area from the peak and an artificial baseline extended from the baseline without transition, there is no error induced by putting in and taking out sample. Thus, the error for transition enthalpy is smaller than $\pm 0.5\%$. In order to decrease measuring error for C_p, the enthalpy method can be used to measure C_p (Cezairliyan, 1988). For this kind of measurement, a step-wise heating is necessary. With each heating scan, a new quasi-balance state is established on the time for the thermal arrest. Each single measurement for one C_p value corresponds to an enthalpy measurement. The time for heating up between two temperature steps may be fixed in these measurements as 25% of the total one. Thus an average B value is one fourth of the original B value. With the same method an empty crucible is measured. The measured curve with the sample minus the measured curve with empty crucible. Then the area is integrated and divided by a single scanning temperature interval, such as 10 K. The obtained value is the averaged C_p at the middle temperature of temperature range during the heating.

10.3.4 Interface Enthalpy of Nanostructured Materials

When the glass is heated above a certain temperature, glass-nanocrystal transition appears. Through a thermal analysis, the crystallization temperature of the glass (T_x), the heat capacities of liquid (C_p^l), the coarse-grained polycrystals (C_p^c), and the melting enthalpy of the coarse-grained polycrystals ($H_m(\infty)$) can be measured. With these thermodynamic data, the crystallization enthalpy of liquid ($H^{l-c}(T)$) may be calculated by the following (Lu, et al., 1993; Lu, 1996):

$$H^{l-c}(T) = H_m(\infty) + \int_{T_m}^{T_x} (C_p^c - C_p^l) dT \qquad (10.8)$$

The transition enthalpy of glass-nanocrystal transition $H^{l-n}(T)$ can be directly measured by heating the glass through the glass-nanocrystal transition temperature T_x. $H^{l-n}(T)$ can also be calculated with Eq. (10.8) when $C_p^c - C_p^l$ is replaced by $C_p^n - C_p^l$.

$$H(T)^{l-n} = H_m(\infty) + \int_{T_m}^{T_x} (C_p^n - C_p^l) dT \qquad (10.9)$$

The difference of H^{l-n} and H^{l-c} divided by the interface area shows the interface enthalpy of the nanostructured materials (H_{GB})

$$H_{GB} = (H^{l-n} - H^{l-c})/(gV_m/r) \qquad (10.10)$$

where g is a geometrical factor of the grains, V_m denotes the molar volume, and r is the averaged grain radius.

To measure T_x and H^{l-n}, it is suggested that the sample is first heated to $T_x - 150$ K with $B = 40$ K min^{-1}, then is cooled to ambient temperature with $B = 200$ K min^{-1}. This procedure lets the glass be fully relaxed and T_x be correctly measured. Then the sample is heated through T_x with $B = 40$ K min^{-1}. The measurement is finished at $T = T_x + 100$ K (Jiang, et al., 1997b).

To measure C_p^l, the procedure is a little complicated. C_p^l should be measured over the full temperature range of T_g (glass transition temperature) and $T_m(\infty)$ (Jiang, et al., 1997b). However, because the undercooled liquid exists only briefly owing to its metastable nature, the temperature range for C_p^l measurement near T_g is limited (for example, for a glass with a high glass forming ability, this temperature range is about 30 K). In order to increase the measuring temperature range, different B values must be utilized to heat the glass over T_g. When B is increased by a factor of 2, T_g increases by 3-4 K. The undercooled liquid remained at about $T_g + 30$ K. Thus, the measured C_p^l was partly in another temperature range. With this method, the temperature range for C_p^l measurement near T_g can be doubled. Together with the measured data of C_p^l near $T_m(\infty)$, C_p^l temperature function in the full temperature range between T_g and $T_m(\infty)$ can be fitted.

Another way to measure H_{GB} is to consider the grain growth of the nanostructured materials. Upon heating the nanostructured materials, grain growth process will occur accompanied by a decrease of the total grain boundary area, of which the thermal effect is large enough to be detected by DSC when the initial grain size is small (Lu and Dong, 1997). During the growth, the total interface area of the materials decreases and the internal energy of the materials releases, which leads to an exothermic peak in the heating trace of DSC. Assuming the total grain boundary area in the nanostructured materials per unit area as $A = 6r^2/(r^3D) = 6/(rD)$, where D is the mass density of the grain boundary. When the grains grow from r_1 to r_2, the grain boundary area will be changed by

$$\Delta A = 6(1/r_2 - 1/r_1)/D \qquad (10.11)$$

With the thermal energy during the growth of the nanostructured materials released Q, H_{GB} is calculated as follows:

$$H_{GB} = Q/\Delta A \qquad (10.12)$$

10.4 Data Interpretation

With the measured thermodynamic and kinetic data, some theories can be developed. These theories will deepen our understanding for nanophases and nanostructured materials. In this section, some models for size-dependent thermodynamic functions of nanophases and for the glass-nanocrystal transition will be introduced.

10.4.1 Size-Dependent Melting Thermodynamics of Nanophases

The size-dependent melting temperature $T_m(r)$ is shown as (Jiang, et al., 1997a; 1998a)

$$T_m(r)/T_m(\infty) = \exp[-(\alpha-1)/(r/r_0-1)] \qquad (10.13)$$

where α is a material constant, r_0 shows a critical radius at which all atoms of a nanocrystal are located on its surface. For low dimensional metallic crystals, r_0 depends on their dimension d and atomic diameter h, the corresponding relationship is

$$r_0 = (3-d)h \qquad (10.14)$$

where $d=0$ for nanoparticles, $d=1$ for nanowires and $d=2$ for thin films. When $d=0$, r means the normal radius. When $d=1$, r denotes the radius of the nanowire. If $d=2$, r shows the half thickness of a thin film. Since a crystal is characterized by its long-range order, the smallest metallic nanocrystal should have at least half of the atoms located within the nanocrystal. Hence, the smallest r is $2r_0$ (Jiang, et al., 1997a; 1998a).

For organic crystals, the smallest unit is not an atom, but a molecule. Since the shape of an organic molecule is usually not spherical, h as a mean diameter of a molecule is defined as (Jiang, et al., 1999a)

$$h = \frac{1}{3}\sum_{i=1}^{3} h_i \qquad (10.15)$$

where the subscript i is taken from 1 to 3 denoting x-, y-, and z-axis of a molecule, h_i is the length of a molecule along the corresponding axis direction. For organic molecules not having any regular shapes, the direction having the largest size is taken as x-axis, and the possible smallest size is defined as another axis. Note that different choices of the axes lead to little difference in h, which changes $H_m(r)$ and $S_m(r)$ functions little, especially when r/h is large enough (for instance, usually $r > 2$ nm in experiments and $h < 0.5$ nm for most organic crystals).

When the nanophases have free surfaces, α has the following form (Jiang, et al., 1997a; 1998a):

$$\alpha = 2S_m(\infty)/(3R) + 1 \tag{10.16}$$

where R is the ideal gas constant.

If the nanophases with a melting temperature of $T_m(\infty)$ and an atomic diameter of h_m are embedded in a matrix whose melting temperature and atomic diameter are $T_M(\infty) > T_m(\infty)$ and h_M, respectively, under the condition that the interface between the nanophases and the matrix is coherent, its α reads (Jiang, et al., 1999b)

$$\alpha = \{[h_M^2/h_m^2]T_m(\infty)/T_M(\infty) + 1\}/2 \tag{10.17}$$

Combining Eqs. (10.13)—(10.17), the size-dependent melting temperatures of all different types of low dimensional crystals are obtained. Note that most phenomenological models for $T_m(r)$, regardless of their initial assumptions, predict a change of $T_m(r)$ following the relation $T_m(r)/T_m(\infty) \sim 1 - 1/r$ (Jiang, et al., 1998a). All these theories are in good agreement with the experimental data when $r \geq 10$ nm. Under this condition ($r \geq 10 r_0$ where $r_0 \approx 1$ nm for $d=0$), Eq. (10.13) is the same for the phenomenological model $T_m(r)/T_m(\infty) \sim 1 - 1/r$ in terms of the relation $\exp(-x) \approx 1 - x$. However, these pheno-menological models that fail for smaller nanocrystals and do not have the lowest limit of $T_m(r)$ at the critical size of r_0, can not explain a dimension-dependent $T_m(r)$ and has no idea for the size-dependent overheating of nanocrystals. Against that, the above equations can illustrate all of the above phenomena. With the above equations, the size-dependent melting entropy $S_m(r)$ and enthalpy $H_m(r)$ have been deduced and are given by (Jiang and Shi, 1998)

$$S_m(r)/S_m(\infty) = 1 - 1/(r/r_0 - 1) \tag{10.18}$$
$$H_m(r)/H_m(\infty) = [T_m(r)/T_m(\infty)][1 - 1/(r/r_0 - 1)] \tag{10.19}$$

where $S_m(\infty)$ and $H_m(\infty)$ are the bulk values of $S_m(r)$ and $H_m(r)$, respectively. Equations (10.18) and (10.19) are valid for both metallic and organic nanophases based on the consideration that the vibrational melting entropy is dominant on the overall melting entropy. It is well known that the transition entropy consists, at least, of three contributions: positional, vibrational and electronic component. If the type of chemical connection does not vary during the transition from a solid to a liquid, the electronic component is negligible. Apparently, the positional component is also small for organic compounds and metals. Thus, the melting for metallic and organic crystals is mainly vibrational in nature. However, for a semiconductor, the contribution for the electronic component of the overall entropy is large. Thus, when the above equations are utilized for semiconductors, the melting entropy must be replaced by its vibrational part (Jiang and Shi, 1998b).

10.4.2 Glass-nanocrystal Transition Thermodynamics

The glass-nanocrystal transition can be generally calculated as (Lu, 1995a; 1996),

$$H^{g-n}(T) = H^{g-n}(T_x) + \int_{T_m}^{T_x} (C_p^n - C_p^l) dT \quad (10.20)$$

$$S^{g-n}(T) = S^{g-n}(0) + \int_0^T (C_p^n - C_p^l)/T dT \quad (10.21)$$

$$G^{g-n}(T) = H^{g-n}(T) - TS^{g-n}(T) \quad (10.22)$$

where $G^{g-n}(T)$, $H^{g-n}(T)$ and $S^{g-n}(T)$ are the temperature-dependent Gibbs free transition energy, transition enthalpy and transition entropy for glass-nanocrystal transition. All amounts in the equations are measurable on DSC except $S^{g-n}(0)$, which should be close to zero according to the third law of thermodynamics. Thus, the thermodynamic functions for glass-nanocrystals are measurable.

The formed nanostructured materials are metastable. This metastability can be interpreted as follows: Assuming that the nanocrystals can be divided as two independent phases of crystals and interfaces and the interaction between the crystals and interfaces is negligible, $G^{g-n}(T)$ is read as

$$G^{g-n}(T) = (1-f)G^c(T) + fG^i(T) - G^g(T) \quad (10.23)$$

where f is the atomic fraction of the interface in the nanostructured materials. $G^c(T)$, $G^i(T)$ and $G^g(T)$ are the Gibbs free energy of crystals, interfaces and glasses, respectively. When $G^{g-n}(T) = 0$, f becomes the maximum $f_{max}(T)$,

$$f_{max}(T) = [G^g(T) - G^c(T)]/[G^i(T) - G^c(T)] \quad (10.24)$$

If the interfaces have a certain thickness β, f_{max} is proportional the grain radius r. Thus, there exists a critical temperature-dependent $r^*(T)$ value of nanostructured materials where the system remains metastable, that is,

$$r^*(T) = \beta f_{max}(T) \quad (10.25)$$

10.5 Examples of Applications

Figure 10.5 to Fig. 10.7 (Jackson, 1990; Jiang, et al., 1998a; 1999a) show the melting temperature of metallic nanoparticles, organic nanowires and metallic thin films as a function of r. As the size of the nanophases decreases, the melting temperatures decrease. This decrease becomes evident when $r<10$ nm. The solid lines in the figures are the predicted functions of Eqs. (10.13) and (10.16).

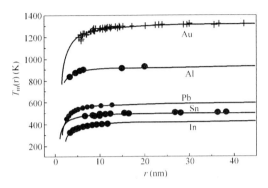

Figure 10.5 Size-dependent melting temperatures of Au, Al, Pb, Sn and In nanoparticles. The solid lines denote the model prediction of Eqs. (10.13) and (10.16)

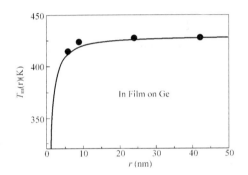

Figure 10.6 Size-dependent melting temperature of In thin film on Ge substrate. The solid line denotes the model prediction of Eqs. (10.13) and (10.16)

Figure 10.8 (Jiang, et al., 1999b) gives the size-dependent melting temperature of In nanoparticles embedded in Al matrix. The necessary condition for this kind of superoverheating is that the interface between the nanoparticles and the matrix is coherent (Jiang, et al., 1999b; Sheng, et al., 1996a; 1996b). It is clear that the melting temperature of nanocrystals increases as the size decreases. This result is also predicted in terms of Eqs. (10.13) and (10.17) as shown in the figure.

Figure 10.9 and Fig. 10.10 present the size-dependent melting enthalpy and entropy of metallic Al and organic benzene nanocrystals. It is observed that both of the melting enthalpy and melting entropy decrease as the size

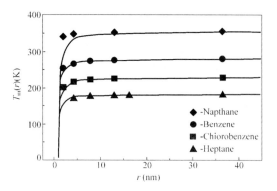

Figure 10.7 Size-dependent melting temperature of organic nanowires of napthane, benzene, chlorobenzene and heptane. The solid lines denote the model prediction of Eqs. (10.13) and (10.16)

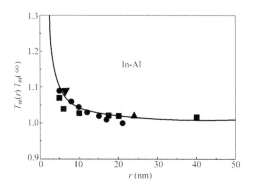

Figure 10.8 Size-dependent superheating of In nanocrystals embedded in Al matrix. The solid line denotes the model prediction of Eqs. (10.13) and (10.17). The symbols ▼, ▲, ■, and ● are experimental results from different authors

decreases. The melting enthalpy drops more strongly than the melting entropy since $H_m(r) = S_m(r) T_m(r)$, where $T_m(r)$ decreases too as the size of nanophases decreases. The solid lines in the figures are the predicted values of Eqs. (10.19) and (10.20).

Figure 10.11 shows the temperature-dependent transition enthalpy of glass-nanocrystal transition of the Ni-P alloy (Lu, et al., 1993). As shown in the figure, the transition enthalpy decreases as temperature increases due to the decrease of grain size. This result is consistent with the theoretical prediction of Eq. (10.21). The reason is that the heat capacities of nanocrystals are evidently higher than the glass in this alloy.

Figure 10.12 (Lu, et al., 1993) gives $f_{max}(T)$ and $r^*(T)$ values for glass-nanocrystal transition of Ni-P alloys as a function of the annealing temperature. The solid lines are the theoretical prediction of Eqs. (10.25)

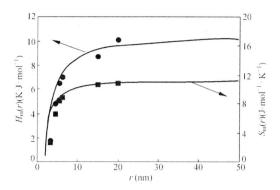

Figure 10.9 Size-dependent melting enthalpy and melting entropy of Al nanocrystals. The solid lines denote the model prediction of Eqs. (10.18) and (10.19).

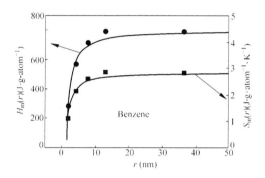

Figure 10.10 Size-dependent melting enthalpy and melting entropy of benzene. The solid lines denote the model prediction of Eqs. (10.18) and (10.19)

and (10.26). It is clear that before a certain annealing temperature is reached, the grain size decreases and atomic fraction of interfaces increases. This is due to the thermodynamic equilibrium among glasses and nanostructured materials.

JMA measurements for the kinetics of glass-nanocrystals transitions for Ni-P alloy are shown in Fig. 10.13 and Fig. 10.14 (Lu, et al., 1994; Lu, 1996). According to Fig. 10.13, at first, transition is present only at the defects of amorphous alloys, such as surface due to the value of mechanism constant $n = 1.3$, later bulk sample begins to transform ($n = 3.5$). At the end, growth of crystallite arises ($n = 0.7$). Figure 10.14 shows that E_x for the isothermal glass-nanocrystal transition is a little different. For the early stage, $E_x = 145$ J/g-atom. For the middle and last stages, $E_x = 245$ J/g-atom. The result shows that the defect crystallization is easier than the crystallization and growth of the bulk sample.

Figure 10.15 (Ye and Lu, 1998) presents the measurement of E_x for

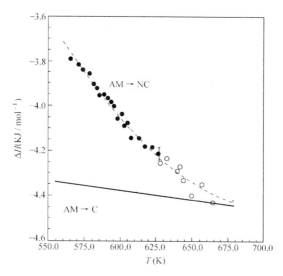

Figure 10.11 Measured transition enthalpy of glass-nanocrystal transition of Ni-P alloys as a function of the annealing temperature. The solid line is the theoretical prediction of Eq. (10.20). T_a in the figure is the annealing temperature

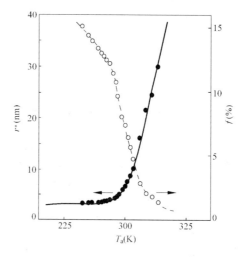

Figure 10.12 Measured $f_{max}(T)$ and $r^*(T)$ values for glass-nanocrystal transition of Ni-P alloys as a function of the annealing temperature T_a. The solid lines are the theoretical prediction of Eqs. (10.24) and (10.25)

element Se in terms of the Kissinger method for anisothermal glass-nanocrystal transition. It is found that when the pressure increases, E_x increases. This phenomenon comes from the fact that when nuclei in the amorphous Se arise, the specific volume of the nuclei increases due to the appearance of glass/

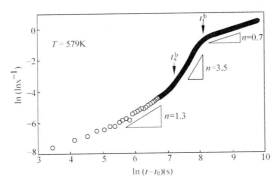

Figure 10.13 Measurement of mechanism exponent of Ni-P alloys at 579 K, where t_s^b is the starting time for the bulk nanocrystallization process and t_f^b the finishing time for bulk nanocrystallization process

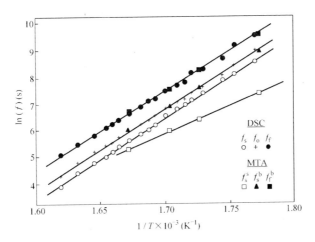

Figure 10.14 Arrhenius plots for the calculation of the activation energies for the nanocrystallization process by using different sets of the specific times measured in the DSC and the MTA (magnetothermal analysis) experiments (t_s^s is the starting time for the surface nanocrystallization process)

nuclei interface (Lu, 1995b). Thus, the pressure makes it difficult to nucleate from the glass.

Figure 10.16 (Lu, et al., 1994) gives the Kissinger plot for the measurement of E_x for the grain growth of the nanostructured materials of $HfNi_5$. The obtained E_x value is comparable to that of the volume self-diffusion of Hf, but is evidently larger than those of the grain boundary self-diffusion of both elements. This implies that the growth of the $HfNi_5$ nanophases is essentially dominated by volume diffusion rather than the interface diffusion, which is the usual case of grain growth in conventional polycrystals.

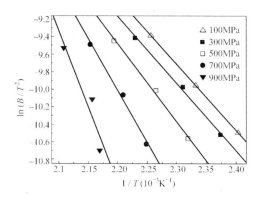

Figure 10.15 Determination of E_x for anisothermal glass-nanocrystal transition of element Se with Kissinger method

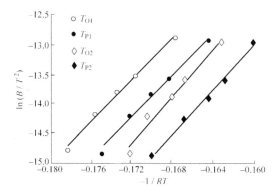

Figure 10.16 Determination of E_x for nanocrystal growth of $HfNi_5$ and the precipitation of Ni from a $HfNi_5$ nanostructured material. On the DSC trace, there are two exothermic peaks. The first one corresponds to the grain growth of $HfNi_5$, and the second to the precipitation of Ni. T_{o1}, T_{p1}, T_{o2} and T_{p2} shown are the onset temperature and the peak temperature for the first and the second peaks, respectively

Figure 10.17 (Sun, et al., 1996) shows the comparison among the heat capacities of nanostructured materials, bulk crystals and glasses for Se. The nanostructured crystals have much higher heat capacities than the coarse-grained crystals due to the effect of a large number of interfaces. This result implies that the entropy of nanostructured materials is larger than the coarse-grained crystals since the vibrational and positional entropies of crystals are closely related to the existence of interfaces.

Figure 10.18 (Jiang, et al., 1997a) presents a DSC measurement of C_p^c and C_p^l values of ZrAlNiCuCo alloys. Clearly the heat capacities of the undercooled liquid of the glass forming alloys increase as temperature decreases, which decreases the Gibbs free energy difference between the

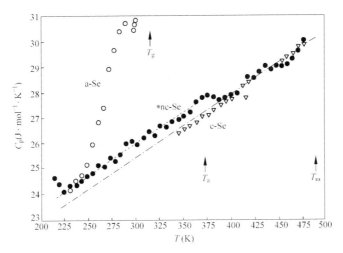

Figure 10.17 Heat capacities of the glassy Se (a-Se), the nanostructured Se (nc-Se) and the coarse-grained polycrystalline Se (c-Se)

crystal and the liquid and increases the glass forming ability of the alloy. At the glass transition temperature, the C_p^l value reaches about 60 J/g-atom, which is comparable with other good glass forming alloys.

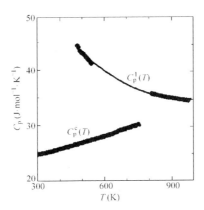

Figure 10.18 Heat capacities of the undercooled liquid ($C_p^l(T)$) and that of the coarse-grained polycrystalline crystals ($C_p^c(T)$) for ZrAlNiCuCo alloys.

Figure 10.19 (Sun, et al., 1997) gives the calculated interface enthalpy of nanostructured Se. It is found that nanostructured materials have lower interface enthalpy than bulk crystals due to the appearance of coherent interfaces, which guarantees the existence of the metastable nanostructured materials.

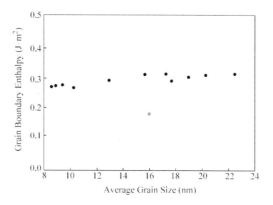

Figure 10.19 Calculated interface energy of nanostructured Se in terms of Eq. (10.10) as a function of averaged grain size

10.6 Limitations

DTA/DSC is a useful instrument to measure many thermodynamic and kinetic parameters of nanophases and nanostructured materials, which include size-dependent melting temperature, size-dependent melting entropy, size-dependent heat capacities, surface (interface) energy and activation energy for glass-nanocrystal transition, as well the activation energy for nanocrystal growth. However, there exists a large measuring error on these parameters since sensitivity of the present commercial DSC instruments is not enough, while a self-made calorimeter can only be utilized in some specific cases. All of these difficulties may be solved only when some more sensitive commercial calorimeter with the sensitivity of a microcalorimeter is manufactured.

Since the nanocrystals are metastable, growth can easily arise during measuring the thermodynamic and kinetic parameters by heating the samples in the DSC if the nanocrystals are not isolated from each other. The release of heat during the growth of nanocrystals leads to unreal data of heat capacities of nanocrystals. The growth of nanocrystals also leads to the result that the measured data belong to nanocrystals in different sizes. Thus, a good method to isolate nanocrystals should be further studied.

As seen from Eq. (10.11), the total area change of grain boundary will be smaller when the initial grain size is larger. This will significantly decrease the heat effect in the DSC measurement. Therefore, for nanostructured materials with relatively large grain sizes (say larger than 50 nm), the grain growth process could not be detected accurately by using DSC. In addition, during the DSC measurements of the nanostructured materials, strain release occurs frequently, of which the heat effect may overlap with that of the grain growth. This brings more difficulties in determining the grain growth kinetics accurately in nanostructured materials. For the glass-nanocrystal transition, the nucleation process is always accompanied by a weak thermal effect that is difficult to determine by DSC. Therefore, crystal nucleation kinetics from glasses determined from DSC measurements should be analyzed with care.

The heating rate limitation is another concern for many investigators. The highest controllable heating rate in the present-day DSC is below 500 K/min. This is, however, not high enough for some cases when a kinetic process is to be seen at an extremely high heating rate (say, for example, in order to separate the overlaps of several different processes).

10.7 Prospects

As development of calorimetric study on nanophases and nanostructured materials, a more complete understanding of the thermodynamic and the kinetic aspects for nanophases and nanostructured materials could be carried out. These works will help us to understand the thermal stability of nanophases and nanostructured materials that is a very important property for the application of nanocrystals due to the metastable nature of nanophases and nanostructured materials.

It is known that there are three different types of materials according to their industrial application, which are structural, functional and intelligent materials. Two former materials have been widely utilized in the industry. Although nanophases as functional and even intelligent materials, especially on information technology industry, have been manufactured by chemical, physical and mechanical methods, these methods are essentially not suitable to produce bulk structural materials due to their high cost. A possible way to produce the bulk nanostructured materials as structural materials is through the bulk glass-nanocrystal transition. However, a quantitative thermodynamic and kinetic description for this kind of transition is still scarce. These descriptions should concentrate on the effect of transition temperature, transition time and alloy element amount on grain size of nanostructured materials.

In the theoretical aspects, the understanding of melting and solidification for nanocrystals will give us a chance to find an analytical function for the size-dependent surface/interface energy, which is the key for the classic melting/solidification theory. This is because the size of nuclei is just in the nanometer range. If the Gibbs free energy of nanophases is known, the difference of the

Gibbs free energies between the nanophases and the bulk crystals denotes the surface energy of the nanophases.

Although melting is a natural phenomenon, its physical nature is still in question. One melting theory is the Lindemann criterion. Although the criterion in nature is a kinetic melting one, it has given a thermodynamic melting temperature. A simple interpretation of the criterion is that the melting temperature is a kinetic melting temperature for surface melting which leads to thermodynamic melting of the atoms within the crystals. The melting of nanophases will supply a bridge relating the macroscopic and microscopic melting, such as the appearance of surface melting. Further investigations on the melting mechanism of nanocrystals and the superheating should be strengthened in terms of experimental and theoretical, as well as computer, simulations.

References

Cezairliyan A.. Cindas Data Series on Materials Properties. Volumes I-2, Specific Heat of Solids, eds. C. Y. Ho. Hemisphere Publishing Corporation, New York (1988)
Gleiter H.. Prog. Mater. Sci.. **33**, 223(1988)
Jackson C. L., G. B. McKenna. J. Chem. Phys.. **93**, 9002(1990)
Jiang Q., J. C. Li, J. Tong. Mat. Sci. Eng.. A **196**, 165(1995)
Jiang Q., N. Aya, F. G. Shi. Appl. Phys.. A **64**, 627(1997a)
Jiang Q., M. Zhao, X. Y. Xu. Phil. Mag.. B **76**, 1(1997b)
Jiang Q., H. Y. Tong, D. T. Hsu, K. Okuyama, F. G. Shi. Thin Solid Films. 312, 357 and references therein (1998a)
Jiang Q., F. G. Shi. Mater. Lett.. **37**, 79(1998b)
Jiang Q., X. L. Wei, M. Zhao. Thermal analysis of metals and alloys. In: Handbook of thermal analysis. ed.. by T. Hatakeyama and Zhenhai Liu. John Wiley & Sons Ltd., Chichester (1998b) pp. 112—124.
Jiang Q., H. X. Shi, M. Zhao. J. Chem. Phys.. in press and references therein (1999a)
Jiang Q., Z. Zhang, J. C. Li. Chem. Phys. Lett.. (1999b) (submitted)
Lai S. L., G. Ramanath, L. H. Allen, P. Infante, Z. Ma. Appl. Phys. Lett.. **67**, 1229(1995)
Lai S. L., J. Y. Guo, V. Petrova, G. Ramanath, L. H. Allen. Phys. Rev. Lett.. **77**, 99(1996)
Lu K, R. Luck, B. Predel. Scripta Metall. Mater.. **23**, 1387(1993)
Lu K, R. Luck, B. Predel. Acta. Metall. Mater.. **42**, 2303(1994)

Lu K. Phys. Rev.. B **51**, 18(1995a)

Lu K., Z. F. Dong, I. Bakonyi, A. Czibaki. Acta Metall. Mater.. **43**, 2641 (1995b)

Lu K.. Mat. Sci. Eng.. R**16**, 161(1996)

Lu K. and Z.F. Dong. J. Mater. Sci. Tech.. **13**, 491(1997)

Speyer R F. Thermal analysis of materials. Marcel Dekker, Inc., New York (1994)

Sheng H. W., G. Ren, L. M. Peng, Z. Q. Hu and K, Lu. Phil. Mag. Lett.. **73**, 179(1996a)

Sheng H. W., J. Xu, L. G. Yu, X. K. Sun, Z. Q. Hu and K, Lu. J. Mater. Res.. **11**, 2841(1996b)

Sun N. X. and K. Lu. Phys. Rev.. B **54**, 6058(1996)

Sun N. X., K. Lu, Q. Jiang. Phys. Rev.. B **56**, 5885(1997)

Suryanarayana C.. International Mater. Rev.. **40**, 41(1995)

Ye F., K. Lu. Acta Mater.. **46**, 5965(1998)

Index

Abbe's imaging theory, 32
absorption spectroscopie, 174
AgI, 220
anelastic relaxation, 331
atomic force microscopy(AFM), 124
Au, 222
Auger electrons, 103

backscattered electrons, 103
ballistic electron emission microscopy (BEEM), 124
battery electrode, 316, 318
bending modulus, 70
blocking temperature, 255

carbon nanotube, 68
catalytic properties, 283, 293
cds, 220
characteristic X-rays, 103
charge carriers, 220
chemical contrast, 12
chromatic aberration, 103
cold field-emission, 102
composite electrodes, 291
contrast, 103
counter electrode(CE), 301
crystallite sizes, 16
cyclic voltammetry(CV), 300, 316, 317

decahedron, 42
density of states, 14
different scanning calorimeter(DSC), 358
differential thermal analysis(DTA), 358
diffusion boundary layer, 298
dislocations, 352
dynamic, 219
Döppler effect, 265

elastic scattering, 1
elasticity measurements, 328
electrocatalytic properties, 284
electrochemical infiltration, 287
electrochemical permeation method, 311
electrochemical properties, 285, 293
electrochemical quartz crystal microbalance (EQCM), 312, 313, 315

electrochemical self-assembly, 288
electrochemical techniques, 293
electrodeposition, 285, 286, 293, 312
electroless deposition, 286
electron energy-loss spectroscopy(EELS), 77
electron holography, 46
electron wavelength, 103
electronic relaxation, 220
electrophoretic deposition, 285, 293
energy dispersive X-ray microanalysis (EDS), 93
energy-dispersive spectrometers, 119
energy-filtered electron imaging, 85
equivalent circuit, 297

Fe_2O_3, 220
FIB milling, 348, 349
foucault lorentz microscopy, 50
Foucault mode, 278
fresnel lorentz microscopy, 49
Fresnel imaging mode, 278
FTIR spectroscopy, 181
full-width at half-maximum intensity (FWHM), 106

galvanostatic intermittent charge-discharge, 310, 318
galvanostatic intermittent titration technique(GITT), 308, 309
grain size, 9

Hall-Petch relation, 337
heat capacity, 369
heating rate, 365
heterogeneous nanostructure, 210

icosahedron, 42
impedance, 293—300, 319, 320
indentation, 348
inelastic scattering, 1
interaction volumes, 104
intercalated electrode, 309, 319
interface enthalpy, 370
internal friction, 330
isomer shift, 266

kinetics of glass-nanocrystal transition, 366

lattice dynamics, 24
linear sweep voltammetry(LSV), 300, 301
liquid-crystal templating(LCT), 289
lorentz microscopy, 48
low melting temperature nanophases, 364
low-voltage SEM, 110

magnetic contrast, 117
magnetic hyperfine structure, 268
magnetic scattering, 274
magnetocrystalline anisotropy, 253
mapping valence states, 91
mechanical property, 326
melting temperature, 363
mesoporous electrodes, 289
metal multilayer(MMLS), 332
metal nanoparticles, 220
metallic glass, 366
mixed ionic-electronic conductor, 300
modification, 124
multilayer, 327
multislice theory, 35
Mössbauer effect, 263
Mössbauer fraction, 263

nanobalance, 71
nanocomposites, 286, 288
nanocrystalline materials, 24
nanocrystalline, 284
nanodiffraction, 52
nanomaterials, 219
nanoparticles(NPs), 219, 285—289, 291—293
nanophase, 283—286, 291, 293, 305, 317, 319
nanoporous, 284
nanoproes, 286
nanostructured electrode, 283, 285, 293, 316, 317, 319
near edge fine structure, 79
near-field scanning optical microscopy (NSOM), 185, 186
Neel equation, 270
neutron scattering, 1
neutron scattering length, 274

nuclear scattering, 274

PbL_2, 220
PbS, 220
phase contrast, 31
phonons, 16
photocatalytic properties, 284
photoluminescence spectroscopy, 174
photon excitation, 103
photovoltage spectroscopy, 212
plastic deformation, 346
powder microelectrode, 291—293, 301, 316, 317
primary beam, 104

quadrupole splitting, 267
quantum conductance, 68
quantum confinement, 174
quartz crystal microbalance(QCM), 312

raman spectroscopy, 181
reference electrode, 305
Reitveld refinement, 275
resonant frequency, 330

scanning electron microscopy(SEM), 99
scanning probe microscopy(SPM), 124
schottky field-emission, 102
secondary electrons, 103
semiconductor nanoparticles, 220
Si, 220
size-dependent melting temperature, 372
small angle scattering, 21
spatial confinement, 219
spectroscopic diagnosis, 197, 201
spherical aberration, 103
spin valves, 89
stacking faults, 41
Stoner-Wohlfarth theory, 255
superconducting quantum interference device(SQUID), 255
supermodulus effect, 332
superparamagnetism, 254
surface enhanced raman scattering(SERS), 181
surface structures, 124

thermal analysis, 358

thermocouple junctions,360
thermodynamics,359
time-resolved laser spectroscopy,220
TiO_2,220
twinning structure,41

ultrafast spectroscopy,192

wavelength-dispersive spectrometers,118

working electrode(WE),301,314

X-ray scattering,1
X-ray spectrometers,118

Young's modulus,327

ZnO,220

Appendix

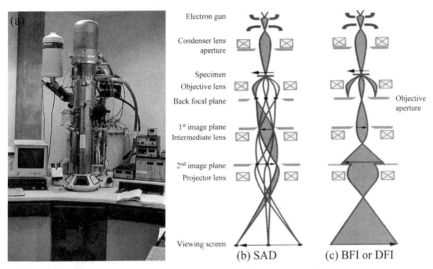

Figure II.1 (a) A high-resolution transmission electron microscope (TEM) is a key tool for nanomaterials research, which can provide not only atomic resolution lattice images but also chemical and electronic information at a spatial resolution better than 1 nm. The most advanced TEM today can reach 0.1 nm image resolution, and 0.2 nm in probe size. (b, c) Optical diagrams of a TEM in selected area diffraction (SAD) mode and bright-field imaging (BFI) and dark-field imaging (DFI) mode.

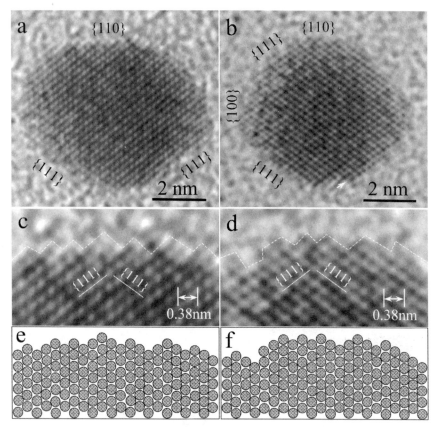

Figure II.2 Atomic structures of nanocrystals can be imaged directly by TEM, which has been used since late 1970s for imaging of surfaces. (a, b) are high-resolution TEM images of FePt nanocrystals oriented along [110], showing the profile images of the {111}, {100} and {110} facets. (c, d) Enlarged atomic structures of the {110} surface, showing the "missing-row" reconstruction, and (e, f) are the corresponding atomic structure models, respectively (Z. R. Dai, S. H. Sun, and Z. L. Wang, Surface Sci., 505 (2002) 325).

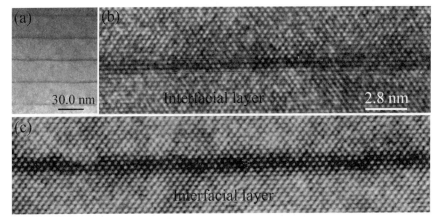

Figure II.3 Growth of heterostructures is an important technique for fabrication of quantum devices for optoelectronic applications. Examining the interfaces in these materials can only be carried out by high-resolution TEM. (a) TEM image of the GaAs superlattice structure formed by periodically exposing GaAs (001) to Sb_2 during the molecular beam epitaxy. (b, c) High-resolution TEM images recorded from the region with Sb exposure, showing that anion exchange is limited to about 4 atomic layers (Courtesy of Dr. Yongqian Wang, Dr. Z. L. Wang, and Dr. April Brown, Georgia Institute of Technology).

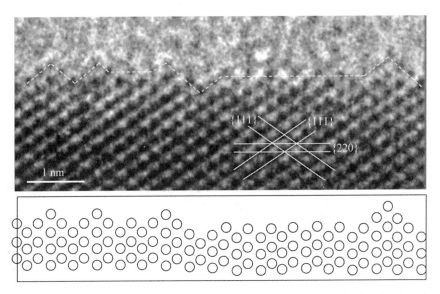

Figure II.4 Many unique properties of nanocrystals are due to the high ratio of surface to volume atoms. Atoms on nanocrystal surfaces some times show reconstruction to minimize surface energy. Shown in this figure is a profile TEM image from Au (110) surface that shows the reconstructed surface. Each dot represents a row of atoms along the electron beam direction (Courtesy of Dr. Z. L. Wang, Georgia Institute of Technology).

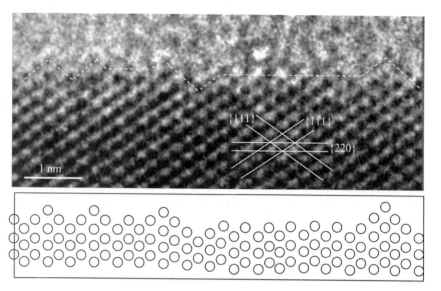

Figure II.5 Nanocrystals are not stable at relatively high temperature, and several particles tend to coalesce, resulting in the formation of a larger particle. Shown in this figure is a coalesced FePt particle that exhibits an irregular shape and twinned structure. The chemical ordering can be directly imaged in the middle section of the picture due to the layered arrangement of the Fe and Pt atoms in the tetragonal structure (Courtesy of Dr. Zurong Dai and Z. L. Wang, Georgia Institute of Technology).

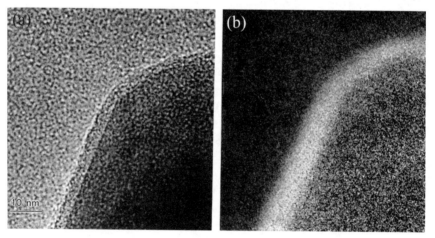

Figure II.6 Chemical imaging is an important technique for studying nanomaterials. Using the electrons that have been inelastically scattered by a characteristic process from the atoms in the sample, high-spatial chemical images can be provided. (a) A conventional TEM image of a TiO_2 particle coated with a 5 nm amorphous SiO_2 layer. (b) A chemical image formed by the electrons that have excited the Si-L edge, showing the distribution of silicon atoms at the surface (Courtesy of Dr. Z. L. Wang, Georgia Institute of Technology).

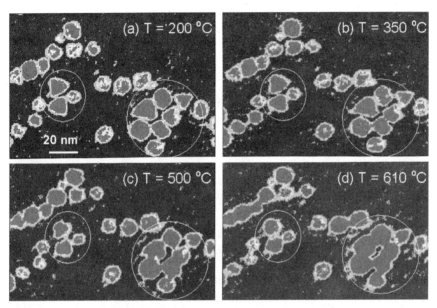

Figure II.7 The melting temperature of nanocrystals is much lower than the melting point of the bulk due to the increased number of surface atoms. Shown here are a series of in-situ TEM images of Pt nanocrystals recorded at different temperatures. The as-synthesized nanocrystals exhibits cubic and tetrahedral shapes, but these shapes cannot be distinguished when the temperature reached 500 °C, around which the surface melting occurred. The nanocrystals showed coalescing at 610 °C, which is much lower than the melting point of 1772 °C for Pt (Courtesy of Dr. Z. L. Wang, Georgia Institute of Technology).

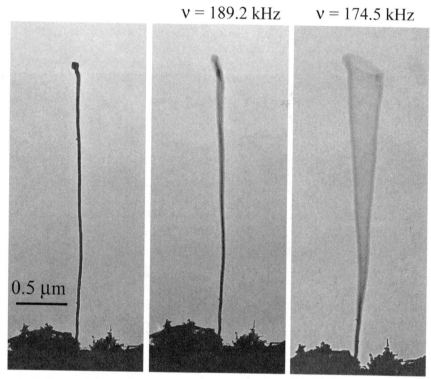

Figure II.8 Characterizing the physical properties of individual nanostructures is rather challenging because of the difficulty in manipulating objects of small size. It is essential to directly image the as-measured object so that the structure and property can be correlated. Using a principle of introducing scanning tunneling microscope in TEM, the mechanical properties of nanowire materials have been measured based on the resonance phenomenon induced by an externally applied alternating voltage (P. Poncharal, Z. L. Wang, D. Ugarte, and W. A. de Heer, Science, 283 (1999) 1513; R. P. Gao, Z. L. Wang, Z. G. Bai, W. de Heer, L. Dai, and M. Gao, Phys. Rev. Lett., 85 (2000) 622). Shown here is the mechanical resonance of a silicon nanowire. Due to the asymmetric shape of the particle at the end of the nanowire, two distinct resonance frequencies have been observed, which correspond to the resonance in the direction parallel and perpendicular to the observation direction, respectively. This is likely to be the most reliable technique for measuring the dynamic modulus of nanowire materials.

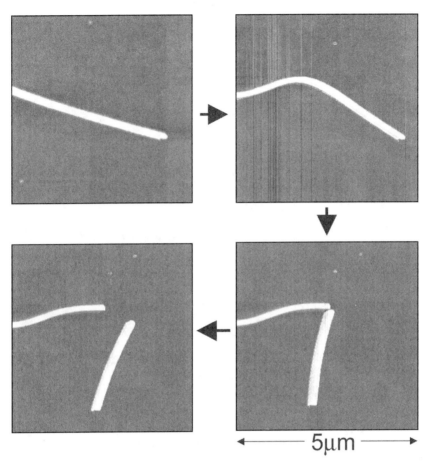

Figure II.9 Nanobelts of semiconducting oxides have been grown by a solid-vapor process (Z. W. Pan, Z. R. Dai, and Z. L. Wang, Science 291 (2001) 1947). The nanobelts are ideal objects for building nano-size sensors and optoelectronic devices. A reduced size of the nanobelt greatly improves its mechanical flexibility. Using an atomic force microscope, the shapes of the nanobelt can be manipulated, and the fracture occurs at extremely large deformation angle (Courtesy of M. Arnold, Dr. P. Avouris, and Dr. Z. L. Wang).

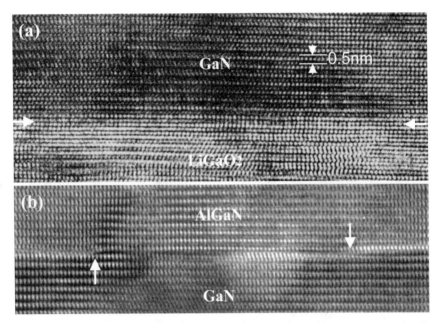

Figure II.10 Thin films are two-dimensional nanostructures. The defect structure in the film is important in determining its electronic properties. This figure gives two high-resolution cross sectional TEM images recorded from the interfaces of GaN on $LiGaO_2$ and AlGaN on GaN, showing interface steps and dislocations (Courtesy of Dr. Zurong Dai and Dr. Z. L. Wang, Georgia Institute of Technology).